Springer Series in
Computational
Mathematics

9

Editorial Board

R.L. Graham, Murray Hill
J. Stoer, Würzburg
R. Varga, Cleveland

Z. Ditzian V. Totik

Moduli of Smoothness

Springer-Verlag
New York Berlin Heidelberg
London Paris Tokyo

Z. Ditzian
Department of Mathematics
University of Alberta
Edmonton, Alberta T6G 2G1
Canada

V. Totik
Bolyai Institute
Attila Jozsef University
6720 Szeged
Hungary

AMS Classification: 41XX, 26 A 16, 46 E 35

Library of Congress Cataloging-in-Publication Data
Ditzian, Zeev
  Moduli of smoothness.
  (Springer series in computational mathematics; v. 9)
  Bibliography: p.
  Includes index.
  1. Smoothness of functions. 2. Moduli theory.
I. Totik, V. II. Title. III. Series.
QA355.D57   1987   515   87-12992

© 1987 by Springer-Verlag New York Inc.
All rights reserved. This work may not be translated or copied in whole or in part without the written permission of the publisher (Springer-Verlag, 175 Fifth Avenue, New York, New York 10010, USA), except for brief excerpts in connection with reviews or scholarly analysis. Use in connection with any form of information storage and retrieval, electronic adaptation, computer software, or by similar or dissimilar methodology now known or hereafter developed is forbidden.
The use of general descriptive names, trade names, trademarks, etc. in this publication, even if the former are not especially identified, is not to be taken as a sign that such names, as understood by the Trade Marks and Merchandise Marks Act, may accordingly be used freely by anyone.

Typeset by Asco Trade Typesetting Ltd., Hong Kong.
Printed and bound by R.R. Donnelley and Sons, Harrisonburg, Virginia.
Printed in the United States of America.

9 8 7 6 5 4 3 2 1

ISBN 0-387-96536-X Springer-Verlag New York Berlin Heidelberg
ISBN 3-540-96536-X Springer-Verlag Berlin Heidelberg New York

PREFACE

The subject of this book is the introduction and application of a new measure for smoothness of functions. Though we have both previously published some articles in this direction, the results given here are new.

Much of the work was done in the summer of 1984 in Edmonton when we consolidated earlier ideas and worked out most of the details of the text. It took another year and a half to improve and polish many of the theorems.

We express our gratitude to Paul Nevai and Richard Varga for their encouragement. We thank NSERC of Canada for its valuable support. We also thank Christine Fischer and Laura Heiland for their careful typing of our manuscript.

Z. Ditzian
V. Totik

CONTENTS

Introduction.. 1

PART I. THE MODULUS OF SMOOTHNESS

Chapter 1. Preliminaries 7

1.1. Notations 7
1.2. Discussion of Some Conditions on $\varphi(x)$ 8
1.3. Examples of Various Step-Weight Functions $\varphi(x)$ 9

Chapter 2. The K-Functional and the Modulus of Continuity 10

2.1. The Equivalence Theorem................................... 10
2.2. The Upper Estimate, $K_{r,\varphi}(f,t^r)_p \leq M\omega_\varphi^r(f,t)_p$, Case I 12
2.3. The Upper Estimate of the K-Functional, The Other Cases 16
2.4. The Lower Estimate for the K-Functional 20

Chapter 3. K-Functionals and Moduli of Smoothness, Other Forms . 24

3.1. A Modified K-Functional 24
3.2. Forward and Backward Differences.......................... 26
3.3. Main-Part Modulus of Smoothness........................... 28
3.4. Computation of Our Modulus for Some Functions 34

Chapter 4. Properties of $\omega_\varphi^r(f,t)_p$ 36

4.1. Extending the Basic Properties of the Classical Moduli ... 36
4.2. Optimal Rate of $\omega_\varphi^r(f,t)$ 40
4.3. Marchaud Inequality....................................... 43

Chapter 5. More General Step-Weight Functions φ 46

5.1. Logarithmic-Type Weights and Internal Zeros 46
5.2. The Necessity of the Finite Overlapping Condition 47
5.3. Growth Order of Type x^β with Arbitrary β 49

Chapter 6. Weighted Moduli of Smoothness...................... 55

6.1. Weighted Moduli of Smoothness and Weighted K-Functionals 55
6.2. The Weighted Main-Part Modulus 59
6.3. Smoothness Properties of Derivatives 62
6.4. Marchaud Inequality for Weighted Main-Part Moduli 67
6.5. Connection with Ordinary Weighted Moduli..................... 71

PART II. APPLICATIONS

Chapter 7. Algebraic Polynomial Approximation 77

7.1. Background.. 77
7.2. Best Polynomial Approximation 79
7.3. Asymptotic Behavior of Derivatives of Best Approximating Polynomials. 84
7.4. Error Bounds for Gaussian Quadrature......................... 87

Chapter 8. Weighted Best Polynomial Approximation 90

8.1. Some Concepts and Description of the Weight 90
8.2. Best Weighted Algebraic Polynomial Approximation 94
8.3. Derivatives of the Optimal Polynomials........................ 98
8.4. Proof of Some Crucial Inequalities for $w \in J_p^*$ 100
8.5. Applications, Calculations, and Specific Examples 109

Chapter 9. Exponential-Type or Bernstein-Type Operators 112

9.1. Background and Notations, Positive Operators on $C(D)$ 112
9.2. Operators on $L_p(D)$, Higher Degree of Smoothness 115
9.3. Direct and Converse Results 117
9.4. The Bernstein-Type Inequality $\|\varphi^{2r}L_n^{(2r)}f\|_p \leq Mn^r\|f\|_p$ 124
9.5. Rate of Convergence for Smooth Functions...................... 134
9.6. Estimate of $\|L_n(R_{2r}(f, \cdot, x), x)\|_{L_p(E_n)}$ 140
9.7. The Estimate $\|\varphi(x)^{2r}L_n^{(2r)}(f)\|_{L_p} \leq M\|\varphi^{2r}f^{(2r)}\|_p$ 152

Chapter 10. Weighted Approximations by Exponential-Type Operators 158

10.1. The Direct and Inverse Result 158
10.2. The Boundedness of the Operators in Weighted Norm.............. 161
10.3. Bernstein-Type Inequality 165
10.4. The Estimate $\|w\varphi^2 L_n^{(2)}(f)\| \leq C(\|w\varphi^2 f^{(2)}\| + \|f\|)$ 168
10.5. The Estimate of $L_n f - f$ for Smooth Functions................... 168
10.6. The Saturation Result 175

Chapter 11. Weighted Polynomial Approximation in $L_p(R)$ 180

11.1. Introduction 180
11.2. The Equivalence Result 181
11.3. The Direct and Converse Results 184
11.4. Proof of the Equivalence Result 186
11.5. Comparisons and Generalizations 195

Chapter 12. Polynomial Approximation in Several Variables 197

12.1. Approximation on Cubes 197
12.2. Approximation on Polytopes 201

Chapter 13. Comparisons and Conclusions 211

13.1. Comparison with Similar Expressions 211
13.2. The Integral Modulus of Smoothness of Ivanov and Sendov 212
13.3. Moduli Generated by Multipliers and Integral Transforms 213
13.4. A Modulus Introduced by Potapov 214
13.5. Hoeffding's Result 215
13.6. Conclusion 216

Appendix 217

A. The Analogue of Definition 5.3.1 217
B. The Definition of the Weighted Modulus of Smoothness on $(0, 1)$. 218

References 219

List of Symbols 226

Index 227

INTRODUCTION

The aim of this book is to introduce and apply a new "natural" modulus of smoothness. This will be a measure of smoothness that will provide us with a better tool to deal with the rate of best approximation, inverse theorems and imbedding theorems. The classical modulus of smoothness

$$\omega^r(f,t) = \sup_{0<h\le t} \|\Delta_h^r f\|$$

has proved to be very useful for problems of the above type. However, in recent years some shortcomings of the classical modulus and the need for a new modulus for measuring smoothness have become evident. In answer to this we suggest our modulus given by

$$\omega_\varphi^r(f,t)_p \equiv \sup_{0<h\le t} \|\Delta_{h\varphi}^r f\|_{L_p}, \tag{1}$$

where the function $\varphi(x)$ and the interval in question will be related to the problem at hand. We should point out that a vital feature of (1) is that the increment $h\varphi(x)$ varies with x. For $\varphi(x) \equiv 1$, (1) is reduced to the classical modulus.

We now outline some recent developments and problems about best polynomial approximation, some special operators on $[0,1]$ or R_+, and characterization of K-functionals, and indicate how natural and simple it is to express the solution to these problems when we use (1). These developments and the possibility of solving many of the natural problems related to them were the motivating force behind this book.

It was observed by S.M. Nikolski that for functions in $C[-1,1]$ the rate of polynomial approximation implied by a condition on $\omega^r(f,t)$ ($\omega^r(f,t) = O(t^\alpha)$ for example) is faster toward the end points of $[-1,1]$. In fact, it was proved that $\omega^r(f,t) = O(t^\alpha)$ $0 < \alpha < r$ (in $C[-1,1]$) is equivalent to the existence of a sequence $\{P_n\}$, of polynomials of degree n, that satisfies

$$|f(x) - P_n(x)| \le K(n^{-1}\sqrt{1-x^2} + n^{-2})^\alpha, \qquad x\in[-1,1] \qquad (2)$$

(see Brudnyi [1], Dzjadyk [1], Freud [1], and Timan [1–2]). It was pointed out by V.P. Motornii [1] and R. DeVore [2] that this type of characterization of $\omega^r(f,t)$ would not be valid in $L_p[-1,1]$, $p < \infty$, that is

$$\inf_{P\in\Pi_n} \|(\sqrt{1-x^2} + n^{-1})^{-\alpha}(f - P_n)(x)\|_{L_p[-1,1]} = O(n^{-\alpha})$$

does not characterize $\omega^r(f,t)_p = O(t^\alpha)$. We believe that the popularity of the Timan-type estimate (2) diverted attention from the natural, and in our opinion more important, problem of characterizing $E_n(f)_p = O(n^{-\alpha})$, where $E_n(f)_p$ is the best nth degree polynomial approximation to f in $L_p[-1,1]$, using smoothness properties of f. This problem was solved by K.G. Ivanov in [1–7], using a modification of a modulus developed and studied extensively by several Bulgarian mathematicians. The problem was also treated by P.L. Butzer, R.L. Stens, and M. Wehrens [2] and M.K. Potapov [2–4] who obtained a solution via generalized translations such as

$$(\tau_h f)(x) = \frac{1}{\pi} \int_{-1}^{1} f(xh + u\sqrt{1-x^2}\sqrt{1-h^2})(1-u^2)^{-1/2}\,du$$

for $x, h \in [-1,1]$

to define the second modulus of smoothness, and proceeded from there.

Now, however, with the aid of (1) we will prove for $0 < \alpha < r$, $1 \le p \le \infty$ and $\varphi(x) = (1-x^2)^{1/2}$ that

$$E_n(f)_p = O(n^{-\alpha}) \quad \text{if and only if} \quad \omega_\varphi^r(f,t)_p = O(t^\alpha). \qquad (3)$$

This result will be extended to include weighted polynomial approximation with Jacobi and other weights. The main advantage is having a simple instrument, that is $\omega_\varphi^r(f,t)_p$, for characterizing the particular class of functions for which more smoothness is required inside the interval than near the points ± 1. Moreover, as we will see, for many functions the behavior of $\omega_\varphi^r(f,t)_p$ is easily computable.

The second set of problems that illustrates the need for the new modulus is that of various classical linear approximation processes on a finite interval or on the half line. For the special but important operator given by the Bernstein polynomials $B_n(f,x)$, H. Berens and G.G. Lorentz [1] and R. DeVore [1, Chapter 8] proved for for $\alpha \le 2$,

$$\omega^2(f,h) = O(h^\alpha) \quad \Leftrightarrow \quad |B_n(f,x) - f(x)| \le M\left(\frac{x(1-x)}{n}\right)^{\alpha/2}.$$

While elegant, results of this type cannot be used for combinations of Bernstein polynomials that would relate their convergence to higher levels of smoothness (Ditzian [1, p. 279]), and the method is inappropriate for the characterization of

Introduction 3

$$\|B_n f - f\|_{C[0,1]} = O(n^{-\alpha}). \tag{4}$$

For this latter and more natural characterization problem it was shown (see Ditzian [1], [3]) that for $\alpha \le 1$, (4) is equivalent to

$$\omega_\varphi^2(f,t)_p = O(t^{2\alpha}), \qquad \varphi(x) = (x(1-x))^{1/2}, \tag{5}$$

with $p = \infty$. The main deficiency of relating the rate of convergence to the classical modulus of continuity manifests itself most clearly when dealing with the L_p norm. For the L_p variant of the Bernstein polynomials, that is, the Kantorovich polynomials $B_n^*(f,x)$, a characterization of the class of functions satisfying

$$\|B_n^*(f) - f\|_{L_p[0,1]} = O(n^{-\alpha})$$

was established (Totik [4], [5], [7], and [8]), and in fact it is (5). From these studies it became obvious that a new modulus is needed to solve some basic problems in approximation theory. In Chapters 9 and 10 we will show the relation between the rate of convergence of many operators and their combinations, with the modulus in (1).

The third problem that illustrates the importance of (1) is related to interpolation spaces and is in fact the problem of characterization of K-functionals. The K-functional for an L_p space and a corresponding weighted Sobolev space is given by

$$K_{r,\varphi}(f,t^r)_p \equiv \inf_{g^{(r-1)} \in \text{A.C.}_{\text{loc}}} \{\|f - g\|_{L_p(I)} + t^r \|\varphi^r g^{(r)}\|_{L_p(I)}\}. \tag{6}$$

The characterization of the K-functionals which were introduced by J. Peetre for investigation of interpolation spaces between two Banach spaces and which is a fascinating activity in itself, turned out to be very helpful in approximation theory as predicted by J. Peetre [1]. The present K-functional for some class of functions φ was already investigated by Z. Ditzian [3], where a characterization was given as well as a conjecture for a simpler one. That conjecture was settled for $r = 1$ and 2 by V. Totik [7], [14] and will be settled here for all r. It is known that for $\varphi(x) \equiv 1$ the K-functional is equivalent to the regular modulus of smoothness (Johnen and Scherer [1]), but a zero of φ (a singularity) has an inherent effect on the characterization. The study of the K-functional of (6) will be an important part of this book and is essentially the second face of our modulus.

At this point the reader may say "The expressions $\omega_\varphi^r(f,t)_p$, $K_{r,\varphi}(f,t^r)_p$, $E_n(f)_p$, and $\|f - B_n^*(f)\|_p$ may be quite simple and their relations nice, but given a function f, how do any (or all) of these expressions behave?" For many functions f we will be able to compute (see Sections 3.4 and 8.5) the behavior of these expressions quite easily (in a few lines) using theorems on the main-part modulus (to be introduced in Chapter 3) and thus answer the above question. For example, if $f(x) = x^\delta |\log 2x|^\gamma$ for $x \in [0,1]$ and $\delta > -(1/p)$, then for $\varphi(x) = \sqrt{x(1-x)}$ and r large enough ($r > 2\delta + (2/p)$) we have

$$K_{r,\varphi}(f,t^r)_p \sim \omega_\varphi^r(f,t)_p \sim \begin{cases} t^{2\delta+(2/p)}(\log 1/t)^\gamma & \text{if } \delta \neq 0, 1, \dots \\ t^{2\delta+(2/p)}(\log 1/t)^{\gamma-1} & \text{if } \delta = 0, 1, \dots \text{ but } \gamma \neq 0, \end{cases}$$

and therefore for

$$E_n(f)_{L_p[0,1]} \equiv \inf_{P \in \Pi_n} \|f - P\|_{L_p[0,1]}$$

we obtain

$$E_n(f)_{L_p[0,1]} \sim \begin{cases} n^{-2\delta-(2/p)}(\log n)^\gamma & \text{if } \delta \neq 0, 1, \dots \\ n^{-2\delta-(2/p)}(\log n)^{\gamma-1} & \text{if } \delta = 0, 1, \dots \text{ but } \gamma \neq 0. \end{cases}$$

For f as above and $-(1/p) < \delta < 1 - (1/p)$ such computation implies

$$E_{[\sqrt{n}]}(f)_{L_p[0,1]} \sim \|B_n^*(f) - f\|_{L_p[0,1]} \sim \omega_\varphi^2(f, 1/\sqrt{n})_p.$$

The text consists of two parts and thirteen chapters. The investigation in Part I will include: the equivalence relation of ω_φ^r with the K-functional (6) (Chapter 2), the introduction of the main-part modulus and its relation to ω_φ^r (Chapter 3), the extension of all important properties of the classical modulus ω^r to ω_φ^r (Chapters 4 and 6) and the discussion of weighted moduli (Chapter 6). Part II is devoted to applications. These will include: the characterization of best polynomial approximation on $[-1, 1]$ with or without a weight function (Chapters 7 and 8), the characterization of the rate of convergence of various operators (Chapters 9 and 10), the characterization of best weighted polynomial approximation on R (Chapter 11), and the characterization of best polynomial approximation on simple polytopes (Chapter 12).

In the last chapter we shall discuss and compare our modulus with related expressions, and we hope that its relative simplicity and the possibility to actually calculate it for many functions will convince the reader that this is the "correct" measure of smoothness on an interval. (It includes, of course, the classical modulus of continuity as a special case.)

It is our goal in this book to stay closer to applications rather than to investigate an abstract modulus. We believe that both parts of our book are equally important. It is through the different applications that reasons for general properties and for conditions under which they are proved are illuminated.

An index (l.m.n) for a formula or l.m.n for a theorem or remark will indicate that it belongs to Chapter l, Section m in that chapter and is the nth such (formula etc.) in that section. Section l.m will indicate Section m of Chapter l.

A list of symbols and an index are given at the end.

This book was written as a research paper, but it has outgrown this description in size. However, it has retained the characteristics of a research work and all the results presented are new.

PART I
THE MODULUS OF SMOOTHNESS

CHAPTER 1
PRELIMINARIES

1.1. Notations

In this chapter we will introduce some of the notations and the technical setup of our book. The rth symmetric difference of a function f is given by

$$\Delta_h^1 f(x) = \Delta_h f(x) = f\left(x + \frac{h}{2}\right) - f\left(x - \frac{h}{2}\right),$$
$$\Delta_h^r f(x) = \Delta_h(\Delta_h^{r-1} f(x)), \tag{1.1.1}$$

or equivalently by

$$\Delta_h^r f(x) = \sum_{k=0}^{r} (-1)^k \binom{r}{k} f(x + r(h/2) - kh). \tag{1.1.2}$$

The forward and backward rth differences are defined similarly by

$$\vec{\Delta}_h^r f(x) = \sum_{k=0}^{r} (-1)^k \binom{r}{k} f(x + (r-k)h) \tag{1.1.3}$$

and

$$\overleftarrow{\Delta}_h^r f(x) = \sum_{k=0}^{r} (-1)^k \binom{r}{k} f(x - kh) \tag{1.1.4}$$

respectively.

From this point on, when we discuss a space of functions on $D = (a, b)$ (possibly infinite) and either of the points $x \pm rh/2$ does not belong to (a, b) we write $\Delta_h^r f(x) = 0$. We use the same convention for $\vec{\Delta}_h^r f(x)$ and $\overleftarrow{\Delta}_h^r f(x)$, but here these expressions will be defined as 0 when $[x, x + rh]$ or $[x - rh, x]$ are not inside the interval (a, b) respectively.

Since a linear substitution maps (a, b) onto $I = (0, 1)$, $R_+ = (0, \infty)$, or $R = (-\infty, \infty)$, it is enough to consider these invervals. In most cases we will treat only the case R_+ as it will exhibit the problems caused by a finite endpoint and by the nonfiniteness of the interval. The spaces which we will consider are $L_p(D)$ $1 \le p \le \infty$ and $C(D)$, $C(D) \subset L_\infty(D)$ where D is I, R_+ or R. We will use the abbreviation L_p for $L_p(I)$, $L_p(R_+)$, or $L_p(R)$ whenever the domain is clear from the context.

We call ψ_1 and ψ_2 of the same order: $\psi_1 \sim \psi_2$, if and only if there exists a positive constant M such that $M^{-1}\psi_1 \le \psi_2 \le M\psi_1$. Whenever we say there exists a constant, we will mean a strictly positive constant. As a rule the constants will be independent of the functions and variables.

A property is said to hold locally in D if for every $x \in D$ there is some neighborhood of x in which it is satisfied.

1.2. Discussion of Some Conditions on $\varphi(x)$

The function $\varphi(x)$ in the investigation of our moduli of smoothness

$$\omega_\varphi^r(f, t)_p \equiv \sup_{0 < h \le t} \|\Delta_{h\varphi(x)}^r f(x)\|_{L_p(D)}$$

is defined for $x \in D$ (recall $D = (a, b)$, where $a = -\infty$ or $a = 0$ and $b = 1$ or $b = \infty$) and will be assumed, except in Chapter 5, to satisfy the following conditions

I. $\varphi \sim 1$ locally, that is, for every proper finite subinterval $[a', b'] \subset D$ there exists a constant $M \equiv M(a', b') > 0$ such that $M^{-1} \le \varphi(x) \le M$ for $x \in [a', b']$.

II. There are two numbers $\beta(a)$ and $\beta(b)$ (where $a = -\infty$ or $a = 0$, $b = 1$ or $b = \infty$) satisfying $\beta(0) \ge 0$, $\beta(1) \ge 0$ and $\beta(\pm\infty) \le 1$ for which

$$\varphi(x) \sim \begin{cases} |x|^{\beta(a)} & \text{as } x \to a+ \quad (a = 0 \text{ or } a = -\infty) \\ x^{\beta(\infty)} & \text{as } x \to \infty \quad (b = \infty) \\ (1-x)^{\beta(1)} & \text{as } x \to 1- \quad (b = 1). \end{cases}$$

III. $\varphi(x)$ is measurable and there exist constants M_0 and h_0 such that for each $0 < h \le h_0$ and every finite interval $E \subset D$

$$\text{meas}\{x : x \pm h\varphi(x) \in E, x \in D\} \le M_0 \text{ meas } E.$$

It should be emphasized that if $\beta(0) \ge 0$, $\beta(1) \ge 0$ or $\beta(\pm\infty) \le 1$ in II are not satisfied, then the behavior of the type described in II is incompatible with Condition III.

In Condition I we excluded zeros of φ inside D as a matter of convenience, as otherwise we would have to split D into subdomains in the interior of which Condition I is satisfied. Condition II will be relaxed in Chapter 5 to include logarithmic singularities. However, Condition III, which probably looks the least natural at first glance, is crucial and an example will be given in Section 5.2, showing that in the absence of Condition III $\omega_\varphi^r(f,h)_p$ can be unbounded or can have no meaning at all, while the corresponding K-functional is well-behaved. Sometimes, we will refer to Condition III as the "finite overlapping condition." Condition III guarantees for $0 < h < h_0$ the measurability of $f(x \pm h\varphi(x))$ from the measurability of f as well as the inequality

$$\int_{\{x:\, x \pm h\varphi(x) \in (a,b)\}} |f(x \pm h\varphi(x))|\, dx \leq K \int_a^b |f(x)|\, dx \quad \text{for } f \in L_1(a,b). \quad (1.2.1)$$

For $C(D)$ Condition III is not necessary.

1.3. Examples of Various Step-Weight Functions $\varphi(x)$

In the applications we will usually encounter the following step-weight functions φ:

$$\varphi(x) = (1 - x^2)^{1/2} \quad \text{for } D = (-1, 1),$$

$$\varphi(x) = (x(1-x))^{1/2} \quad \text{and} \quad \varphi(x) = \sqrt{x}(1-x) \quad \text{for } D = (0,1),$$

$$\varphi(x) = x^{1/2}, \quad \varphi(x) = (x(1+x))^{1/2} \quad \text{and} \quad \varphi(x) = x \quad \text{for } D = (0, \infty).$$

These are related to polynomial and operator approximation (see Chapters 7–10). We can get a somewhat more general example with the aid of a boundedly differentiable function ψ that is bounded away from zero and infinity as follows: $\psi \in C^2(a,b)$, where $(a,b) = D$, and

$$\varphi(x) = \begin{cases} |x|^{\beta(a)} \psi(x) & \text{as } x \to a+ \quad (a = 0 \text{ or } a = -\infty) \\ (1-x)^{\beta(1)} \psi(x) & \text{as } x \to 1- \quad (b = 1) \\ x^{\beta(\infty)} \psi(x) & \text{as } x \to \infty \quad (b = \infty). \end{cases}$$

Of course, we must have here $\beta(0) \geq 0$, $\beta(1) \geq 0$, and $\beta(\pm\infty) \leq 1$. We now mention a discontinuous step-weight function φ on R_+. Let φ_1 be any weight function mentioned above and let $\varphi(x) = \varphi_1(t_k)$ for $x \in [t_{k+1}, t_k]$, $k = 0$, $\pm 1, \ldots$, where the monotone sequence $\{t_k\}_{k=-\infty}^\infty$ satisfies the conditions

$$t_k \sim t_{k+1} \quad (t_k - t_{k+1}) \sim t_k.$$

The first characterization of the K-functional for $p < \infty$ of the interpolation pair $L_p(a,b)$ and a weighted Sobolev space used the modulus $\omega_\varphi^r(f,t)_p$ with weight φ of this type (see Ditzian [3]).

CHAPTER 2
THE *K*-FUNCTIONAL AND THE MODULUS OF CONTINUITY

After the introduction and preliminaries we are now in the "heart of the matter." In this chapter the equivalence between our modulus of continuity and a certain Peetre *K*-functional will be proved. This will be the starting point for almost every other chapter, some in content and some in technique. The connection with the *K*-functional is important for establishing properties of the modulus as well as for demonstrating its applicability.

2.1. The Equivalence Theorem

For a positive integer r the K-functional of the pair of spaces $L_p(a,b)$, $1 \le p \le \infty$, and a corresponding weighted Sobolev space with weight function φ^r is given by

$$K_{r,\varphi}(f,t^r)_p = \inf_g \{\|f - g\|_p + t^r \|\varphi^r g^{(r)}\|_p; g^{(r-1)} \in \text{A.C.}_{\cdot\text{loc}}\}, \quad (2.1.1)$$

where $g^{(r-1)} \in \text{A.C.}_{\cdot\text{loc}}$ means that g is $r-1$ times differentiable and $g^{(r-1)}$ is absolutely continuous in every closed finite interval $[c,d]$ such that $[c,d] \subset (a,b) = D$.

For various K-functionals probably the most important problem is that of characterizing their behavior using structural properties of the functions. This was done for $K_{r,\varphi}(f,t^r)_p$ of (2.1.1) by Z. Ditzian [3], with less general φ and more complicated characterization. Here for the characterization we will use the modulus of smoothness

$$\omega_\varphi^r(f,t)_p = \sup_{0 < h \le t} \|\Delta_{h\varphi}^r f\|_p \quad (2.1.2)$$

2.1. The Equivalence Theorem

with the same parameters r, φ, p, and t as in (2.1.1). This will yield a simpler description of the K-functional and settle positively the conjecture of Z. Ditzian [3] which was partially solved by V. Totik [7].

We found out that the conjecture was also settled by X.L. Zhou [1] who used a nice identity of a combinatorial nature to derive the result from the paper in which it was posed (Ditzian [3]). The equivalence theorem below, however, is more general and contains no second term which makes it neater and more useful for applications.

Remarks. (a) (2.1.1) and (2.1.2) f is assumed to be locally in L_p; this assumption is made implicitly throughout the paper.

(b) We recall that $\Delta^r_{h\varphi(x)} f(x) = 0$ if $x + rh\varphi(x)/2$ or $x - rh\varphi(x)/2$ is not in the domain D of $L_p(D)$.

It is clear that the K-functional (2.1.1) is defined for a wider class of φ than $\omega^r_\varphi(f, t)_p$. For the r-modulus of smoothness (2.1.2) one has to guarantee that $\Delta^r_{h\varphi(x)} f(x)$ given by

$$\Delta^r_{h\varphi(x)} f(x) \equiv \sum_{k=0}^{r} (-1)^k \binom{r}{k} f\left(x + \left(\frac{r}{2} - k\right) h\varphi(x)\right) \quad (2.1.3)$$

is measurable, and then that $\|\Delta^r_{h\varphi} f\|_p$ is finite, at least for $h < h_0$. (It may happen that $\|\Delta^r_{h\varphi} f\|_p$ is not finite while $f \in L_p$ and φ are measurable.) Conditions I, II, and III on φ given in Section 1.2 will ensure both measurability and finiteness of the norm for $h \le h_1$ ($h_1 = h_0/r$ where h_0 is given in III of Section 1.2). Often, and with no loss of generality, we will assume that $\Delta^r_{h\varphi} f$ is measurable and of finite L_p norm for $0 < h \le 1$. (This represents a modification in φ, but no real change in the proofs.) The main result of this chapter and one of the main results of the whole book is the following.

Theorem 2.1.1. *Suppose φ satisfies Conditions I, II, and III of Section 1.2, r is a positive integer, $f \in L_p(D)$ locally where $D = (0, 1)$, R_+, or R, and $1 \le p \le \infty$. Then*

$$M^{-1} \omega^r_\varphi(f, t)_p \le K_{r,\varphi}(f, t^r)_p \le M \omega^r_\varphi(f, t)_p, \quad 0 < t \le t_0 \quad (2.1.4)$$

for some constants $M > 0$ and t_0.

This result (and its proof) is valid also if $C(D)$ (the space of bounded continuous functions on D) replaces $L_\infty(D)$. In this case D may also be $[0, 1]$, $(0, 1]$, $[0, 1)$, or $[0, \infty)$.

The two inequalities given in (2.1.4) have separate and distinctly different proofs. In Sections 2.2 and 2.3 we will prove the upper estimate and in Section 2.4 the lower estimate for the K-functional.

2.2. The Upper Estimate, $K_{r,\varphi}(f,t^r)_p \leq M\omega^r_\varphi(f,t)_p$, Case I

In this section we will lay the foundation for the proof of the upper estimate in all cases, but will explicitly prove it only for $D = R_+$, $\beta(0) \geq 1$ (cf. Section 1.2) and $1 \leq p < \infty$, which we call Case I.

Thus, let $D = R_+$, $\beta(0) \geq 1$ and $1 \leq p < \infty$. We will need the following simple lemma.

Lemma 2.2.1. *Suppose φ and φ_1 satisfy Conditions I–III of Section 1.2 and for $x \in (\xi_1, \xi_2)$ $0 < A_1^{-1}\varphi(x) \leq \varphi_1(x) \leq \varphi(x)$. Then we have for $t > 0$*

$$\int_0^t \int_{\xi_1}^{\xi_2} |\Delta^r_{\tau\varphi_1(x)} f(x)|^p \, dx \, d\tau \leq A_1 \int_0^t \int_{\xi_1}^{\xi_2} |\Delta^r_{\tau\varphi(x)} f(x)|^p \, dx \, d\tau \qquad (2.2.1)$$

and also

$$\int_0^t \int_{\xi_1}^{\xi_2} |\bar{\Delta}^r_{\tau\varphi_1(x)} f(x)|^p \, dx \, d\tau \leq A_1 \int_0^t \int_{\xi_1}^{\xi_2} |\bar{\Delta}^r_{\tau\varphi(x)} f(x)|^p \, dx \, d\tau, \qquad (2.2.2)$$

where $\bar{\Delta}$ denotes either forward or backward differences.

PROOF. We observe that

$$\int_0^t |\Delta^r_{\tau\varphi_1(x)} f(x)|^p \, d\tau = \frac{1}{\varphi_1(x)} \int_0^{t\varphi_1(x)} |\Delta^r_u f(x)|^p \, du \leq \frac{A_1}{\varphi(x)} \int_0^{t\varphi(x)} |\Delta^r_u f(x)|^p \, du$$

$$= A_1 \int_0^t |\Delta^r_{\tau\varphi(x)} f(x)|^p \, d\tau$$

and apply Fubini's theorem. The same proof is valid for (2.2.2). $\square$

We will be aided by the functions $\psi_k(x)$ given by

$$\psi_k(x) \equiv \psi(4^k x), \quad k = 0, \pm 1, \ldots, \quad \text{where}$$

$$\psi(x) = \begin{cases} 1, & x \leq 1, \\ 0, & 3 \leq x, \end{cases} \quad \psi(x) \in C^\infty(R). \qquad (2.2.3)$$

In fact we can choose ψ decreasing. We define $\omega^{*r}_\varphi(f,t)_p$ by

$$\omega^{*r}_\varphi(f,t)_p \equiv \left\{ \frac{1}{t} \int_0^t \int_0^\infty |\Delta^r_{\tau\varphi(x)} f(x)|^p \, dx \, d\tau \right\}^{1/p}, \qquad (2.2.4)$$

and prove

$$K_{r,\varphi}(f,t^r)_p \leq M\omega^{*r}_\varphi(f,t)_p \leq M\omega^r_\varphi(f,t)_p. \qquad (2.2.5)$$

As

$$\omega^{*r}_\varphi(f,t)_p \leq \left(\frac{1}{t} \int_0^t \{\omega^r_\varphi(f,\tau)_p\}^p \, d\tau \right)^{1/p} \leq \omega^r_\varphi(f,t)_p,$$

2.2. The Upper Estimate, $K_{r,\varphi}(f,t^r)_p \leq M\omega^r_\varphi(f,t)_p$, Case I

only the first inequality of (2.2.5) has to be proved. We define

$$f_\tau(x) = r^r \int_0^{1/r} \cdots \int_0^{1/r} \left(\sum_{l=1}^r (-1)^{l+1}\binom{r}{l}f(x + l\tau(u_1 + \cdots + u_r))\right)du_1\cdots du_r \quad (2.2.6)$$

and

$$g_\tau(x) = \sum_{k=-\infty}^\infty f_{\tau c\varphi(4^{-k})}\psi_{k-1}(x)(1-\psi_k(x)), \quad (2.2.7)$$

where $c > 0$ is small enough to satisfy

$$c\varphi(4^{-k}) < \inf[\varphi(x) | 4^{-k-1} \leq x \leq 4^{-k+3}] \quad (2.2.8)$$

and

$$\frac{r}{2}c\varphi(4^{-k}) \leq 4^{-k+1} \quad (2.2.9)$$

for all k. The inequality (2.2.9) can be satisfied as $\beta(0) \geq 1$ and $\beta(\infty) \leq 1$ here, which implies $\varphi(x) = O(x)$ for $x \to 0+$ and $x \to \infty$. The function $G_t(x)$ given by

$$G_t(x) = \frac{2}{t}\int_{t/2}^t g_\tau(x)\,d\tau$$

will provide a constructive proof for

$$K_{r,\varphi}(f,t^r)_p \leq M\omega^{*r}_\varphi(f,t)_p,$$

namely we will show that

(a) $\|f - G_t\|_p \leq \dfrac{M}{2}\omega^{*r}_\varphi(f,t)_p$ and (b) $\|\varphi^r G_t^{(r)}\|_p \leq \dfrac{M}{2}t^{-r}\omega^{*r}_\varphi(f,t)_p.$

With the notation

$$F_{t,k}(x) = \frac{2}{t}\int_{t/2}^t f_{\tau c\varphi(4^{-k})}(x)\,d\tau,$$

we have

$$G_t(x) = \sum_{k=-\infty}^\infty \psi_{k-1}(x)(1-\psi_k(x))F_{t,k}(x).$$

To prove (a) we write

$$I_k = \int_{4^{-k}}^{4^{-k+2}} |f(x) - F_{t,k}(x)|^p\,dx = \int_{4^{-k}}^{4^{-k+2}} \left|\frac{2}{t}\int_{t/2}^t (f(x) - f_{\tau c\varphi(4^{-k})}(x))\,d\tau\right|^p dx$$

$$= \int_{4^{-k}}^{4^{-k+2}} \left|\frac{2}{t}\int_{t/2}^t r^r \int_0^{1/r}\cdots\int_0^{1/r} \vec{\Delta}^r_{\tau c\varphi(4^{-k})(u_1+\cdots+u_r)}f(x)\,du_1\cdots du_r\,d\tau\right|^p dx.$$

14 2. The K-Functional and the Modulus of Continuity

We set $u = u_1 + \cdots + u_r$, observe that for any u the $r - 1$-dimensional volume of $\{u = u_1 + \cdots + u_r\} \cap \{0 \leq u_i \leq 1/r, 1 \leq i \leq r\}$ is bounded by a constant $B(r)$, and write $B_r \equiv B(r)r^r$. We now have

$$I_k \leq B_r \int_{4^{-k}}^{4^{-k+2}} \frac{2}{t} \int_{t/2}^{t} \int_0^1 |\vec{\Delta}^r_{\tau u c \varphi(4^{-k})} f(x)|^p \, du \, d\tau \, dx$$

$$\leq 2B_r \int_{4^{-k}}^{4^{-k+2}} \frac{1}{t} \int_0^t |\vec{\Delta}^r_{w c \varphi(4^{-k})} f(x)|^p \, dw \, dx$$

$$= 2B_r \int_{4^{-k}}^{4^{-k+2}} \frac{1}{t} \int_0^t |\Delta^r_{w c \varphi(4^{-k})} f(x + \frac{r}{2} w c \varphi(4^{-k}))|^p \, dw \, dx.$$

Changing the order of integration, and recalling that the choice of c in (2.2.9) implies $(r/2)c\varphi(4^{-k}) \leq 4^{-k+1}$, we enlarge the domain of integration of x to compensate for dropping $+(r/2)wc\varphi(4^{-k})$, and obtain by Lemma 2.2.1

$$\int_{4^{-k}}^{4^{-k+2}} |f(x) - F_{t,k}(x)|^p \, dx \leq 2B_r \frac{1}{t} \int_0^t \int_{4^{-k}}^{4^{-k+3}} |\Delta^r_{w c \varphi(4^{-k})} f(x)|^p \, dx \, dw$$
$$\leq B \frac{1}{t} \int_0^t \int_{4^{-k}}^{4^{-k+3}} |\Delta^r_{\tau\varphi(x)} f(x)|^p \, dx \, d\tau. \tag{2.2.10}$$

Since for every x the product $\psi_{k-1}(x)(1 - \psi_k(x))$ can differ from zero for at most two k, $0 \leq \psi_{k-1}(x)(1 - \psi_k(x)) \leq 1$ and $\sum_k \psi_{k-1}(x)(1 - \psi_k(x)) = 1$ for all x, and $\psi_{k-1}(x)(1 - \psi_k(x)) \neq 0$ only in $[4^{-k}, 3 \cdot 4^{-k+1}]$, we have

$$\|f - G_t\|_p^p \leq 2^p \sum_{k=-\infty}^{\infty} \int_0^{\infty} (\psi_{k-1}(x)(1 - \psi_k(x)))^p |f(x) - F_{t,k}(x)|^p \, dx$$

$$\leq 2^p \sum_{k=-\infty}^{\infty} \int_{4^{-k}}^{4^{-k+2}} |f(x) - F_{t,k}(x)|^p \, dx$$

$$\leq 2^p B \left\{ \sum_{k=-\infty}^{\infty} \frac{1}{t} \int_0^t \int_{4^{-k}}^{4^{-k+3}} |\Delta^r_{\tau\varphi(x)} f(x)|^p \, dx \, d\tau \right\} \leq 6B 2^p \{\omega_\varphi^{*r}(f, t)\}^p,$$

from which (a) follows.

We will now prove (b). The definition of $f_\tau(x)$ in (2.2.6) implies

$$\frac{d^r}{dx^r} f_\tau(x) = r^r \sum_{l=1}^{r} (-1)^{l+1} \binom{r}{l} (l\tau)^{-r} \vec{\Delta}^r_{l\tau/r} f(x),$$

and therefore, for

$$F_{t,k}(x) = \frac{2}{t} \int_{t/2}^{t} f_{\tau c \varphi(4^{-k})}(x) \, d\tau$$

we have

2.2. The Upper Estimate, $K_{r,\varphi}(f,t^r)_p \le M\omega_\varphi^r(f,t)_p$, Case I

$$\int_{4^{-k}}^{4^{-k+2}} \left| \frac{d^r}{dx^r} F_{t,k}(x) \right|^p dx$$

$$\le Ct^{-rp}(\varphi(4^{-k}))^{-rp} \sum_{l=1}^{r} \frac{2}{t} \int_{t/2}^{t} \int_{4^{-k}}^{4^{-k+2}} |\vec{\Delta}_{lt c\varphi(4^{-k})/r}^r f(x)|^p \, dx \, d\tau$$

$$\le Ct^{-rp}(\varphi(4^{-k}))^{-rp} \sum_{l=1}^{r} \frac{2}{t} \int_{t/2}^{t} \int_{4^{-k}}^{4^{-k+3}} |\Delta_{lt c\varphi(4^{-k})/r}^r f(x)|^p \, dx \, d\tau.$$

As $lt c\varphi(4^{-k})/r \le c\varphi(4^{-k}) \le 4^{-k+1}$, we may use Lemma 2.2.1 with $\varphi_1(x) = lc\varphi(4^{-k})/r$ for which $C_1^{-1}\varphi(x) \le \varphi_1(x) \le \varphi(x)$ in $[4^{-k}, 4^{-k+3}]$ (see (2.2.8)) and obtain

$$\int_{4^{-k}}^{4^{-k-2}} \left| \frac{d^r}{dx^r} F_{t,k}(x) \right|^p dx \tag{2.2.11}$$

$$\le C_1 t^{-rp}(\varphi(4^{-k}))^{-rp} \frac{1}{t} \int_0^t \int_{4^{-k}}^{4^{-k+3}} |\Delta_{\tau\varphi(x)}^r f(x)|^p \, dx \, d\tau.$$

In the sum representing $G_t(x)$ at most two terms can be different from 0, and in fact

$$G_t(x) = F_{t,k}(x) + \psi_k(x)(F_{t,k+1}(x) - F_{t,k}(x)) \quad \text{for } 4^{-k} \le x \le 4^{-k+1} \tag{2.2.12}$$

and therefore for $4^{-k} \le x \le 4^{-k+1}$,

$$G_t^{(r)}(x) = F_{t,k}^{(r)}(x) + \sum_{j=0}^{r} \binom{r}{j} \psi_k^{(j)}(x)(F_{t,k+1}(x) - F_{t,k}(x))^{(r-j)}. \tag{2.2.13}$$

We now use for $0 < j < r$ the inequality

$$\|g^{(j)}\|_{L_p[c,d]} \le M[(d-c)^{-j}\|g(x)\|_{L_p[c,d]} + (d-c)^{r-j}\|g^{(r)}(x)\|_{L_p[c,d]}] \tag{2.2.14}$$

with M depending only on r (see Ditzian [3, p. 310]) with $c = 4^{-k}$, $d = 4^{-k+1}$, and $g(x) = F_{t,k+1}(x) - F_{t,k}(x)$, and observe that $|\psi_k^{(j)}(x)| \le M_1 4^{kj}$ to obtain

$$t^{rp} \int_{4^{-k}}^{4^{-k+1}} |\varphi(x)^r G_t^{(r)}(x)|^p \, dx$$

$$\le M_2 t^{rp}(\varphi(4^{-k}))^{rp} \Bigg\{ \int_{4^{-k}}^{4^{-k+1}} |F_{t,k}^{(r)}(x)|^p \, dx$$

$$+ \sum_{j=0}^{r} 4^{kjp} 4^{k(r-j)p} \int_{4^{-k}}^{4^{-k+1}} |F_{t,k+1}(x) - F_{t,k}(x)|^p \, dx$$

$$+ \sum_{j=0}^{r} 4^{kjp} 4^{-kjp} \int_{4^{-k}}^{4^{-k+1}} [|F_{t,k+1}^{(r)}(x)|^p + |F_{t,k}^{(r)}(x)|^p] \, dx \Bigg\}.$$

Using (2.2.10), we have

$$\int_{4^{-k}}^{4^{-k+1}} |F_{t,k+1}(x) - F_{t,k}(x)|^p \, dx$$

$$\leq 2^p \left\{ \int_{4^{-k}}^{4^{-k+1}} |f(x) - F_{t,k+1}(x)|^p \, dx + \int_{4^{-k}}^{4^{-k+1}} |f(x) - F_{t,k}(x)|^p \, dx \right\}$$

$$\leq M_3 \frac{1}{t} \int_0^t \int_{4^{-k}}^{4^{-k+4}} |\Delta^r_{\tau\varphi(x)} f(x)|^p \, dx \, d\tau.$$

The above estimates, together with (2.2.11), yield

$$t^{rp} \int_{4^{-k}}^{4^{-k+1}} |\varphi(x)^r G_t^{(r)}(x)|^p \, dx$$

$$\leq C[1 + (t\varphi(4^{-k})4^k)^{rp}] \frac{1}{t} \int_0^t \int_{4^{-k}}^{4^{-k+4}} |\Delta^r_{\tau\varphi(x)} f(x)|^p \, dx \, d\tau.$$

Since in the case treated here $t\varphi(4^{-k})4^k$ is uniformly bounded for $0 \leq t \leq 1$ and $k = 0, \pm 1, \ldots$, we have

$$t^{rp} \int_{4^{-k}}^{4^{-k+1}} |\varphi(x)^r G_t^{(r)}(x)|^p \, dx \leq C \frac{1}{t} \int_0^t \int_{4^{-k}}^{4^{-k+4}} |\Delta^r_{\tau\varphi(x)} f(x)|^p \, dx \, d\tau. \quad (2.2.15)$$

From (2.2.15) we can easily derive (b), that is, $\|\varphi^r G_t^{(r)}\| \leq (M/2) t^{-r} \omega_\varphi^{*r}(f, t)$, which completes the proof of Case I. $\square$

2.3. The Upper Estimate of the K-Functional, The Other Cases

We still have to deal with the estimate $K_{r,\varphi}(f, t^r)_p \leq M \omega_\varphi^r(f, t)_p$ in several other cases. These cases are:

Case II. $D = R_+$, f is locally in $L_p(R_+)$, $1 \leq p < \infty$, and $\beta(0) < 1$.
Case III. $D = (0,1)$, f is locally in $L_p(0,1)$, $1 \leq p < \infty$, $\beta(0) \geq 0$ and $\beta(1) \geq 0$.
Case IV. $D = R$, f is locally in $L_p(R)$, $1 \leq p < \infty$, $\beta(-\infty) \leq 1$ and $\beta(\infty) \leq 1$.
Case V. D, f, and β are as in Cases I–IV but with $p = \infty$.

Of course many of the obstacles were already encountered in proving the estimate for Case I. Most of the remaining difficulties will be treated in proving Case II and only a relatively short discussion will be dedicated to the other cases.

Case II. $f \in L_p(R_+)$ $1 \leq p < \infty$, and $\beta \equiv \beta(0) < 1 (\beta(\infty) \leq 1)$. Unlike Case I, here $x - (r/2)h\varphi(x)$ becomes negative as $x \to 0+$, in which case the construction of Section 2.2 fails. We recall that $A_1 x^\beta \leq \varphi(x) \leq A x^\beta$, $\beta = \beta(0)$, for $0 \leq x \leq 1/2$ and define

2.3. The Upper Estimate of the K-Functional, The Other Cases

$$\omega_\varphi^{*r}(f,C,t)_p \equiv \left\{\frac{1}{t}\int_0^t \int_{Ct^*}^\infty |\Delta_{\tau\varphi(x)}^r f(x)|^p \, dx \, d\tau\right\}^{1/p}$$
$$+ \left\{\frac{1}{t^*}\int_{Lt^*/4r}^{Lt^*}\int_0^{Nt^*} |\Delta_u^r f(x)|^p \, dx \, du\right\}^{1/p}, \quad (2.3.1)$$

where $C \geq 1$ (arbitrary), $t^* = (rA)^{1/1-\beta}t^{1/1-\beta}$, $L = (A_1/Ar8^\beta)^{1/1-\beta}$, and $N = 12C + r$. We will complete the proof of the upper estimate of the K-functional when we show that

$$K_{r,\varphi}(f,t^r)_p \leq M\omega_\varphi^{*r}(f,C,t)_p \quad (2.3.2)$$

and then that

$$\omega_\varphi^{*r}(f,C,t)_p \leq M\omega_\varphi^{*r}(f,t)_p, \quad (2.3.3)$$

where

$$\omega_\varphi^{*r}(f,t)_p \equiv \left\{\frac{1}{t}\int_0^t \int_0^\infty |\Delta_{\tau\varphi(x)}^r f(x)|^p \, dx \, d\tau\right\}^{1/p},$$

as defined by (2.2.4) in Section 2.2.

We now explain heuristically the choices of t^*, C, N and L in (2.3.1). To separate the behavior near 0 (second term) from that away from 0 (first term), we chose $t^* = (Ar)^{1/1-\beta}t^{1/1-\beta}$ which implies for $0 < h \leq t < t_0$ and $x > Ct^* \geq t^*$ the inequality $x - (r/2)h\varphi(x) \geq x/2$. (For the present proof $C = 1$ is sufficient; the use of different C will be helpful later in Section 3.3.) The choice of N is necessary for the proof of (2.3.2) so that there is sufficient overlapping between (Ct^*, ∞) and $(0, Nt^*)$. The choice of L (so that for $x > Lt^*/8$ we have $Lt^*/\varphi(x) \leq t$) will be used to prove (2.3.3). With different L the second term on the right of (2.3.1) would have been bounded by $M\omega_\varphi^{*r}(f, Bt)$ (and that difficulty should and could have been overcome later). Finally we mention that any choice $t^* = Bt^{1/1-\beta}$ with appropriate modification of L would work both here and later.

We will prove (2.3.2) first. We define $l \equiv l(Ct^*)$ by

$$l = \max\{k; 4^{-k} \geq Ct^*\}. \quad (2.3.4)$$

In the definition of

$$F_{t,k}(x) \equiv \frac{2}{t}\int_{t/2}^t f_{\tau c \varphi(4^{-k})}(x) \, d\tau$$

(where f_τ is given by (2.2.6)) c is restricted by (2.2.8) and (instead of (2.2.9)) by

$$\frac{r}{2}tc\varphi(4^{-k}) \leq 4^{-k} \quad \text{for } k \leq l. \quad (2.3.5)$$

This latter inequality is possible because of the choice of l. We define G_t with the aid of $\psi_k(x)$ given in (2.2.3) as

$$G_t(x) = \sum_{k=-\infty}^{l} \psi_{k-1}(x)(1-\psi_k(x))F_{t,k}(x)$$
$$+ \psi_l(x)\frac{2}{Lt^*}\int_{Lt^*/2}^{Lt^*} \tilde{f}_v(x)\,dv \quad (2.3.6)$$
$$\equiv G_{t,1}(x) + G_{t,2}(x),$$

where $\tilde{f}_v(x)$ is given by

$$\tilde{f}_v(x) = (2r)^r \int_{1/2r}^{1/r} \cdots \int_{1/2r}^{1/r} \sum_{j=1}^{r}(-1)^{j+1}\binom{r}{j} \quad (2.3.7)$$
$$\times f(x+jv(u_1+\cdots+u_r))\,du_1\cdots du_r.$$

Using the same calculation as in Section 2.2 (where we have to use (2.3.5) instead of (2.2.9)), we obtain

$$\|f - G_{t,1}\|_{L_p(3\cdot 4^{-l},\infty)} \leq M\left\{\frac{1}{t}\int_0^t \int_{Ct^*}^{\infty} |\Delta^r_{\tau\varphi(x)}f(x)|^p\,dx\,d\tau\right\}^{1/p} \quad (2.3.8)$$

$$t^r\|\varphi^r G_{t,1}^{(r)}\|_{L_p(3\cdot 4^{-l},\infty)} \leq M\left\{\frac{1}{t}\int_0^t \int_{Ct^*}^{\infty} |\Delta^r_{\tau\varphi(x)}f(x)|^p\,dx\,d\tau\right\}^{1/p}. \quad (2.3.9)$$

We also have the estimate for

$$\|f - F_{t,l}\|_{L_p[4^{-l},4^{-l+2}]}^p \quad \text{and} \quad \left\|\left(\frac{d}{dx}\right)^r F_{t,l}(x)\right\|_{L_p[4^{-l},4^{-l+2}]}^p$$

given in (2.2.10) and (2.2.11) with $k = l$, respectively. Near zero we now estimate

$$\int_0^{3\cdot 4^{-l}}\left|f(x) - \frac{2}{Lt^*}\int_{Lt^*/2}^{Lt^*}\tilde{f}_v(x)\,dv\right|^p dx$$

$$\leq \int_0^{3\cdot 4^{-l}} \frac{2}{Lt^*}\int_{Lt^*/2}^{Lt^*}(2r)^r \int_{1/2r}^{1/r}\cdots\int_{1/2r}^{1/r}|\vec{\Delta}^r_{v(u_1+\cdots+u_r)}f(x)|^p\,du_1\cdots du_r\,dv\,dx$$

$$\leq M\int_0^{3\cdot 4^{-l}}\frac{2}{Lt^*}\int_{Lt^*/2}^{Lt^*}\int_{1/2}^{1}|\vec{\Delta}^r_{vu}f(x)|^p\,du\,dv\,dx$$

$$\leq M_1\int_0^{3\cdot 4^{-l}}\frac{1}{t^*}\int_{Lt^*/4}^{Lt^*}|\vec{\Delta}^r_w f(x)|^p\,dw\,dx$$

$$\leq M_1\int_0^{(12C+r)t^*}\frac{1}{t^*}\int_{Lt^*/4}^{Lt^*}|\Delta^r_w f(x)|^p\,dw\,dx.$$

We can also write

2.3. The Upper Estimate of the K-Functional, The Other Cases

$$t^{rp} \int_0^{3\cdot 4^{-1}} (\varphi(x))^{rp} \left| \left(\frac{d}{dx}\right)^r \frac{2}{Lt^*} \int_{Lt^*/2}^{Lt^*} \tilde{f}_v(x) \, dv \right|^p dx$$

$$\leq M t^{rp} \varphi(Ct^*)^{rp} \int_0^{3\cdot 4^{-1}} \sum_{j=1}^r \frac{2}{Lt^*} \int_{Lt^*/2}^{Lt^*} v^{-rp} \left| \vec{\Delta}_{jv/2r}^r f\left(x + \frac{jv}{2}\right) \right|^p dv \, dx \quad (2.3.10)$$

$$\leq M_1 t^{rp} \varphi(Ct^*)^{rp} (t^*)^{-rp} \int_0^{3\cdot 4^{-1}+rLt^*} \frac{2}{Lt^*} \int_{Lt^*/4r}^{Lt^*} |\Delta_v^r f(x)|^p \, dv \, dx$$

$$\leq M_2 \int_0^{Nt^*} \frac{1}{t^*} \int_{Lt^*/4r}^{Lt^*} |\Delta_v^r f(x)|^p \, dv \, dx,$$

where, in the last step, $t\varphi(t^*) \sim t^*$, $3 \cdot 4^{-1} \leq 12Ct^*$, $L \leq 1$ and $N = 12C + r$ were used. To conclude the proof of (2.3.2), we need the procedure used at the end of Section 2.2 (see (2.2.12), (2.2.13), and (2.2.14)) to deal with $G_t(x)$ in $[4^{-1}, 3 \cdot 4^{-1}]$ which is represented there by

$$G_t(x) = F_{t,l}(x) + \psi_l(x) \left\{ \left(\frac{2}{Lt^*} \int_{Lt^*/2}^{Lt^*} \tilde{f}_v(x) \, dv \right) - F_{t,l}(x) \right\};$$

but the necessary estimates have already been proved (including (2.2.10) and (2.2.11) for $F_{t,l}(x)$).

It remains to show (2.3.3) to complete the proof of Case II. Using $A_1 x^\beta \leq \varphi(x) \leq A x^\beta$ for $0 \leq x \leq 1/2$ and $Lt^*/\varphi(x) \leq t$ for $1 \geq x > Lt^*/8$, which follows from

$$\frac{Lt^*}{\varphi(x)} \leq \frac{Lt^*}{A_1 x^\beta} \leq \frac{(Lt^*)^{1-\beta}}{A_1} 8^\beta \leq 8^\beta A_1^{-1} L^{1-\beta} (rA) t$$

with our choice $L = (A_1/Ar8^\beta)^{1/1-\beta}$, we can write (recalling $\Delta_u^r f(x) = 0$ if $x < ru/2$)

$$\frac{1}{t^*} \int_{Lt^*/4r}^{Lt^*} \int_0^{Nt^*} |\Delta_u^r f(x)|^p \, dx \, du = \frac{1}{t^*} \int_{Lt^*/4r}^{Lt^*} \int_{ru/2}^{Nt^*} |\Delta_u^r f(x)|^p \, dx \, du$$

$$\leq \int_{Lt^*/8}^{Nt^*} \frac{1}{t^*} \int_{Lt^*/4r}^{Lt^*} |\Delta_u^r f(x)|^p \, du \, dx$$

$$= M \int_{Lt^*/8}^{Nt^*} \frac{\varphi(x)}{t^*} \int_{Lt^*/4r\varphi(x)}^{Lt^*/\varphi(x)} |\Delta_{\tau\varphi(x)}^r f(x)|^p \, d\tau \, dx$$

$$\leq M_1 \int_{Lt^*/8}^{Nt^*} \frac{1}{(t^*)^{1-\beta}} \int_0^t |\Delta_{\tau\varphi(x)}^r f(x)|^p \, d\tau \, dx$$

$$\leq M_2 \int_0^\infty \frac{1}{t} \int_0^t |\Delta_{\tau\varphi(x)}^r f(x)|^p \, d\tau \, dx.$$

Case III. To prove the upper estimate for Case III we will not need to introduce any new idea but some minor changes; for instance, instead of

$t_k = 4^k$, $k = 0, \pm 1, \ldots$, we choose any decreasing sequence satisfying

$$t_k - t_{k+1} \sim t_k \sim t_{k+1} \quad \text{for } k \geq 0;$$
$$t_k - t_{k+1} \sim 1 - t_k \sim 1 - t_{k+1} \quad \text{for } k < 0. \qquad (2.3.11)$$

For $D = (-1, 1)$, which corresponds to the important problem of best polynomial approximation, we could choose, for example, $t_k = -1 + 4^{-k}$ for $k > 0$, $1 - 4^k$ for $k < 0$, and define $\psi_k(x) = \psi(4^k(x + 1))$ for $k > 0$, $\psi_k(x) = \psi(4^{-k}(1 - x))$ for $k < 0$, and $\psi_0(x) = 1 - \psi_1(x) - \psi_{-1}(x)$, where $\psi(x)$ is given in (2.2.3). A linear transformation from $(-1, 1)$ to $(0, 1)$ (or direct choice) will yield an analogous construction for $D = (0, 1)$. With these $\psi_k(x)$ and t^* defined appropriately (when $\beta(0) < 1$ or $\beta(1) < 1$ or both), we can follow Case I or Case II to prove Case III.

Case IV. Dealing with $L_p(R)$, we will again have a change in $\psi_k(x)$; for negative k it remains as in (2.2.3), for $k > 0$ $\psi_k(x) = \psi_{-k}(-x)$, and $\psi_0(x) = 1 - \psi_1(x) - \psi_{-1}(x)$. The rest of the proof follows Case I and in fact this is the easiest situation.

Case V. In all the above cases we used $p < \infty$. This was not a real restriction, but simply a matter of convenience so that we can raise the various expressions to the power p. For $p = \infty$ or $C(D)$, we take the supremum of the expressions in question on $[4^{-k}, 4^{-k+2}]$ for Cases I and II and on $[0, 3 \cdot 4^{-l}]$ for Case II near 0. This was done explicitly in Ditzian [3,§3] for the case treated there and it is easier than the L_p case above. We could have stated after every step in the proofs of Cases I and II what would happen if $p = \infty$, but those proofs were long and complicated enough as they were.

2.4. The Lower Estimate for the K-Functional

In this section we shall show $\omega_\varphi^r(f, t) \leq M K_{r,\varphi}(f, t^r)_p$ for some $t_0 > 0$ and $0 < t < t_0$. We will consider only the case $L_p(R_+)$ as we will encounter and overcome all the obstacles already in that case. Let g_t be chosen so that

$$\|f - g_t\|_p \leq 2K_{r,\varphi}(f, t^r)_p \quad \text{and} \quad t^r\|\varphi^r g_t^{(r)}\|_p \leq 2K_{r,\varphi}(f, t^r)_p \qquad (2.4.1)$$

are satisfied. We estimate $\omega_\varphi^r(f, t)$ using

$$\omega_\varphi^r(f, t)_p \leq \omega_\varphi^r(f - g_t, t)_p + \omega_\varphi^r(g_t, t)_p. \qquad (2.4.2)$$

It is enough to show that for small t and $0 < h \leq t$

$$\|\Delta_{h\varphi}^r(f - g_t)\|_p \leq C\|f - g_t\|_p, \qquad (2.4.3)$$

and that

2.4. The Lower Estimate for the K-Functional

$$\|\Delta^r_{h\varphi} g_t\|_p \leq Ch^r \|\varphi^r g_t^{(r)}\|_p, \qquad (2.4.4)$$

where C is independent of f, h, and t. To prove (2.4.3) we just have to show for $v = (r/2) - j$ $j = 0, \ldots, r$ that

$$\int |f(x + vh\varphi(x))|^p\, dx \leq M \int |f(x)|^p\, dx$$

with the understanding that if $x + vh\varphi(x) \notin R_+$, then $f(x + vh\varphi(x)) = 0$. However, the last inequality is implied by the finite overlapping property, that is, Condition III of Section 1.2. (see (1.2.1)).

For $x \geq rh\varphi(x)/2$, $D = R_+$ we have

$$|\Delta^r_{h\varphi(x)} g_t(x)| = \left| \int_{-h\varphi(x)/2}^{h\varphi(x)/2} \cdots \int_{-h\varphi(x)/2}^{h\varphi(x)/2} g_t^{(r)}(x + u_1 + \cdots + u_r)\, du_1 \cdots du_r \right|$$

$$\leq M \int_{-rh\varphi(x)/2}^{rh\varphi(x)/2} ((rh\varphi(x)/2) - |u|)^{r-1} |g_t^{(r)}(x + u)|\, du, \qquad (2.4.5)$$

otherwise the left hand side is (defined to be) zero. We shall now show that for $x > (r/2)h\varphi(x) \equiv \eta\varphi(x)$ one has

$$\int_{-\eta\varphi(x)}^{\eta\varphi(x)} [(\eta\varphi(x) - |u|)^{r-1}/\varphi(x + u)^r]\, du \leq M\eta^r. \qquad (2.4.6)$$

Since we have always $(\eta\varphi(x) - |u|) \leq \eta\varphi(x)$, if

$$\frac{x}{2} \leq x - \frac{r}{2} h\varphi(x) \leq x + \frac{r}{2} h\varphi(x) \leq 2x,$$

one can write

$$\int_{-\eta\varphi(x)}^{\eta\varphi(x)} \frac{(\eta\varphi(x) - |u|)^{r-1}}{\varphi(x + u)^r}\, du \leq \eta^{r-1} \int_{-\eta\varphi(x)}^{\eta\varphi(x)} \frac{\varphi(x)^{r-1}}{(\min_{x/2 \leq v \leq 2x} \varphi(v))^r}\, du$$

$$\leq \eta^{r-1} \eta\varphi(x) \varphi(x)^{r-1} \left(M \frac{1}{\varphi(x)^r} \right) \leq M\eta^r.$$

Since we always assume $x > (r/2)h\varphi(x) = \eta\varphi(x)$, we have $x + \eta\varphi(x) < 2x$. Therefore, to use the above estimate we require only $x \geq rh\varphi(x) = 2\eta\varphi(x)$. If $\beta(0) \geq 1$ and $x > 0$ or if $\beta(0) < 1$ and $x \geq 1/2$, the inequality $x \geq 2\eta\varphi(x)$ is satisfied for $0 < h \leq h_0$ with some fixed h_0. To examine the remaining case, i.e., $\beta \equiv \beta(0) < 1$ and $x \in [0, 1/2]$, we recall that $A_1 x^\beta \leq \varphi(x) \leq A x^\beta$ in that interval. If $(rA)^{1/1-\beta} h^{1/1-\beta} \leq x \leq 1/2$, we have again $x - \eta\varphi(x) \geq x/2$. That is, $x - \eta\varphi(x) < x/2$ occurs only for

$$(rA_1/2)^{1/1-\beta} h^{1/1-\beta} < x < (rA)^{1/1-\beta} h^{1/1-\beta},$$

in which case we have $x \sim h^{1/1-\beta}$ and therefore

$$\int_{-\eta\varphi(x)}^{0} + \int_{0}^{\eta\varphi(x)} \frac{(\eta\varphi(x) - |u|)^{r-1}}{\varphi(x+u)^r} du$$

$$\leq M \int_{0}^{x} \frac{v^{r-1}}{\varphi(v)^r} dv + M(\eta\varphi(x))^{r-1} \int_{x}^{x+\eta\varphi(x)} \frac{dv}{\varphi(v)^r}$$

$$\leq M \left\{ \int_{0}^{x} v^{r-\beta r-1} dv \right\} + (\eta\varphi(x))^r \max_{x < v < 2x} \{\varphi(v)^{-r}\}$$

$$\leq M_1 \{(h^{1/1-\beta})^{r(1-\beta)} + \eta^r\} \leq M_2 \eta^r.$$

To estimate $\|\Delta_{h\varphi}^r g_t\|_p$ we now write for $1 \leq p < \infty$, using the Hölder inequality and the above estimate,

$\|\Delta_{h\varphi}^r g_t\|_p^p$

$$\leq M \int_{x \geq rh\varphi(x)/2} \left\{ \int_{-rh\varphi(x)/2}^{rh\varphi(x)/2} \frac{((rh\varphi(x)/2) - |u|)^{r-1}}{\varphi(x+u)^r} \right.$$

$$\times \left. [|\varphi(x+u)^r g_t^{(r)}(x+u)|] du \right\}^p dx$$

$$\leq M_1 (h^r)^{p-1} \int_{x \geq rh\varphi(x)/2} \left\{ \int_{-rh\varphi(x)/2}^{rh\varphi(x)/2} \frac{((rh\varphi(x)/2) - |u|)^{r-1}}{\varphi(x+u)^r} \right.$$

$$\times \left. |\varphi(x+u)^r g_t^{(r)}(x+u)|^p du \right\} dx$$

$$= M_1 h^{r(p-1)} \int_{0}^{\infty} \varphi(v)^{r(p-1)} |g_t^{(r)}(v)|^p \left\{ \int_{E(v,h)} ((rh\varphi(x)/2) - |x-v|)^{r-1} dx \right\} dv$$

$$\equiv M_1 h^{r(p-1)} \int_{0}^{\infty} \varphi(v)^{r(p-1)} |g_t^{(r)}(v)|^p I(v,h) dv,$$

where

$$I(v,h) = \int_{E(v,h)} ((rh\varphi(x)/2) - |x-v|)^{r-1} dx$$

and

$$E(v,h) = \left\{ x; x \geq \frac{r}{2} h\varphi(x), |x-v| \leq \frac{r}{2} h\varphi(x) \right\}.$$

We will now show that $I(v,h) \leq M_2 h^r \varphi(v)^r$ and that will complete the proof of (2.4.4). We split the integral $I(v,h)$ into two integrals $I_1(v,h)$ and $I_2(v,h)$ over the domains $E_1(v,h) = E(v,h) \cap \{x; v/2 \leq x \leq 2v\}$ and $E_2(v,h) = E(v,h) \cap E_1(v,h)^c$, respectively. To estimate $I_1(v,h)$, we write

$$I_1(v,h) \leq \max_{x \in E_1(v,h)} \left(\frac{r}{2} h\varphi(x) \right)^{r-1} \text{meas } E_1(v,h).$$

2.4. The Lower Estimate for the K-Functional

Clearly

$$\max_{x \in E_1(v,h)} \varphi(x) \le \max_{v/2 \le x \le 2v} \varphi(x) \le C\varphi(v)$$

and hence

$$\operatorname{meas} E_1(v,h) = \operatorname{meas}\left\{x; \frac{v}{2} < x < 2v, 0 \le x - \frac{r}{2}h\varphi(x) \le v \le x + \frac{r}{2}h\varphi(x)\right\}$$

$$\le \operatorname{meas}\left\{x; x - \frac{r}{2}hC\varphi(v) \le v \le x + \frac{r}{2}hC\varphi(v)\right\} = rhC\varphi(v),$$

and therefore $I_1(v,h) \le Mh^r\varphi(v)^r$. Since $x \in E(v,h)$ implies $x \ge v/2$, we have that $x \in E_2(v,h)$ implies $x - rh\varphi(x)/2 < v < x/2$, and therefore, as discussed earlier, $x \le (rA)^{1/1-\beta}h^{1/1-\beta}$; otherwise (at least for $h \le h_0$) $E_2(v,h)$ is empty. Now on $E_2(v,h)$ we have $x - (r/2)h\varphi(x) \le v < x/2$ and

$$I_2(v,h) = \int_{E_2(v,h)} ((rh\varphi(x)/2) - |x-v|)^{r-1}\,dx \le \int_{E_2(v,h)} v^{r-1}\,dx$$

$$= v^{r-1}\operatorname{meas} E_2(v,h) \le v^{r-1}\operatorname{meas}\left\{x: 0 < x - \frac{rh}{2}\varphi(x) < v\right\} \le Mv^r$$

$$\le M_1\varphi(v)^r v^{r(1-\beta)} \le M_2(\varphi(v))^r x^{r(1-\beta)} \le M_3 h^r \varphi(v)^r,$$

where for the third inequality we used Condition III of φ from Section 1.2. For $p = \infty$ the result is an immediate consequence of (2.4.5) and (2.4.6).

CHAPTER 3
K-FUNCTIONALS AND MODULI OF SMOOTHNESS, OTHER FORMS

The modification of the K-functional, use of forward (and backward) differences, and the introduction of the main-part modulus are the subject of this chapter. As a result of Theorem 3.3.2 about the main-part modulus, we will show in Section 3.4 the ease with which one can determine the behavior of $\omega_\varphi^r(f,t)_p$ for many functions.

3.1. A Modified K-Functional

In applications and in particular for polynomials of best approximation we will need the following theorem.

Theorem 3.1.1. *Suppose* $f \in L_p[0,1]$, $1 \le p \le \infty$, $\varphi(x) = \sqrt{x(1-x)}$ *and* $r \ge 1$, *and* $\bar{K}_{r,\varphi}(f,t^r)_p$ *is given by*

$$\bar{K}_{r,\varphi}(f,t^r)_p \equiv \inf_{g^{(r-1)} \in \text{A.C.}} \{\|f-g\|_p + t^r\|\varphi^r g^{(r)}\|_p + t^{2r}\|g^{(r)}\|_p\}. \quad (3.1.1)$$

Then there exist constants M *and* t_0 *such that*

$$M^{-1}\omega_\varphi^r(f,t)_p \le \bar{K}_{r,\varphi}(f,t^r)_p \le M\omega_\varphi^r(f,t)_p \quad \text{for } 0 < t < t_0. \quad (3.1.2)$$

Somewhat more general and not harder to prove (but not endowed with many additional applications) is the following:

Theorem 3.1.2. *Suppose* $f \in L_p[0,1]$, *for* $\varphi(x)$ *we have* $\beta = \beta(0) = \beta(1) < 1$ (*see Section 1.2*) *and* $\bar{K}_{r,\varphi}(f,t^r)_p$ *is given by*

$$\bar{K}_{r,\varphi}(f,t^r)_p = \inf_{g^{(r-1)} \in \text{A.C.}} \{\|f-g\|_p + t^r\|\varphi^r g^{(r)}\|_p + t^{r/1-\beta}\|g^{(r)}\|_p\}. \quad (3.1.3)$$

3.1. A Modified K-Functional   25

Then there exist constants M and t_0 such that

$$M^{-1}\omega_\varphi^r(f,t)_p \leq \bar{K}_{r,\varphi}(f,t^r)_p \leq M\omega_\varphi^r(f,t)_p \quad \text{for } 0 < t < t_0. \tag{3.1.4}$$

In fact, analogues for $L_p(R_+)$ are also easy to state and prove.

PROOF OF THEOREMS 3.1.1 AND 3.1.2. We need to prove only the second inequality in (3.1.2) and (3.1.4) since $\bar{K}_{r,\varphi}(f,t^r)_p \geq K_{r,\varphi}(f,t^r)_p$. We can split the p-power of the third term of $\bar{K}_{r,\varphi}(f,t^r)$ as

$$\|g^{(r)}\|_p^p = \|g^{(r)}\|_{L_p[0,t^*]}^p + \|g^{(r)}\|_{L_p[t^*, 1-t^*]}^p + \|g^{(r)}\|_{L_p[1-t^*, 1]}^p,$$

where $t^* = Bt^{1/1-\beta}$. (We may write $t^* = r^2t^2$ for Theorem 3.1.1 and $t^* = (Ar)^{1/1-\beta}t^{1/1-\beta}$ with A satisfying $\varphi(x) \leq A(x(1-x))^\beta$ for Theorem 3.1.2.) We now have

$$t^{r/(1-\beta)}\|g^{(r)}\|_{L_p[t^*, 1-t^*]} \leq Mt^r\|\varphi^r g^{(r)}\|_{L_p[0,1]}.$$

We will complete the proof when we show that the G_t that was constructed in Section 2.3 satisfies

$$t^{r/(1-\beta)}\|G_t^{(r)}\|_{L_p[0,t^*]} \leq M_1\omega_\varphi^r(f,t)_p$$

(and, of course, a similar situation will occur near 1 when we follow the procedure of Case III in Section 2.3). For $0 \leq x \leq t^*$ we have to deal only with $G_{t,2}(x)$ of (2.3.6), and in fact the proof follows (2.3.10) almost word for word; that is, we write with L, N, C, and l of Section 2.3

$$t^{rp/1-\beta}\|G_{t,2}^{(r)}(x)\|_{L_p[0,t^*]}^p$$

$$\leq t^{rp/1-\beta}\|G_{t,2}^{(r)}(x)\|_{L_p[0,4^{-1}]}^p$$

$$= t^{rp/1-\beta}\int_0^{4^{-1}}\left|\left(\frac{d}{dx}\right)^r\frac{2}{Lt^*}\int_{Lt^*/2}^{Lt^*}\tilde{f}_v(x)\,dv\right|^p dx$$

$$\leq Mt^{rp/1-\beta}\int_0^{4^{-1}}\sum_{j=1}^r\frac{2}{Lt^*}\int_{Lt^*/2}^{Lt^*}v^{-rp}\left|\vec{\Delta}_{jv/2r}^r f\left(x+\frac{jv}{2}\right)\right|^p dv\,dx$$

$$\leq M_1 t^{rp/1-\beta}t^{*-rp}\int_0^{4^{-1}+rLt^*}\frac{2}{Lt^*}\int_{Lt^*/4r}^{Lt^*}|\Delta_v^r f(x)|^p\,dv\,dx$$

$$\leq M_2\int_0^{Nt^*}\frac{1}{t^*}\int_{Lt^*/4r}^{Lt^*}|\Delta_v^r f(x)|^p\,dv\,dx.$$

To complete the proof we recall now that it was proved at the end of Section 2.3, Case II, that the right hand side of the above is bounded by $\omega_\varphi^r(f,t)_p^p$. □

Corollary 3.1.3. For $L_p[0,1]$, $\beta(0) = \beta(1) = \beta < 1$ we have

$$\omega^r(f,t^{1/1-\beta})_p \leq M\omega_\varphi^r(f,t)_p, \tag{3.1.5}$$

where $\omega^r(f,\eta)$ is the classical modulus of smoothness ($\varphi \equiv 1$).

PROOF. Using (3.1.3) and (3.1.4) we observe that

$$K_{r,1}(f, t^{r/1-\beta})_p \leq \bar{K}_{r,\varphi}(f, t^r)_p \leq M\omega_\varphi^r(f, t)_p,$$

but $K_{r,1}(f, t^{r/1-\beta})_p \sim \omega^r(f, t^{1/1-\beta})_p$ which completes the proof. □

3.2. Forward and Backward Differences

In a recent paper, J. Löfstrom [2] gave an elegant proof for the forward difference analogue of the equivalence theorem of Ditzian [3] in the special case when φ is a step function. Though our theorem used symmetric differences, the method will yield similar results on forward or backward differences as well. The present extension is presented for use in the proof of certain properties of the modulus of continuity and for completeness of the investigation of the K-functional.

For the forward or backward differences, $\bar{\Delta}_h^r$, we define

$$\bar{\omega}_\varphi^r(f, t)_p = \sup_{0 < h \leq t} \|\bar{\Delta}_{h\varphi}^r f\|_p \tag{3.2.1}$$

and

$$\bar{\omega}_\varphi^{*r}(f, t)_p = \left\{\frac{1}{t} \int_0^t \int_D |\bar{\Delta}_{\tau\varphi(x)}^r f(x)|^p \, dx \, d\tau\right\}^{1/p}, \quad 1 \leq p < \infty; \tag{3.2.2}$$

$$\bar{\omega}_\varphi^{*r}(f, t)_\infty \equiv \bar{\omega}_\varphi^r(f, t)_\infty,$$

and can state and prove the following theorem.

Theorem 3.2.1. *For $K_{r,\varphi}(f, t^r)_p$ given in (2.1.1) we have*

$$\bar{\omega}_\varphi^r(f, t)_p \sim K_{r,\varphi}(f, t^r)_p \sim \bar{\omega}_r^{*r}(f, t)_p \quad \text{for } 0 < t < t_0. \tag{3.2.3}$$

PROOF. We only need to deal with forward differences, as the case of backward differences is similar. It is enough to show

$$\bar{\omega}_\varphi^r(f, t)_p \leq K_{r,\varphi}(f, t^r)_p \tag{3.2.4}$$

and

$$K_{r,\varphi}(f, t^r)_p \leq K\bar{\omega}_\varphi^{*r}(f, t). \tag{3.2.5}$$

We will deal with two cases.

Case I. $L_p(D)$ with $D = R_+$, $D = R$ or $D = (0, 1)$ and $\beta(1) \geq 1$ (cf. Section 1.2). (If we were dealing with backward differences, Case I would cover $D = R$, $D = R_+$ with $\beta(0) \geq 1$, and $D = (0, 1)$ with $\beta(0) \geq 1$.) In this case the proof is similar to Case I of Theorem 2.1.1 because for $x \in D$ and h sufficiently small (but depending only on D and φ and not on x), $x + rh\varphi(x) \in D$.

3.2. Forward and Backward Differences

Case II. $D = (0, 1)$ and $\beta = \beta(1) < 1$ (for backward differences the corresponding situation is $D = (0, 1)$ with $\beta(0) < 1$ and $D = R_+$ with $\beta(0) < 1$). We have, by assumption, $A^{-1}(1-x)^\beta < \varphi(x) < A(1-x)^\beta$ for $1/2 \le x < 1$. Obviously, the behavior near 0 could be treated as in Case I. When $x > 1/2$ the same type of modified modulus of continuity that worked in Case II, i.e., in Section 2.3 (taking into account that 1 and not 0 is the point in need of special care) will be effective here. That is, we write for $1 \le p < \infty$

$$\vec{\omega}_\varphi^{*r}(f, C, t)_p = \left\{\frac{1}{t}\int_0^t \int_0^{1-C t^*} |\vec{\Delta}_{\tau\varphi(x)}^r f(x)|^p \, dx \, d\tau \right\}^{1/p} \\ + \left\{\frac{1}{t^*}\int_{Lt^*/4r}^{Lt^*} \int_{1-Nt^*}^1 |\Delta_u^r f(x)|^p \, dx \, du \right\}^{1/p}, \quad (3.2.6)$$

where $t^* = (rA)^{1/1-\beta} t^{1/1-\beta}$, $C \ge 1$, $N = 12C + r$, and L small enough (t^* is chosen so that $x + r(h/2)\varphi(x) < 1 - (1-x)/2$ for $0 < h \le t$ and $x < 1 - Ct^*$). For the estimate (3.2.5) we need the inequality

$$K_{r,\varphi}(f, t^r)_p \le \vec{\omega}_\varphi^{*r}(f, C, t)_p \le \vec{\omega}_\varphi^{*r}(f, t)_p. \quad (3.2.7)$$

We concentrate here only on the second inequality of (3.2.7), as the first inequality can be verified by the method of Section 2.3. The estimate of the main part (first part on the right of (3.2.6)) follows Section 2.3. The second part of (3.2.6) is estimated by

$$\frac{1}{t^*}\int_{Lt^*/4r}^{Lt^*}\int_{1-Nt^*}^1 |\Delta_u^r f(x)|^p \, dx \, du$$

$$= \frac{1}{t^*}\int_{Lt^*/4r}^{Lt^*}\int_{1-Nt^*}^{1-ru/2} |\Delta_u^r f(x)|^p \, dx \, du$$

$$\le \frac{1}{t^*}\int_{Lt^*/4r}^{Lt^*}\int_{1-(N+r)t^*}^{1-ru} |\vec{\Delta}_u^r f(x)|^p \, dx \, du$$

$$\le M \int_{1-(N+r)t^*}^{1-Lt^*/4} \frac{1}{t\varphi(x)} \int_0^{t\varphi(x)} |\vec{\Delta}_u^r f(x)|^p \, du \, dx$$

$$\le M \int_0^1 \frac{1}{t} \int_0^t |\vec{\Delta}_{\tau\varphi(x)}^r f(x)|^p \, d\tau \, dx \le M(\vec{\omega}_\varphi^{*r}(f, t))^p$$

(L is chosen so that $Lt^* \le t\varphi(x)$ for $x < 1 - (1/8)Lt^*$).

The proof of (3.2.4) follows Section 2.4 with minor modifications. That is, we write

$$|\vec{\Delta}_{h\varphi(x)}^r g(x)| \le M \int_0^{rh\varphi(x)} \left(rh\varphi(x)/2 - \left|\frac{r}{2}h\varphi(x) - u\right|\right)^{r-1} |g^{(r)}(x + u)| \, du$$

and, using steps that are very similar to those in Section 2.4, we prove that

$$\int_0^{rh\varphi(x)} \frac{\left(rh\varphi(x)/2 - \left|\frac{r}{2}h\varphi(x) - u\right|\right)^{r-1}}{\varphi(x+u)^r} du \leq Mh^r$$

(for x such that $x + rh\varphi(x) \leq 1$ if $D = (0,1)$) and

$$\int_{\substack{x \leq v \leq x+rh\varphi(x) \\ x, x+rh\varphi(x) \in D}} \left(rh\varphi(x)/2 - \left|\frac{r}{2}h\varphi(x) + x - v\right|\right)^{r-1} dx \leq Kh^r(\varphi(v))^r. \quad \square$$

3.3. Main-Part Modulus of Smoothness

The discussion here will be restricted to f in $L_p(0,1)$ or $L_p(R_+)$ where $\beta(c)$ (cf. Section 1.2) related to φ satisfies $\beta(0) < 1$ for $D = R_+$ and either $\beta(0) < 1$ or $\beta(1) < 1$ or both for $D = (0,1)$. In earlier discussions (Ditzian [3], Totik [6]) the modulus of smoothness equivalent to the K-functional had two terms (in most cases), a "main term" which was supported by a subinterval that tended to the whole of D in a way that depended on t and an additional term governing the situation near the endpoint (or points) of D. The same situation occurred in Chapter 2 when we defined the auxiliary modulus $\omega_\varphi^{*r}(f, C, t)_p$ in (2.3.1). The disadvantage of $\omega_\varphi^r(f, t)$ (see (2.1.2)) is that it can be too sensitive to the values of the function near the endpoints. Moreover, the exact domain in which the expression is treated, that is, $\{x | x \pm rh\varphi(x)/2 \in D\}$, is difficult to calculate. The main-part modulus of smoothness to be introduced below will use a domain which is smaller, and in fact, bounded away from the endpoint(s), and chosen in such a way that most of the "essential behavior" is preserved. Consequently, both the domain and the expression of the main-part modulus itself will be substantially easier to compute. We will formally define the main-part modulus of smoothness here and show that it is equivalent to the modulus we discussed earlier if its behavior is $O(t^\alpha)$ for instance. However, the main-part modulus is not always equivalent to our modulus, as will be shown later in this section.

Definition 3.3.1. The main-part modulus of smoothness is given for $L_p(R_+)$ by

$$\Omega_\varphi^r(C, f, t)_p = \sup_{0 < h \leq t} \|\Delta_{h\varphi}^r f\|_{L_p(Ch^*, \infty)}, \quad \Omega_\varphi^r(1, f, t)_p \equiv \Omega_\varphi^r(f, t)_p, \quad (3.3.1)$$

where $A_1 x^{\beta(0)} \leq \varphi(x) \leq A x^{\beta(0)}$ for $0 < x < 1/2$, $h^* = (Ar)^{1/1-\beta(0)} h^{1/1-\beta(0)}$ and $\beta(0) < 1$. (If $\beta(0) \geq 1$, then replace Ch^* by 0). The main-part modulus for $L_p(0,1)$ is given by

$$\Omega_\varphi^r(C, f, t)_p = \sup_{0 \leq h \leq t} \|\Delta_{h\varphi}^r f\|_{L_p(Ch_0^*, 1 - Ch_1^*)}, \quad \Omega_\varphi^r(1, f, t)_p \equiv \Omega_\varphi^r(f, t)_p, \quad (3.3.2)$$

where $A_1 x^{\beta(0)} \leq \varphi(x) \leq A x^{\beta(0)}$ for $0 < x \leq 1/2$, $\beta(0) < 1$ and $h_0^* = (Ar)^{1/1-\beta(0)} h^{1/1-\beta(0)}$; $B_1(1-x)^{\beta(1)} \leq \varphi(x) \leq B(1-x)^{\beta(1)}$ for $1/2 \leq x < 1$,

$\beta(1) < 1$ and $h_1^* = (Br)^{1/1-\beta(1)} h^{1/1-\beta(1)}$. (If $\beta(0) \geq 1$, Ch_0^* is replaced by 0, and if $\beta(1) \geq 1$, $1 - Ch_1^*$ is replaced by 1.)

This definition depends on the choice of A and B but this will not cause any problem.

We are now ready to state and prove the relation between $\Omega_\varphi^r(C,f,t)_p$ and $\omega_\varphi^r(f,t)_p$ (cf. Totik [7, Theorem 2]).

Theorem 3.3.2. *Suppose $f \in L_p[a,b]$ where $(a,b) = D$, $\Omega_\varphi^r(C,f,t)_p$ is given by Definition 3.3.1 and $\omega_\varphi^r(f,t)_p$ is given by (2.1.2). Then we have for any $C \geq 0$ and some $M = M(C)$*

$$\Omega_\varphi^r(C,f,t)_p \leq \omega_\varphi^r(f,t)_p \leq M \int_0^t (\Omega_\varphi^r(C,f,\tau)_p/\tau)\,d\tau \qquad (0 < t < t_0). \quad (3.3.3)$$

Remark 3.3.3. For very small C (near 0) $\omega_\varphi^r(f,t)_p = \Omega_\varphi^r(C,f,t)_p$ and therefore the inequality is interesting only for bigger C. If $\varphi(x) = \sqrt{x(1-x)}$ ($A = 1$) for instance, any $C < 1/4$ would satisfy $\omega_\varphi^r(f,t)_p = \Omega_\varphi^r(C,f,t)_p$ for t small enough.

Corollary 3.3.4. *For $\alpha > 0$, $\omega_\varphi^r(f,t)_p = O(t^\alpha)$ and $\Omega_\varphi^r(C,f,t)_p = O(t^\alpha)$ are equivalent, and also $\omega_\varphi^r(f,t)_p \sim t^\alpha$ and $\Omega_\varphi^r(C,f,t)_p \sim t^\alpha$ are equivalent.*

For the interval $[-1,1]$ and for $\varphi(x) = (1-x^2)^{1/2}$, that will be used for results relating to best polynomial approximation, Corollary 3.3.4 reads:

Corollary 3.3.5. *For $\alpha > 0$ and $\varphi(x) = (1-x^2)^{1/2}$ $\omega_\varphi^r(f,t)_p = O(t^\alpha)$ and*

$$\|\Delta_{h\varphi}^r f\|_{L_p[-1+r^2h^2, 1-r^2h^2]} = O(h^\alpha)$$

are equivalent and also $\omega_\varphi^r(f,t)_p \sim t^\alpha$ and

$$\|\Delta_{h\varphi}^r f\|_{L_p[-1+r^2h^2, 1-r^2h^2]} \sim h^\alpha$$

are equivalent.

The advantages of the main-part modulus are best illustrated by these corollaries. In Corollary 3.3.5, for example, we see that the domain of integration $(-1 + r^2h^2, 1 - r^2h^2)$ replaces (for the situation at hand) the more cumbersome domain $((-1 + (rh/2)^2)/(1 + (rh/2)^2), (1 - (rh/2)^2)/(1 + (rh/2)^2))$ which appears in the definition of $\omega_\varphi^r(f,t)$. Furthermore it is not only the expression of the domain that is more cumbersome but the estimate of $\Delta_{h\varphi}^r f$ and its integration on that domain can cause much difficulty even for quite a simple function f.

PROOF OF THEOREM 3.3.2. Since for $C_1 < C_2$, $\Omega_\varphi^r(C_1,f,t)_p \geq \Omega_\varphi^r(C_2,f,t)_p$, we may assume $C \geq 1$ (or as large as we wish). We will deal only with $L_p(R_+)$, $1 \leq p < \infty$, since the other cases are similar. For the sake of brevity we sup-

30   3. K-Functionals and Moduli of Smoothness, Other Forms

press in part the parameters φ, r, and p and write $\Omega^r_\varphi(C,f,t)_p \equiv \Omega(C,f,t)$ etc. If $\beta(0) \geq 1$, then ω and Ω coincide, so we may assume $0 \leq \beta = \beta(0) < 1$. Let

$$\Omega^*(C,f,t) = \left\{ \frac{1}{t} \int_0^t \int_{Ct^*}^\infty |\Delta^r_{\tau\varphi(x)} f(x)|^p \, dx \, d\tau \right\}^{1/p}, \qquad (3.3.4)$$

and

$$V(t) = V(C,f,t) = \left\{ \frac{1}{t^*} \int_{Lt^*/4r}^{Lt^*} \int_0^{(12C+r)t^*} |\Delta^r_u f(x)|^p \, dx \, du \right\}^{1/p}, \qquad (3.3.5)$$

where $L = (r8^\beta A_1/A)^{1/1-\beta}$ as in Section 2.3. We will show that

$$V(t) = V(C,f,t) \leq M \int_0^t (\Omega^*(C,f,\tau)/\tau) \, d\tau \qquad (3.3.6)$$

which will be the main part of the proof. Using (3.3.6), the main estimate of Section 2.4 and formulas (2.3.1) and (2.3.2), we have

$$\omega^r_\varphi(f,t)_p \leq MK_{r,\varphi}(f,t^r)_p \leq M_2 \omega^{*r}_\varphi(f, C2^{1/1-\beta}, t)_p$$
$$= M_2[\Omega^*(C2^{1/1-\beta},f,t) + V(C2^{1/1-\beta},f,t)]$$
$$\leq M_3 \left\{ \int_0^t (\Omega^*(C2^{1/1-\beta},f,\tau)/\tau) \, d\tau + \Omega^*(C2^{1/1-\beta},f,t) \right\}.$$

Observing that for $C_1 < C_2$ $\Omega^*(C_1,f,t) \geq \Omega^*(C_2,f,t)$, that $\Omega^*(C,f,t) \leq \Omega(C,f,t)$ and that for $t \leq \xi \leq 2t$

$$\Omega^*(C2^{1/1-\beta},f,t) \leq \sup_{\tau \leq t} \left\{ \int_{C2^{1/1-\beta}t^*}^\infty |\Delta^r_{\tau\varphi(x)} f(x)|^p \, dx \right\}^{1/p}$$
$$\leq \sup_{\tau \leq \xi} \left\{ \int_{C\xi^*}^\infty |\Delta^r_{\tau\varphi(x)} f(x)|^p \, dx \right\}^{1/p} \leq \Omega^r_\varphi(C,f,\xi)$$

as $C\xi^* \leq C2^{1/1-\beta}t^*$ and $\xi > t$, we have

$$\omega^r_\varphi(f,t)_p \leq 2M_3 \int_0^{2t} (\Omega^r_\varphi(C,f,\tau)_p/\tau) \, d\tau.$$

Now as

$$\omega^r_\varphi(f,t)_p \sim K_{r,\varphi}(f,t^r)_p \sim K_{r,\varphi}(f,(t/2)^r)_p \sim \omega^r_\varphi\left(f, \frac{t}{2}\right)_p,$$

we have

$$\omega^r_\varphi(f,t)_p \leq M_4 \int_0^t (\Omega^r_\varphi(C,f,\tau)_p/\tau) \, d\tau.$$

It remains to prove (3.3.6) for which it is enough to show that

$$V(t) \leq V(t/2^{1-\beta}) + R\Omega^*(C,f,tB) \qquad (3.3.7)$$

for some R depending on β and r but not on f and t, and some $B < (1/2)^{1-\beta}$.

3.3. Main-Part Modulus of Smoothness

In fact, using $\lim_{m \to \infty} V(t/2^{m(1-\beta)}) = 0$ and (3.3.7), we have

$$V(t) \leq R \sum_{m=0}^{\infty} \Omega^*(C, f, Bt/2^{m(1-\beta)}) \leq R_1 \int_0^t (\Omega_\varphi^r(C, f, \tau)_p/\tau) \, d\tau.$$

To complete the proof we now have to show (3.3.7). We write (see Timan [2, p. 103, (5)])

$$\Delta_u^r f(x) = \sum_{v_1=0}^{1} \cdots \sum_{v_r=0}^{1} \Delta_{u/2}^r f\left(x + \left(v_1 + \cdots + v_r - \frac{r}{2}\right) u/2\right)$$

$$\equiv \sum_{j=0}^{r} a_j \Delta_{u/2}^r f\left(x + \left(j - \frac{r}{2}\right) u/2\right),$$

where the a_j's satisfy $a_0 = 1$, $a_j > 0$ and $\sum_{j=0}^{r} a_j = 2^r$. Consequently, with $N = 12C + r$ we have

$$V(t) = \left\{ \frac{1}{t^*} \int_{Lt^*/4r}^{Lt^*} \int_0^{Nt^*} |\Delta_u^r f(x)|^p \, dx \, du \right\}^{1/p}$$

$$\leq \sum_{j=0}^{r} a_j \left\{ \frac{1}{t^*} \int_{Lt^*/4r}^{Lt^*} \int_{ru/2}^{Nt^*} \left| \Delta_{u/2}^r f\left(x + \left(j - \frac{r}{2}\right) u/2\right) \right|^p dx \, du \right\}^{1/p}$$

$$\equiv \sum_{j=0}^{r} a_j V_j(t).$$

We write for $j > 0$ ($N > rL$)

$$V_j(t) = \left\{ \frac{1}{t^*} \int_{Lt^*/4r}^{Lt^*} \int_{(ru/4)+(ju/2)}^{Nt^*+(ju/2)-(ru/4)} |\Delta_{u/2}^r f(x)|^p \, dx \, du \right\}^{1/p}$$

$$\leq \left\{ \frac{1}{t^*} \int_{Lt^*/4r}^{Lt^*} \int_{(ru/4)+(Lt^*/8r)}^{2Nt^*} |\Delta_{u/2}^r f(x)|^p \, dx \, du \right\}^{1/p} \equiv W(t).$$

For $j = 0$ we have

$$V_0(t) \leq \left\{ \frac{1}{t^*} \int_{Lt^*/4r}^{Lt^*} \int_{ru/4}^{Nt^*/2} |\Delta_{u/2}^r f(x)|^p \, dx \, du \right\}^{1/p}$$

$$+ \left\{ \frac{1}{t^*} \int_{Lt^*/4r}^{Lt^*} \int_{Nt^*/2}^{Nt^*} |\Delta_{u/2}^r f(x)|^p \, dx \, du \right\}^{1/p}$$

$$\leq \left\{ \frac{2}{t^*} \int_{Lt^*/8r}^{Lt^*/2} \int_{r\eta/2}^{Nt^*/2} |\Delta_\eta^r f(x)|^p \, dx \, d\eta \right\}^{1/p} + W(t) = V(t/2^{1-\beta}) + W(t).$$

To complete the proof it is enough to show $2^r W(t) \leq R\Omega(C, f, tB)$ for some $B < 2^{-1+\beta}$ independent of f and t for which we need the following lemma.

Lemma 3.3.6. For $0 < \eta \leq \delta$ and $c \geq b + r\delta$ we have

$$\left\{ \int_a^b |\vec{\Delta}_\eta^r f(x)|^p \, dx \right\}^{1/p} \leq M \inf_{g^{(r-1)} \in A.C.[a,c]} (\|f - g\|_{L_p[a,c]} + \delta^r \|g^{(r)}\|_{L_p[a,c]}),$$

where M depends on p but not on f, a, b, c, or δ.

32   3. K-Functionals and Moduli of Smoothness, Other Forms

This was proved many times before; we state it only to emphasize that in the construction the constant M is independent of many of the parameters.

We now write

$$W(t) \le \sup_{Lt^*/4r \le u < Lt^*} \left\{ \int_{(ru/2)+(Lt^*/8r)}^{2Nt^*} |\Delta_{u/2}^r f(x)|^p \, dx \right\}^{1/p}$$

$$\le \sup_{0 < u \le Lt^*} \left\{ \int_{Lt^*/8r}^{2Nt^*} |\vec{\Delta}_{u/2}^r f(x)|^p \, dx \right\}^{1/p}$$

$$\le M \inf \Big\{ \|f - g\|_{L_p[Lt^*/8r, 3Nt^*]} + (Lt^*)^r \|g^{(r)}\|_{L_p[Lt^*/8r, 3Nt^*]};$$

$$g^{(r-1)} \in \text{A.C.} \left[\frac{Lt^*}{8r}, 3Nt^*\right] \Big\}$$

$$\le M_1 \inf\{\|f-g\|_{L_p[Lt^*/8r, 3Nt^*]} + t^r \|\varphi^r g^{(r)}\|_{L_p[Lt^*/8r, 3Nt^*]}; \quad g^{(r-1)} \in \text{A.C.}\}$$

$$\le M_1 \inf\{\|f-g\|_{L_p[Lt^*/8r, \infty]} + t^r \|\varphi^r g^{(r)}\|_{L_p[Lt^*/8r, \infty]};$$

$$g^{(r-1)} \in \text{A.C.}[Lt^*/8r, b), \forall b < \infty\} \equiv M_1 K(t),$$

where M_1 is independent of t. To prove the last but one inequality we note that for x in the interval $[Lt^*/8r, 3Nt^*]$ we have, recalling $t^* \sim t^{1/1-\beta}$, $B_1 t^{\beta/1-\beta} \le \varphi(x) \le B_2 t^{\beta/1-\beta}$ in that interval and therefore $(t^*/t) \sim \varphi(x)$ there. The construction of $G_{\zeta,1}$ in Section 2.3 with ζ chosen such that

$$C\zeta^* = \frac{Lt^*}{8r} \frac{1}{12}, \quad \zeta = t\left(\frac{L}{Cr\,96}\right)^{1/1-\beta} = t \cdot B$$

($L < 1$, $C \ge 1$, $r \ge 1$ and hence $B < (1/2)^{1/1-\beta}$) implies $K(t) \le M\Omega^*(C, f, tB)$. (This follows from the fact that for the l of (2.3.4) in the construction of $G_{\zeta,1}$ we have $3 \cdot 4^{-l} \le 12Ch^* = Lt^*/8r$ and from the use of (2.3.8) and (2.3.9) with the corresponding ζ.)

For $p = \infty$ the proof is very similar. □

Remark 3.3.7. In Theorem 3.3.2.

$$\tilde{\Omega}_\varphi^r(C, f, t)_p \equiv \sup_{h \le t} \|\Delta_{h\varphi}^r f\|_{L_p[Ct^*, \infty)}$$

can replace $\Omega_\varphi^r(C, f, t)$ and this stronger theorem was proved first, but we believe that the concept of $\Omega_\varphi^r(C, f, t)_p$ is more natural.

We already noted that for small C $\Omega_\varphi^r(C, t, f) = \omega_\varphi^r(f, t)$. This clearly happens for $D = R_+$, $A_1 x^\beta \le \varphi(x) \le A x^\beta$ for $0 < x < 1/2$ and $C < (A_1/2A)^{1/1-\beta}$. For bigger C we have the following result.

Theorem 3.3.8. Suppose $L < C_1 < C_2$ for appropriate $L = L(\varphi)$. Then for some M independent of f and t, we have

$$\Omega_\varphi^r(C_2, f, t) \le \Omega_\varphi^r(C_1, f, t) \le M\Omega_\varphi^r(C_2, f, t). \tag{3.3.8}$$

3.3. Main-Part Modulus of Smoothness

This result will be proved in Theorem 6.2.1 as part of a more general result, and therefore, to avoid duplication, we do not prove it here.

The above discussion seems to imply that $\Omega_\varphi^r(f,t)_p$ is of the same order of magnitude as $\omega_\varphi^r(f,t)_p$. While, as we have already seen, this is correct in many cases, which we hope justifies the name "main-part modulus," it is not always true.

Theorem 3.3.9. Let $1 \leq p \leq \infty$, $r = 2$ and $\varphi(x) = \sqrt{x}$. Then for some $f \in L_p(R_+)$ $\omega_\varphi^2(f,t)_p \neq O(\Omega_\varphi^2(f,t)_p)$.

PROOF. For $p = \infty$ we choose $f(x) = (\log x)^{-1}$ for $0 < x \leq 1/2$, $f(x) = 0$ for $x > 1$ and $f(x) \in C^2$. Choosing $x_0 = t^2$, we can write

$$\omega_\varphi^2(f,t)_\infty \geq |f(x_0 - t\varphi(x_0)) - 2f(x_0) + f(x_0 + t\varphi(x_0))|$$
$$= |-2(\log t^2)^{-1} + (\log 2t^2)^{-1}| \sim |\log t|^{-1}.$$

On the other hand we also have

$$\Omega_\varphi^2(f,t)_\infty \leq \sup_{0 \leq h < t} \sup_{x > 4h^2} |\Delta_{h\sqrt{x}}^2 f(x)| \sim \sup_{\substack{h < t \\ x > 4h^2}} (h\sqrt{x})^2 f''(\xi_x),$$

where $\xi_x \in (x - h\sqrt{x}, x + h\sqrt{x})$ and therefore for $x > 4h^2$ $\xi_x \in (x/2, 2x)$, which implies $\Omega_\varphi^2(f,t) \sim (\log t)^{-2}$.

For $1 \leq p < \infty$ we will construct f so that for any M there exists t such that

$$\omega_\varphi^2(f,t)_p \geq M \sup_{h \leq t} \|\Delta_{h\varphi}^2 f\|_{L_p[4h^2, \infty)}.$$

(In the present situation $t^* = (2t)^2$.) For a function $\Phi(x) \in C^2(R)$ satisfying $\Phi(x) = 1$ for $|x| \leq 1/2$, $\Phi(x) = 0$ for $|x| \geq 1$ and $0 \leq \Phi(x) \leq 1$ otherwise, we define $f_k(x) = M_k \Phi((x - x_k)/\eta_k)$ and $f(x) = \sum_{k=1}^\infty f_k(x)$ with certain sequences $\{M_k\}$, $\{\eta_k\}$, and $\{x_k\}$ to be chosen below. Obviously $|f_k''(x)| \leq K M_k \eta_k^{-2}$ and $|f_k(x)| \leq M_k$. We will choose now x_k, η_k, and M_k inductively. Let $x_1 = 1$, for given x_k we choose $\eta_k = x_k/3$ and $M_k^p = k^{-2} \eta_k^{-1}$; given x_k, η_k, and M_k for $k < m$, we choose x_m so that

$$x_m \sum_{k=1}^{m-1} \eta_k^{1/p} (M_k/\eta_k^2) < m^{-2/p} \quad \text{and} \quad 0 < x_m < x_{m-1}/48.$$

We estimate $\omega_\varphi^2(f,t)_p$ for $t = 2\sqrt{x_m}$. For $x \in [4x_m, 16x_m]$ one has $x - t\sqrt{x} = x - 2\sqrt{x_m}\sqrt{x} \in [0, 8x_m]$ and $x + t\sqrt{x} \in [8x_m, 24x_m]$, but since $f(x) = 0$ in $(4x_m, 24x_m)$, we have for $t = 2\sqrt{x_m}$

$$\omega_{\sqrt{x}}^2(f,t)_p \geq \left\{ \int_{t^2}^\infty |\Delta_{t\sqrt{x}}^2 f(x)|^p \, dx \right\}^{1/p} \geq \left\{ \int_{4x_m}^{16x_m} |\Delta_{t\sqrt{x}}^2 f(x)|^p \, dx \right\}^{1/p}$$
$$= \left\{ \int_{4x_m}^{16x_m} |f(x - t\sqrt{x})|^p \, dx \right\}^{1/p}.$$

Since $m\{x; x - t\sqrt{x} \in I\} > m(I)$ for $t > 0$, we have

$$\omega^2_{\sqrt{x}}(f,t)_p \geq \left\{ \int_{4x_m}^{16x_m} |f(x - t\sqrt{x})|^p \, dx \right\}^{1/p} \geq \left\{ \sum_{k=m}^{\infty} \eta_k M_k^p \right\}^{1/p} \geq m^{-1/p}.$$

For $0 < h < t$ there exists v, $v \geq m$, such that $2\sqrt{x_{v+1}} < h \leq 2\sqrt{x_v}$. For $x > 4h^2$ we have $x - h\sqrt{x} \geq 2h^2 \geq 8x_{v+1} \geq x_{v+1} + \eta_{v+1}$ and therefore

$$\left\{ \int_{4h^2}^{\infty} |\Delta^2_{h\sqrt{x}} f(x)|^p \, dx \right\}^{1/p} \leq \sum_{k=1}^{v} \|\Delta^2_{h\sqrt{x}} f_k(x)\|_p \leq K_1 \left(\sum_{k=1}^{v-1} h^2 \|f_k''\|_p + \|f_v\|_p \right)$$

$$\leq K_2 \left(4x_v \sum_{k=1}^{v-1} \eta_k^{1/p} M_k \cdot \eta_k^{-2} + M_v \eta_v^{1/p} \right)$$

$$\leq K_3 v^{-2/p} \leq K_3 m^{-2/p},$$

where K_i, $i = 1, 2, 3$ are independent of v and m. This concludes the proof for $1 \leq p < \infty$. $\square$

We could have shown the result of Theorem 3.3.9 for other r as well.

3.4. Computation of Our Modulus for Some Functions

In this section we will compute the order of magnitude of our modulus for several functions and utilize the main-part modulus and the results of Section 3.3. We hope these examples will illustrate the advantages of the main-part modulus of smoothness, as well as the computability of $\omega^r_\varphi(f,t)$.

1. $D = (0,1)$, $\varphi(x) = x^{1/2}$ and $f(x) = x^\rho$, $-1/p < \rho < 1 - (1/p)$.

$$\Omega^2_\varphi(f,t)_\infty = \Omega^2_\varphi(1, f, t)_\infty = \sup_{h \leq t} \sup_{x \geq 4t^2} |\Delta^2_{h\sqrt{x}} f(x)|$$

$$= \sup_{h \leq t} \sup_{x \geq 4h^2} h^2 x f''(\xi_x) \sim \sup_{\substack{h \leq t \\ x \geq 4h^2}} x h^2 \xi_x^{\rho-2}$$

$$\sim \sup_{\substack{h \leq t \\ x \geq 4h^2}} h^2 x^{\rho-1} \sim t^{2\rho}.$$

Since $\Omega^2_\varphi(f,t)_\infty \sim t^{2\rho}$, we have, using Corollary 3.3.4, $\omega^2_\varphi(f,t)_\infty \sim t^{2\rho}$.

For $p < \infty$ and $\rho \neq 0$ ($-1/p < \rho < 1 - (1/p)$) we have

$$\Omega^2_{\sqrt{x}}(f,t)_p = \sup_{h \leq t} \left\{ \int_{4h^2} |\Delta^2_{h\sqrt{x}} f(x)|^p \, dx \right\}^{1/p} \sim \sup_{h < t} \left\{ \int_{4h^2} |xh^2 x^{\rho-2}|^p \, dx \right\}^{1/p}$$

$$\sim t^{2\rho + (2/p)}.$$

It is easy to calculate that for $\rho > 1 - (1/p)$, $\rho \neq 1$ $\Omega^2_{\sqrt{x}}(f,t)_p \sim t^2$, but for

$$\rho = 1 - \frac{1}{p} < 1, \quad \Omega^2_{\sqrt{x}}(f,t)_p \sim t^2 |\log t|^{1/p}.$$

3.4. Computation of Our Modulus for Some Functions

Using Theorem 3.3.2, we have

$$\omega^2_{\sqrt{x}}(f,t)_p \sim \Omega^2_{\sqrt{x}}(f,t)_p$$

for

$$\Omega^2_{\sqrt{x}}(f,t)_p \sim t^\alpha \quad \text{or} \quad \Omega^2_{\sqrt{x}}(f,t)_p \sim t^2 |\log t|^\beta.$$

More generally, we can deduce the behavior of $\Omega^r_\varphi(f,t)_p$ from that of the rth derivative of f if the latter exists in the interior of D and, e.g., for $D = R_+$, satisfies there $f^{(r)}(x) \sim f^{(r)}(u)$ whenever $x \sim u$. For example, if in the above, $D = R_+$, we have

$$\Omega^r_\varphi(f,t)_p \sim t^r \|\varphi^r f^{(r)}\|_{L_p[t^*, \infty)}.$$

Using this observation, straightforward calculations yield the following.

2. Let $f(x) = x^\delta |\log x/2|^\gamma$, $\delta > -(1/p)$ and $x \in (0,1)$. Then for $r > 2\delta + (2/p)$

$$f^{(r)}(x) \sim \begin{cases} x^{\delta - r} |\log x/2|^\gamma & \text{for } \delta \neq 0, 1, \ldots \\ x^{\delta - r} |\log x/2|^{\gamma - 1} & \text{for } \delta = 0, 1, \ldots \text{ but } \gamma \neq 0, \end{cases}$$

and therefore, for $\varphi(x) = \sqrt{x(1-x)}$ (or $\varphi(x) = \sqrt{x}$)

$$\Omega^r_\varphi(f,t)_p \sim \begin{cases} t^{2\delta + 2/p} |\log t|^\gamma & \text{for } \delta \neq 0, 1, \ldots \\ t^{2\delta + 2/p} |\log t|^{\gamma - 1} & \text{for } \delta = 0, 1, \ldots \text{ but } \gamma \neq 0. \end{cases}$$

Using Theorem 3.3.2, $\omega^r_\varphi(f,t)_p \sim \Omega^r_\varphi(f,t)_p$ for the above behavior of Ω^r_φ. If $r < 2\delta + (2/p)$ then $\Omega^r_\varphi(f,t)_p \sim t^r$.

3. Let $D = R_+$, $f(x) = x^\rho(1+x)^\gamma$ with $\rho > -1/p$ and $\rho + \gamma < -1/p$ and let $\varphi(x) = x^\beta$, where $0 \leq \beta \leq 1$. Then

$$\omega^2_\varphi(f,t)_p \sim \Omega^2_\varphi(f,t)_p \sim \begin{cases} t^2 & \text{if } \rho + 2(\beta - 1) > -1/p \\ t^2 |\log t|^{1/p} & \text{if } \rho + 2(\beta - 1) = -1/p. \\ t^{(\rho p + 1)/p(1-\beta)} & \text{if } \rho + 2(\beta - 1) < -1/p \end{cases}$$

4. Let $f(x) = \log x$, $D = (0,1)$ and $\varphi(x) = x^\beta$ for $0 \leq \beta \leq 1$. Then for $1 \leq p < \infty$

$$\omega^r_\varphi(f,t)_p \sim \Omega^r_\varphi(f,t)_p \sim \begin{cases} t^r & \text{if } r(1-\beta) < 1/p \\ t^r (\log t)^{1/p} & \text{if } r(1-\beta) = 1/p. \\ t^{1/p(1-\beta)} & \text{if } r(1-\beta) > 1/p \end{cases}$$

5. In contrast with example 1, for $D = (0,1)$, $\varphi(x) = x$ and $f(x) = x^\delta$ where $\delta > -(1/p)$ we have $\omega^r_\varphi(f,t)_p \sim t^r$. (Here the main-part modulus is not needed.)

CHAPTER 4
PROPERTIES OF $\omega_\varphi^r(f,t)_p$

In this chapter we will extend many of the properties of the classical modulus of smoothness to $\omega_\varphi^r(f,t)_p$.

4.1. Extending the Basic Properties of the Classical Moduli

The classical rth modulus of smoothness

$$\omega^r(f,t)_p = \sup_{0<h\le t} \|\Delta_h^r f\|_p$$

was investigated extensively. Here we will outline some of the properties of $\omega^r(f,t)$ and discuss their extension to $\omega_\varphi^r(f,t)_p$. Some of the extensions will be proved in the present section while others will be stated and proved in Sections 4.2 and 4.3 and in Chapter 6.

Probably the most basic fact about $\omega^r(f,t)_p$ is
(a) $\lim_{t\to 0} \omega^r(f,t)_p = 0$ for all $f \in L_p(D)$ if $1 \le p < \infty$, or for all $f \in C$ which satisfy

$$\lim_{\substack{x\to c \\ x\in D}} f(x) = L_c,$$

where c is an endpoint of D (possibly $\pm\infty$) and L_c is finite, if $p = \infty$. For $\varphi(x)$ satisfying the restrictions in Section 1.2 and in particular the finite overlapping property (see (1.2.1)) we have

$$\lim_{t\to 0} \omega_\varphi^r(f,t)_p = 0 \quad \text{for all } f \in L_p(D) \quad \text{if} \quad 1 \le p < \infty. \tag{4.1.1}$$

4.1. Extending the Basic Properties of the Classical Moduli

For $p = \infty$, (4.1.1), i.e., $\lim_{t\to 0} \omega^r_\varphi(f,t)_p = 0$, is satisfied if $f \in C(D)$ and for every endpoint c of D (possibly $\pm\infty$) there is a finite L_c for which

$$\lim_{\substack{x \to c \\ x \in D}} f(x) = L_c.$$

To derive (4.1.1) for $1 \le p < \infty$ we observe that for every $\varepsilon > 0$ there exists $g \in C(D)$ of compact support in D such that $\|f - g\|_p \le \varepsilon$. Using (1.2.1) we have $\omega^r_\varphi(f - g, t) \le M\varepsilon$ where M is independent of ε, f and g and since it is obvious that $\omega^r_\varphi(g,t)_p \to 0$ we complete the proof for $p < \infty$. For $p = \infty$ if $L_c = 0$ and $f \in C(D)$ the above proof is valid. If $L_c \ne 0$ we consider $f_1(x) = L_c$ in a fixed neighborhood of c and $f_1 \in C^\infty(D)$ otherwise. Obviously, $\omega^r_\varphi(f_1, t)_\infty \to 0$, $\omega^r_\varphi(f - f_1, t)_\infty \to 0$ and $\omega^r_\varphi(f,t)_\infty \le \omega^r_\varphi(f - f_1)_\infty + \omega^r_\varphi(f_1,t)_\infty$ which completes the proof. □

The following result will be used frequently.

Theorem 4.1.1. *Suppose φ_1 and φ_2 satisfy the conditions of Section 1.2, $f \in L_p(D)$, $1 \le p \le \infty$ and $\varphi_1(x) \le C\varphi_2(x)$ for $x \in D$. Then for some M and $t_0 > 0$ we have*

$$\omega^r_{\varphi_1}(f,t)_p \le M\omega^r_{\varphi_2}(f,t)_p, \qquad 0 < t < t_0. \tag{4.1.2}$$

PROOF. Using Theorem 2.1.1, we have $\omega^r_{\varphi_i}(f,t)_p \sim K_{r,\varphi_i}(f,t^r)_p$. Using the definition of $K_{r,\varphi_i}(f,t^r)_p$, it is clear that

$$K_{r,\varphi_1}(f,t^r)_p \le K_{r,C\varphi_2}(f,t^r)_p \le M_1 K_{r,\varphi_2}(f,t^r)_p$$

which completes the proof. □

The second property of $\omega^r(f,t)_p$ is:

(b) $\omega^r(f,t)_p$ is a nondecreasing function of t.

Examining the definition, we have:

(b)' $\omega^r_\varphi(f,t)_p$ is a nondecreasing function of t.

Another important property of $\omega^r(f,t)$ is the inequality

(c) $\omega^r(f, \lambda t) \le K\lambda^r \omega^r(f,t)$ for $\lambda \ge 1$.

It turns out that (c) carries over to $\omega^r_\varphi(f,t)$ while the combinatorial proof that is used in most texts to prove (c) via $\omega^r(f, nt) \le n^r \omega^r(f,t)$ is not valid for $\omega^r_\varphi(f,t)$. The difficulty is that identities of the form

$$\Delta^r_{nh} f(x) = \sum_{v_1=0}^{n-1} \cdots \sum_{v_1=0}^{n-1} \Delta^r_h f\left(x + \left(v_1 + \cdots + v_r - \frac{nr}{2}\right)h\right)$$

are no longer valid when we replace h by $h\varphi(x)$ and even for $r = 2$ the identity $\Delta^2_h f(x) = \Delta_h(\Delta_h f)(x)$ does not generalize, as $\Delta^2_{h\sqrt{x}} f(x) = f(x - h\sqrt{x}) - 2f(x) + f(x + h\sqrt{x})$ (for $\varphi(x) = \sqrt{x}$) and

$$\Delta_{h\varphi}(\Delta_{h\varphi}f(x)) = \Delta_{h\varphi}\left(f\left(x+\frac{h}{2}\sqrt{x}\right) - f\left(x-\frac{h}{2}\sqrt{x}\right)\right)$$

$$= f\left(x+\frac{h}{2}\sqrt{x}+\frac{h}{2}\sqrt{x+\frac{h}{2}\sqrt{x}}\right)$$

$$- f\left(x+\frac{h}{2}\sqrt{x}-\frac{h}{2}\sqrt{x+\frac{h}{2}\sqrt{x}}\right)$$

$$- f\left(x-\frac{h}{2}\sqrt{x}+\frac{h}{2}\sqrt{x-\frac{h}{2}\sqrt{x}}\right)$$

$$+ f\left(x-\frac{h}{2}\sqrt{x}-\frac{h}{2}\sqrt{x-\frac{h}{2}\sqrt{x}}\right)$$

(recall $\Delta_{h\varphi}f(z) = f(z+(h/2)\varphi(z)) - f(z-(h/2)\varphi(z))$) are different and the second expression is quite complicated. However, we do have the analogue of (c) without the "nice" identities.

Theorem 4.1.2. *For t, $\lambda t \in [0, t_0]$ and $\lambda \geq 1$ we have*

$$\omega_\varphi^r(f, \lambda t)_p \leq M\lambda^r \omega_\varphi^r(f, t)_p. \tag{4.1.3}$$

Here t_0 is the constant from Theorem 2.1.1.

PROOF. Using Theorem 2.1.1, that is $M_1^{-1}\omega_\varphi^r(f,\tau)_p \leq K_{r,\varphi}(f,\tau^r)_p \leq M_1\omega_\varphi^r(f,\tau)_p$ we have

$$\omega_\varphi^r(f, \lambda t)_p \leq M_1 K_{r,\varphi}(f,(\lambda t)^r)_p \leq M_1 \lambda^r K_{r,\varphi}(f, t^r)_p \leq M_1^2 \lambda^r \omega_\varphi^r(f, t)_p. \quad \square$$

Another basic property of the classical modulus of smoothness is

(d) $\omega^{r+1}(f, t) \leq 2\omega^r(f, t)$

which is generalized in the following theorem.

Theorem 4.1.3. *For $0 < t < t_0$ we have $\omega_\varphi^{r+1}(f, t)_p \leq M\omega_\varphi^r(f, t)_p$.*

We note that here again we have a property which is trivial for $\varphi \equiv 1$ while its extension to other φ requires the deep result of Section 2.1.

PROOF. We consider again only $D = R_+$. Actually we prove $\omega_\varphi^{r+1}(f, t)_p \leq M_1 K_{r,\varphi}(f, t^r)_p$. We take g_t as chosen in Section 2.4 in (2.4.1). Obviously (cf. (1.2.1))

$$\|\Delta_{h\varphi}^{r+1}(f - g_t)\|_p \leq M_2 K_{r,\varphi}(f, t^r)_p.$$

The main effort is to show $\|\Delta_{h\varphi}^{r+1} g_t\|_p \leq Ch^r \|\varphi^r g_t^{(r)}\|_p$. Again we follow the

4.1. Extending the Basic Properties of the Classical Moduli

consideration of Section 2.4. For $x \geq (r+1)h\varphi(x)/2$ we now obtain instead of (2.4.5)

$$|\Delta_{h\varphi(x)}^{r+1} g_t(x)| \leq M \int_{-r h\varphi(x)/2}^{r h\varphi(x)/2} ((r h\varphi(x)/2 - |u|)^{r-1} \left[\left| g_t^{(r)}\left(x + u + \frac{h}{2}\varphi(x)\right) \right| \right.$$
$$\left. + \left| g_t^{(r)}\left(x + u - \frac{h}{2}\varphi(x)\right) \right| \right] du. \tag{4.1.4}$$

The rest of the proof, while similar to that of Section 2.4, requires a somewhat different choice of sets and integrals as analogues of $E(v, h)$ and $I(v, h)$ (see Section 2.4), and the exact description of these analogues, being nontrivial, together with some additional details, are given below. Following the proof in Section 2.4, we have to estimate

$$(h^r)^{p-1} \int_{x \geq (r+1)h\varphi(x)/2} \int_{-r h\varphi(x)/2}^{r h\varphi(x)/2} \frac{(r h\varphi(x)/2 - |u|)^{r-1}}{\varphi\left(x - \frac{h}{2}\varphi(x) + u\right)^r}$$
$$\times \left| \varphi\left(x - \frac{h}{2}\varphi(x) + u\right)^r g_t^{(r)}\left(x - \frac{h}{2}\varphi(x) + u\right) \right|^p du\, dx$$
$$= h^{r(p-1)} \int_0^\infty \varphi(v)^{r(p-1)} |g_t^{(r)}(v)|^p$$
$$\times \int_{E'(v,h)} \left(r h\varphi(x)/2 - \left| x - \frac{h}{2}\varphi(x) - v \right| \right)^{r-1} dx\, dv,$$

where

$$E'(v, h) = \left\{ x : x \geq \frac{r+1}{2} h\varphi(x),\ \left| x - \frac{h}{2}\varphi(x) - v \right| < \frac{r}{2} h\varphi(x) \right\}.$$

To this end it is enough to verify that

$$I'(v, h) = \int_{E'(v,h)} \left(r h\varphi(x)/2 - \left| x - \frac{h}{2}\varphi(x) - v \right| \right)^{r-1} dx \leq M h^r \varphi(v)^r.$$

We write $I'(v, h) = I_1'(v, h) + I_2'(v, h)$ where $I_1'(v, h)$ and $I_2'(v, h)$ are the corresponding integrals over the sets

$$E_1'(v, h) = \left\{ x \in E(v, h): \frac{v}{2} \leq x - \frac{h}{2}\varphi(x) \leq v \right\}, \quad E_2'(v, h) = E'(v, h) \setminus E_1'(v, h).$$

Since $x \in E_1'(v, h)$ implies $x \sim v$, the estimate of $I_1'(v, h)$ coincides with that of $I_1(v, h)$ at the end of Section 2.4. $x \in E'(v, h)$ implies $x - (h/2)\varphi(x) \geq v/2$, and hence for $x \in E_2'(v, h)$ we have

$$x - \frac{r+1}{2} h\varphi(x) \leq v \leq x - \frac{h}{2}\varphi(x).$$

Therefore, following the proof at the end of Section 2.4, we obtain

$$I_2'(v,h) \leq \int_{E_2'(v,h)} \left(v - \left(x - \frac{r+1}{2}h\varphi(x)\right)\right)^{r-1} dx$$

$$\leq \int_{\{x:\, 0 \leq x - [(r+1)/2]h\varphi(x) \leq v\}} v^{r-1}\, dx \leq Mv^r \leq Mh^r\varphi(v)^r,$$

which concludes the proof. □

Another inequality about the classical moduli of smoothness is

(e) $\omega^{k+r}(f,t)_p \leq Mt^r\omega^k(f^{(r)},t)_p$

which is valid if $f^{(r)} \in L_p$ and $f^{(r-1)} \in A.C._{\text{loc}}$. Generalizations of this inequality to $\omega_\varphi^r(f,t)_p$ will be dealt with in Chapter 6 which is devoted to weighted moduli of smoothness and not here because of the "nature of the beast".

4.2. Optimal Rate of $\omega_\varphi^r(f,t)$

It is well known for $\omega^r(f,t)_p$ that the optimal rate in which $\omega^r(f,t)_p$ tends to zero is $O(t^r)$. That is, $\omega^r(f,t)_p = o(t^r)$ implies that f is a polynomial of order $r-1$; and for $1 < p \leq \infty$ $\omega^r(f,t)_p = O(t^r)$ if and only if $f^{(r-1)} \in A.C._{\text{loc}}$ and $f^{(r)} \in L_p$ while $\omega^r(f,t)_1 = O(t^r)$ if and only if $f^{(r-2)} \in A.C._{\text{loc}}$ and $f^{(r-1)} \in B.V.(D)$. These are actually small o and big O saturation theorems. Analogues for $\omega_\varphi^r(f,t)_p$ constitutes the following theorem.

Theorem 4.2.1. For $\omega_\varphi^r(f,t)_p$ we have:

(a) $\liminf_{t \to 0+} \omega_\varphi^r(f,t)_p/t^r = 0$ implies that f is a polynomial of degree $r-1$. (f is a polynomial of degree $r-1$ implies $\omega_\varphi^r(f,t) = 0$.)
(b) For $1 < p \leq \infty$ $\omega_\varphi^r(f,t)_p = O(t^r)$ if and only if $f^{(r-1)} \in A.C._{\text{loc}}$ and $\|\varphi^r f^{(r)}\|_p < \infty$.
(c) $\omega_\varphi^r(f,t)_1 = O(t^r)$ if and only if $f^{(r-2)} \in A.C._{\text{loc}}$ and $f^{(r-1)}$ is equivalent to a function v which is locally of bounded variation and $\int_D \varphi(\tau)^r |dv(\tau)| < \infty$ (integration with respect to the variation of v).

Of course, in these statements "f is a polynomial" etc. means that f coincides a.e. with a polynomial etc. We also remark that there is no integrability problem in (c) since a nonnegative function is integrated with respect to a positive measure.

PROOF. To prove (a) we show first that for every t_0 $\omega_\varphi^r(f,t_0) = 0$. Using (4.1.3) with $\lambda t = t_0$, we have

$$\omega_\varphi^r(f,t_0) = \omega_\varphi^r\left(f, \frac{t_0}{t}t\right) \leq Mt_0^r(\omega_\varphi^r(f,t)/t^r);$$

4.2. Optimal Rate of $\omega_\varphi^r(f,t)$

taking limit on t for which the lim inf is achieved, we have $\omega_\varphi^r(f,t_0) = 0$. We also have $\omega_\varphi^r(f,t)_p \sim K_{r,\varphi}(f,t^r)_p$ and therefore $K_{r,\varphi}(f,t^r)_p = 0$. If $K_{r,\varphi}(f,t^r)_p = 0$, there exists a sequence g_n satisfying $\|f - g_n\|_p + t^r\|\varphi^r g_n^{(r)}\| \le t^r 2^{-n}$, and therefore, $f = g_n + \sum_{i=n}^{\infty}(g_{i+1} - g_i)$ and

$$\|\varphi^r f^{(r)}\|_p \le \|\varphi^r g_n^{(r)}\|_p + \sum_{i=n}^{\infty}(\|\varphi^r g_{i+1}^{(r)}\|_p + \|\varphi^r g_i^{(r)}\|_p) \le 4 \cdot 2^{-n}.$$

Therefore, $\|\varphi^r f^{(r)}\| = 0$ which implies $\varphi(x)^r f^{(r)}(x) = 0$ a.e. or f is a polynomial of degree $r-1$.

We now prove (b). If we assume $\|\varphi^r f^{(r)}\|_p < \infty$, then $K_{r,\varphi}(f,t^r)_p \le Mt^r$ and hence $\omega_\varphi^r(f,t)_p \le M_1 t^r$. Assuming $\omega_\varphi^r(f,t)_p \le Mt^r$, using Theorem 4.1.1 on a finite subinterval $[c,d]$ of D with $\varphi_2 = \varphi$ and $\varphi_1 = 1$, we obtain $\omega^r(f,t)_{L_p[c,d]} \le M_2 t^r$ and therefore $f^{(r-1)} \in \text{A.C.}[c,d]$ and $f^{(r)} \in L_p[c,d]$. We have to show that $\|\varphi^r f^{(r)}\|_p \le M$. Since $f^{(r-1)} \in \text{A.C.}_{\text{loc}}$ and $f^{(r)}$ is locally integrable, we have

$$\lim_{h \to 0} |\Delta_{h\varphi(x)}^r f(x)|/h^r = \varphi(x)^r |f^{(r)}(x)| \quad \text{a.e.,}$$

which implies using Fatou's lemma

$$\|\varphi^r f^{(r)}\|_p \le \liminf_{h \to 0+} \|\Delta_{h\varphi}^r f\|_p/h^r \le M < \infty,$$

and this is conclusion (b) of our theorem.

To prove (c) we first assume that $f^{(r-2)} \in \text{A.C.}_{\text{loc}}$, $f^{(r-1)}$ is equivalent to a function v with locally finite variation and $\int_D \varphi(x)^r |dv(x)| < M$. We will show that

$$\|\Delta_{h\varphi}^r f\|_1 \le M_1 h^r \int \varphi(x)^r |dv(x)|. \tag{4.2.1}$$

We can write

$$|\Delta_{h\varphi(x)}^r f(x)| = \left| \int_{-h\varphi(x)/2}^{h\varphi(x)/2} \cdots \int_{-h\varphi(x)/2}^{h\varphi(x)/2} [f^{(r-1)}(x + (h\varphi(x)/2) + u_1 + \cdots + u_{r-1}) \right.$$
$$\left. - f^{(r-1)}(x - (h\varphi(x)/2) + u_1 + \cdots + u_{r-1})] du_1 \cdots du_{r-1} \right|$$
$$\le \left| \int_{-h\varphi(x)/2}^{h\varphi(x)/2} \cdots \int_{-h\varphi(x)/2}^{h\varphi(x)/2} dv(x + u_1 + \cdots + u_{r-1} + u) du_1 \cdots du_{r-1} \right|$$
$$\le M_2 \int_{rh\varphi(x)/2}^{rh\varphi(x)/2} ((rh\varphi(x)/2) - |u|)^{r-1} |dv(x + u)|.$$

Now using the method of Section 2.4, we have

$$\int_a^b |\Delta_{h\varphi(x)}^r f(x)| dx \le M_2 \int I(v,h) |dv(v)|,$$

where

$$I(v,h) = \int_{E(v,h)} ((rh\varphi(x)/2) - |x-v|)^{r-1} dx$$

and $E(v,h) = \{x; x \geq (r/2)h\varphi(x), |x-v| \leq (r/2)h\varphi(x)\}$ and hence the estimate $I(v,h) \leq M_3 h^r \varphi(v)^r$ is satisfied (cf. Section 2.4). Therefore

$$\|\Delta^r_{h\varphi} f\|_{L_1} \leq M_2 M_3 h^r \int \varphi(v)^r |dv(v)|.$$

It remains to show the necessity of Condition (c) of Theorem 4.2.1. We recall that $\omega^r_\varphi(f,t)_1 \leq Mt^r$, implies $\omega^r_{\varphi_1}(f,t)_1 \leq M_1 t^r$ for any φ_1 satisfying Conditions I–III of Section 1.2 such that $\varphi_1 \in C(D)$ and $\varphi \sim \varphi_1$ and we will show that the latter implies $f^{(r-2)} \in$ A.C.$_{\cdot\text{loc}}$ and that $f^{(r-1)}$ is equivalent to a function $v \in$ B.V.$_{\cdot\text{loc}}$ with $\int \varphi_1(x)^r |dv(x)| \leq M_2$. Using $\omega^r_\varphi(f,t)_1 \leq Mt^r$ on any finite subinterval $[c,d] \subset D$, and using Theorem 4.1.1 for the domain $[c,d]$ (rather than D), we have

$$\omega^r_1(f,t)_{L_1[c,d]} \leq M(c,d) \omega^r_\varphi(f,t)_{L_1[c,d]} \leq M(c,d) \omega^r_\varphi(f,t)_1$$

and hence $\omega^r(f,t)_{L_1[c,d]} \leq Ct^r$. Therefore $f^{(r-2)} \in$ A.C.$_{\cdot\text{loc}}$ and $f^{(r-1)}$ is equivalent to a function, say v_*, of bounded variation in any finite closed subinterval of D. Since $K_{r,\varphi_1}(f,t^r) \leq Mt^r$, we have for some function g_t with locally absolutely continuous $(r-1)$-st derivative $\|f - g_t\|_1 \leq 2Mt^r$ and $\|\varphi_1^r g_t^{(r)}\|_1 \leq 2M$. Using weak* convergence in the space of B.V. (dual to C), we have for some $\mu \in$ B.V.$_{\cdot\text{loc}}$ and $\{g_n\}$ a subsequence of g_t, $\|g_n - f\|_1 \to 0$ and $\int \Psi \varphi_1^r g_n^{(r)} = \langle \Psi, \varphi_1^r g_n^{(r)} \rangle \to \int \Psi(x) d\mu(x)$, for all $\Psi \in C(D)$ with support inside D. Here we used the separability of $C[c,d]$ for every $[c,d] \subset D$ to get a single subsequence. We write $d\mu(x) = \varphi_1(x)^r dv(x)$, and, as φ_1 is positive and continuous, we have $v \in$ B.V.$_{\cdot\text{loc}}$. For Ψ belonging to the Schwartz space of test functions, we write (note that Ψ/φ_1^r is in $C(D)$)

$$0 = \lim_{n\to\infty} \langle \Psi^{(r)}, f - g_n \rangle = \lim_{n\to\infty} \{\langle \Psi^{(r)}, f \rangle - \langle \Psi^{(r)}, g_n \rangle\}$$
$$= \langle \Psi, dv_* \rangle - \lim_{n\to\infty} \langle \Psi, g_n^{(r)} \rangle$$
$$= \langle \Psi, dv_* \rangle - \lim_{n\to\infty} \langle \Psi/\varphi_1^r, \varphi_1^r g_n^{(r)} \rangle$$
$$= \langle \Psi, dv_* \rangle - \langle \Psi/\varphi_1^r, d\mu \rangle$$
$$= \langle \Psi, dv_* \rangle - \langle \Psi, dv \rangle,$$

and therefore $dv_* = dv$. However, as a weak* limit,

$$\int \varphi_1(x)^r |dv(x)| = \|\mu\|_{\text{B.V.}} \leq \overline{\lim_n} \|\varphi_1^r g_n^{(r)}\|_1 \leq 2M,$$

and this completes our proof. □

4.3. Marchaud Inequality

One of the most useful results about moduli of continuity is the Marchaud inequality

$$\omega^r(f,t) \le Ct^r \left\{ \int_t^c \frac{\omega^{r+1}(f,u)}{u^{r+1}} du + \|f\| \right\} \tag{4.3.1}$$

which is usually proved using the identity (see for instance Timan [2, p. 105])

$$\vec{\Delta}_{2h}^r f(x) - 2^r \vec{\Delta}_h^r f(x) = \sum_{v=0}^{r-1} \sum_{\mu=v+1}^{r} \binom{r}{\mu} \vec{\Delta}_h^{r+1} f(x+vh). \tag{4.3.2}$$

This identity, while still useful, cannot be employed to estimate $\omega_\varphi^r(f,t)_p$, without additional work. We will prove the following analogue of the Marchaud inequality.

Theorem 4.3.1. *For $\omega_\varphi^r(f,t)_p$ and $K_{r,\varphi}(f,t^r)_p$ we have*

$$\omega_\varphi^r(f,t)_p \le Ct^r \left\{ \int_t^c \frac{\omega_\varphi^{r+1}(f,u)_p}{u^{r+1}} du + \|f\|_p \right\}, \tag{4.3.3}$$

and (analogously)

$$K_{r,\varphi}(f,t^r)_p \le Ct^r \left\{ \int_t^c \frac{K_{r+1,\varphi}(f,u^{r+1})_p}{u^{r+1}} du + \|f\|_p \right\}, \tag{4.3.4}$$

where $c > 0$ is any fixed constant.

PROOF. We will actually prove our result for $D = R_+$ and we first show that this is sufficient. We may, in case $D = (0,1)$, write $f(x) = f(x)\psi(x) + (1 - \psi(x))f(x) \equiv f_1(x) + f_2(x)$ where $\psi(x) = 1$ on $[0, 1/4]$, $\psi(x) = 0$ on $[3/4, \infty)$ ψ is monotone and $\psi \in C^\infty$. With this notation, the use of Theorem 2.1.1 and of inequality (2.2.14) it is not difficult to show that

$$\omega_\varphi^{r+1}(f_i,t)_p \le M(r)[\omega_\varphi^{r+1}(f,t)_p + t^{r+1}\|f\|_p]. \tag{4.3.5}$$

The inequality

$$\omega_\varphi^r(f_i,t)_p \le Ct^r \left\{ \int_t^c \frac{\omega_\varphi^{r+1}(f_i,u)_p}{u^{r+1}} du + \|f_i\|_p \right\}$$

combined with $\omega_\varphi^r(f,t)_p \le \omega_\varphi^r(f_1,t)_p + \omega_\varphi^r(f_2,t)_p$ and (4.3.5) will imply (4.3.3). Since the situation for f_2 is symmetric to that of f_1, and to prove our result for the latter, we are dealing with the case $D = R_+$, we may restrict ourselves to $D = R_+$. A similar technique can be employed to show that the problem for $D = R$ can also be reduced to that on $D = R_+$.

We will use forward differences. This is permitted since, as we have already shown in Sections 2.1 and 3.2, $\omega_\varphi^r(f,t)_p \sim \vec{\omega}_\varphi^r(f,t)_p \sim K_{r,\varphi}(f,t^r)_p$ and $\omega_\varphi^{r+1}(f,t)_p \sim \vec{\omega}_\varphi^{r+1}(f,t)_p \sim K_{r+1,\varphi}(f,t^{r+1})_p$. Morever, since for K-functionals if we take $\varphi_1 \sim \varphi$, we get $K_{j,\varphi}(f, t^j)_p \sim K_{j,\varphi_1}(f, t^j)_p$, we may also assume that φ is increasing on $(0, 1)$. We now use the identity (4.3.2) to get the inequality

$$|\vec{\Delta}^r_{h\varphi(x)} f(x)| \leq 2^{-r} |\vec{\Delta}^r_{2h\varphi(x)} f(x)|$$
$$+ 2^{-r} \sum_{v=0}^{r-1} \sum_{\mu=v+1}^{r} \binom{r}{\mu} |\vec{\Delta}^{r+1}_{h\varphi(x)} f(x + vh\varphi(x))| \qquad (4.3.6)$$

and therefore

$$\overline{\omega}^r_\varphi(f, t)_p = \sup_{0 < h \leq t} \|\vec{\Delta}^r_{h\varphi} f\|_{L_p(R_+)} \leq 2^{-r} \sup_{h \leq 2t} \|\vec{\Delta}^r_{h\varphi} f\|_{L_p(R_+)}$$
$$+ \frac{r}{2} \sup_{v < r} \sup_{0 < h \leq t} \|\vec{\Delta}^{r+1}_{h\varphi(x)} f(x + vh\varphi(x))\|_{L_p(R_+)}.$$

We will show for an M independent of $v < r$, $0 < h \leq t$ and t that

$$\|\vec{\Delta}^{r+1}_{h\varphi(x)} f(x + vh\varphi(x))\|_{L_p(R_+)} \leq M K_{r+1,\varphi}(f, t^{r+1})_p \leq M_1 \overline{\omega}^{r+1}_\varphi(f, t)_p \qquad (4.3.7)$$

and this will imply

$$\overline{\omega}^r_\varphi(f, t) \leq 2^{-r} \overline{\omega}^r_\varphi(f, 2t) + L \overline{\omega}^{r+1}_\varphi(f, t)$$

from which, using the standard technique, one obtains

$$\overline{\omega}^r_\varphi(f, t) \leq L \sum_{l=0}^{m-1} 2^{-lr} \overline{\omega}^{r+1}_\varphi(f, 2^l t) + 2^{-mr} \omega^r_\varphi(f, 2^m t)$$
$$\leq C t^r \left\{ \int_t^c \frac{\overline{\omega}^{r+1}_\varphi(f, u)}{u^{r+1}} du + \|f\|_p \right\} \quad \text{for } \frac{c}{2} < t 2^m \leq c.$$

Recalling that $\overline{\omega}^j_\varphi(f, t) \sim \omega^j_\varphi(f, t)$ (see Theorem 2.1.1 and 3.2.1), it is clear that showing (4.3.7) and in fact the first inequality there will complete the proof. We now choose g_t satisfying

$$\|f - g_t\|_{L_p(R_+)} \leq 2 K_{r+1,\varphi}(f, t^{r+1})_p$$

and

$$t^{r+1} \|\varphi^{r+1} g_t^{(r+1)}\|_{L_p(R_+)} \leq 2 K_{r+1,\varphi}(f, t^{r+1})_p.$$

Using (1.2.1), we have

$$\|\vec{\Delta}^{r+1}_{h\varphi(x)} (f(x + vh\varphi(x)) - g_t(x + vh\varphi(x)))\|_{L_p(R_+)}$$
$$\leq 2^{r+1} \sup_{0 \leq k \leq r+1} \|f(x + (v+k)h\varphi(x)) - g_t(x + (v+k)h\varphi(x))\|_{L_p(R_+)}$$
$$\leq M \|f - g_t\|_{L_p(R_+)}.$$

To estimate $\|\vec{\Delta}^{r+1}_{h\varphi(x)} g_t(x + vh\varphi(x))\|_p$, we write, using the Taylor formula with integral remainder,

$$|\vec{\Delta}^{r+1}_{h\varphi(x)} g_t(x + vh\varphi(x))|$$
$$\leq M \sup_{0 < j \leq r+1} \left| \int_0^{jh\varphi(x)} (jh\varphi(x) - u)^r g_t^{(r+1)}(x + vh\varphi(x) + u) du \right|$$

4.3. Marchaud Inequality

or

$$\|\vec{\Delta}_{h\varphi(x)}^{r+1} g_t(x + vh\varphi(x))\|_p$$
$$\leq M_1 \sup_{0<j\leq r+1} \left\| \int_0^{jh\varphi(x)} (jh\varphi(x) - u)^r |g_t^{(r+1)}(x + vh\varphi(x) + u)| du \right\|_p.$$

We can now estimate for $1 \leq p < \infty$

$$\left\| \int_0^{jh\varphi(x)} (jh\varphi(x) - u)^r |g_t^{(r+1)}(x + vh\varphi(x) + u)| du \right\|_p$$
$$\leq \left\{ \int_0^\infty \left| \int_0^{jh\varphi(x)} (jh\varphi(x))^r |g_t^{(r+1)}(x + vh\varphi(x) + u)| du \right|^p dx \right\}^{1/p} = I_p.$$

For $1 < p < \infty$ we have

$$I_p \leq Mh^{r+1} \left\{ \int_0^\infty \left| \frac{1}{jh\varphi(x)} \int_0^{jh\varphi(x)} [\varphi(x + vh\varphi(x) + u)]^{r+1} \right. \right.$$
$$\left. \left. \times |g_t^{(r+1)}(x + vh\varphi(x) + u)| du \right|^p dx \right\}^{1/p}$$
$$\leq Mh^{r+1} \|M(F)\|_p \leq M_1 h^{r+1} \|F\|_{L_p},$$

where $F(x) = \varphi(x + vh\varphi(x))^{r+1} |g_t^{r+1}(x + vh\varphi(x))|$, $M(F)$ is the maximal function of $F(x)$ and the inequality $\|M(F)\|_p \leq C(p)\|F\|_p$ for $1 < p < \infty$ was used in the last step. By (1.2.1)

$$\|F\|_{L_p(R_+)} \leq M \|\varphi^{r+1} g_t^{(r+1)}\|_{L_p} \leq 2Mt^{-r-1} K_{r+1,\varphi}(f, t^{r+1})_p$$

and this concludes the proof for $1 < p < \infty$. For $p = \infty$ we use the maximal function in the same manner. For $p = 1$ we use Fubini's theorem to write with $y = x + vh\varphi(x)$

$$I_1 \leq Mh^r \int_0^\infty \int_0^{jh\varphi(x)} \varphi(x + vh\varphi(x) + u)^r |g_t^{(r+1)}(x + vh\varphi(x) + u)| du\, dx$$
$$\leq M_1 h^r \int_0^\infty \int_0^{Ch\varphi(y)} \varphi(y + u)^r |g_t^{(r+1)}(y + u)| du\, dy$$
$$\leq M_2 h^r \int_0^\infty \varphi(v)^r |g_t^{(r+1)}(v)| \left\{ \int_0^{C_1 h\varphi(v)} du \right\} dv \leq M_3 h^{r+1} \|\varphi^{r+1} g_t^{(r+1)}\|_{L_1}$$

and this completes the proof. $\square$

CHAPTER 5
MORE GENERAL STEP-WEIGHT FUNCTIONS φ

In Section 1.2 three conditions were imposed on φ which were the basis of all the proofs up to now. In this chapter we will relax Conditions I and II and will show that in a way III cannot be relaxed without changing the definition of $\omega_\varphi^r(f,t)_p$. For $\varphi(x) \sim x^{\beta(0)}$ where $\beta(0) < 0$ and $\varphi(x) \sim x^{\beta(\infty)}$ where $\beta(\infty) > 1$ we will provide such a change.

5.1. Logarithmic-Type Weights and Internal Zeros

We observe that if φ has zeros in D, Condition I of Section 1.2 is not satisfied. If we have the conditions of Section 1.2 in the subintervals at whose endpoints φ is zero, we can define the modulus separately in each such subinterval and the K-functional will be equivalent in some cases to the sum of those moduli. We do not know if this equivalence is valid in general.

Assertion 5.1.1. An examination of Theorem 2.1.1 for $D = R_+$ (other domains yield similar results) shows that Condition II of Section 1.2 can be replaced by the weaker set of conditions:

(i) $\varphi(x) = O(x)$ as $x \to \infty$
(ii) $\varphi(y) \le A\varphi(x)$ for $0 < y \le 2x < 1$ with a constant A independent of y and x.
(iii) If $\overline{\lim}_{x \to 0+} \varphi(x)/x = \infty$, then there exist constants $\beta < 1$ and $A > 1$ such that $\varphi(Ax)/(Ax)^\beta < \varphi(x)/x^\beta$, for $0 < x < 1$.

These conditions will allow $\varphi(x)$ to behave like $x^{\beta_1}|\log x|^{\beta_2}$ or $x^{\beta_1}|\log x|^{\beta_2}|\log|\log x||^{\beta_3}$ etc. with the restriction that if $\beta_1 = 1$, then $\beta_2 \le 0$

and if $\beta_1 = 1$ and $\beta_2 = 0$, then $\beta_3 \leq 0$ etc. All these conditions on β_i are necessary in order that $\omega_\varphi(f,t)_p$ be finite. For example we show:

EXAMPLE 5.1.2. If $\varphi(x) = x|\log x|$ for $0 < x < 1/2$, then there exists a function $f_0 \in L_p(R_+)$ $1 \leq p < \infty$ such that $\omega_\varphi^r(f_0,t)_p = \infty$, for all $0 < t < t_0$.

The example will actually be constructed in Section 5.2 for all φ not satisfying the finite overlapping property, which we will show here is not satisfied by $\varphi(x) = x|\log x|$. This follows from

$$m\{x: x - h\varphi(x) \in (0,\varepsilon)\} \geq \varepsilon/2h,$$

satisfied for $0 < \varepsilon < h$ and ε small enough since the derivative of $x - h\varphi(x)$ is $1 - h|\log x| + h$ for $x \leq 1$ and is equal to h at $e^{-1/h}$.

5.2. The Necessity of the Finite Overlapping Condition

Condition III of Section 1.2 consists essentially of two conditions. The first condition asserts the measurability of $\varphi(x)$. It is clear that if $\varphi(x)$ is not measurable, then there exists a continuous function f for which $\Delta_{h\varphi}^r f$ is not measurable. The second condition, the finite overlapping condition, given by

$$m\{x; x \pm h\varphi(x) \in E, x \in D\} \leq Km(E) \quad \text{for every interval } E \subset D, \quad (5.2.1)$$

where $0 < h < h_0$ is crucial for the definition of $\omega_\varphi^r(f,t)_p$ for $1 \leq p < \infty$. In the absence of (5.2.1) we cannot guarantee the measurability of $\Delta_{h\varphi}^r f$ anymore. The problem is even more deep-rooted as even if $f(x \pm h\varphi(x))$ are measurable we still may encounter a difficulty as is seen from the following theorem.

Theorem 5.2.1. *If $D = R$, φ is measurable and satisfies Condition I of Section 1.2 but (5.2.1) is not satisfied, then there exists a function $f \in L_p(D)$, $1 \leq p < \infty$ for which $\omega_\varphi^r(f,t)_p = \infty$, for all $0 < t < t_0$.*

This theorem, in a slightly changed form, is valid for $D = R_+$ and $D = (0,1)$ as well.

PROOF. If (5.2.1) is not satisfied, we may assume that it fails for $x - h\varphi(x)$. We will demonstrate the existence of f for $r = 2$, but for any other r the result can be proved similarly. We now define

$$\Phi_h(x) \equiv x - h\varphi(x) \quad \text{and} \quad \mu_h(E) \equiv m(\Phi_h^{-1}(E)) \quad \text{for any Borel set } E. \quad (5.2.2)$$

For the proof we need to distinguish the following two cases:

(a) There exists a sequence $h_n \to 0$ such that μ_{h_n} is not absolutely continuous with respect to the Lebesgue measure m.
(b) For every small $|h|$ μ_h is absolutely continuous with respect to m.

In case (a) let $E_n \subseteq D$ be a Borel set of Lebesgue measure zero such that $\mu_{h_n}(E_n) > 0$. With χ_n the characteristic function of E_n the function

$$f(x) \equiv \sum_{n=1}^{\infty} n \max((\mu_{h_n}(E_n))^{-1/p}, 1)\chi_n(x)$$

which is equivalent to zero and satisfies $f(x) \geq 0$ proves our theorem as

$$\omega_\varphi^2(f,h)_p \geq \left\{ \int_{\Phi_{h_n}^{-1}(E_n)} |\Delta_{h_n\varphi(x)}^2 f(x)|^p \, dx \right\}^{1/p} \geq n$$

for all $n \geq n_0(h)$.

To prove our theorem in case (b) we may now assume that μ_h is absolutely continuous with respect to the Lebesgue measure m for every small h. Assuming (5.2.1) does not hold with the negative sign (there) for any $h_0 > 0$ and $K > 0$, there exist a sequence $\{h_n\}$, $0 < h_n < 1/n$ and a sequence of closed intervals I_n in the interior of D such that

$$m(\Phi_{h_n}^{-1}(I_n)) > n^{3p} m(I_n). \tag{5.2.3}$$

We will show that we can choose I_n such that in addition to (5.2.3)

$$m(I_n) < \frac{1}{n} \quad \text{and} \quad \Phi_{h_n}^{-1}(I_n) \cap I_n = \varnothing$$

are also satisfied. Condition I of Section 1.2 implies $h_n \varphi(x) \geq \delta$ for $x \in I_n$, and therefore we have only to show the existence of J_n such that

$$J_n \subset I_n, \quad m(\Phi_{h_n}^{-1}(J_n)) \geq n^{3p} m(J_n) \quad \text{and} \quad m(J_n) \leq \min\left(\frac{1}{n}, \delta\right).$$

We observe that for an interval $F \subset I_n$ either

$$m(\Phi_{h_n}^{-1}(F)) > n^{3p} m(F) \quad \text{or} \quad m(\Phi_{h_n}^{-1}(I_n \backslash F)) \geq n^{3p} m(I_n \backslash F)$$

holds. Splitting I_n into two equal intervals and repeating the process on the subinterval satisfying (5.2.3) and so on a sufficient number of times, we get the desirable J_n, which, with no loss generality, we call I_n. Using the absolute continuity of μ_h, we can construct two intervals I_n'' and I_n', $I_n'' \subset I_n' \subset I_n$ such that $m(\Phi_{h_n}^{-1}(I_n'')) > n^{3p} m(I_n)$ and the endpoints of I_n'' and I_n' are in the interior of I_n' and I_n respectively. Using again the absolute continuity of μ_{h_n}, we can show that for a given I_n there exists an $\varepsilon_n > 0$ such that

$$m(\Phi_{h_n}^{-1}(I_n \backslash E_n) \backslash E_n) \geq n^{3p} m(I_n) \tag{5.2.4}$$

for every Borel set E_n satisfying $m(E_n) < \varepsilon_n$.

To construct f, we first construct the continuous function $f_n(x)$ by

$$f_n(x) = (n^2 (m(I_n))^{1/p})^{-1} \quad \text{for} \quad x \in I_n'',$$
$$f_n(x) = 0 \quad \text{for} \quad x \notin I_n' \quad \text{and} \tag{5.2.5}$$
$$f_n(x) \text{ is linear on } I_n' \backslash I_n''.$$

This definition implies $\|f_n\|_p \leq n^{-2}$. Moreover, if $f_n^* \geq 0$ is any function that coincides with f_n on a set $D \setminus E_n$ with $m(E_n) < \varepsilon_n$, then using (5.2.3), (5.2.4), and (5.2.5), we have

$$\|\Delta_{h_n\varphi}^2 f_n^*\|_p \geq \left\{\int_{\Phi_{h_n}^{-1}(I_n''\setminus E_n)\setminus E_n} n^{-2p}(m(I_n))^{-1}\,dx\right\}^{1/p} \qquad (5.2.6)$$
$$\geq \{n^{-2p} m(\Phi_{h_n}^{-1}(I_n''\setminus E_n)\setminus E_n)/m(I_n)\}^{1/p} > n.$$

The function f_n is Lip 1 and has compact support in (the interior of) D; and therefore, there exists a constant M_n for which

$$\|\Delta_{\tau\varphi}^2 f_n\|_p \leq M_n \tau^{1/p} \quad \text{for } 0 < \tau < 1. \qquad (5.2.7)$$

We now construct sequences $\{n_k\}$ and $\{h_{n_k}\}$ such that

$$\sum_{k=m+1}^{\infty} n_k^{-1} < \varepsilon_{n_m}, \quad \sum_{k=1}^{m-1} M_{n_k} h_{n_m}^{1/p} < 1, \quad m = 2, 3, \ldots \qquad (5.2.8)$$

and define

$$f(x) \equiv \sum_{k=1}^{\infty} f_{n_k}(x). \qquad (5.2.9)$$

The function f is in L_p as $\|f_{n_k}(x)\|_p \leq n_k^{-2}$. We define

$$f_{n_m}^*(x) \equiv f_{n_m}(x) + \left(\sum_{k=m+1}^{\infty} f_{n_k}(x)\right),$$

and using (5.2.5), (5.2.6), (5.2.7), and (5.2.8), we have for $h = h_{n_m}$

$$\|\Delta_{h\varphi}^2 f\|_p \geq \|\Delta_{h\varphi}^2 f_{n_m}^*\|_p - \sum_{k=1}^{m-1} M_{n_k} h^{1/p} \geq \|\Delta_{h\varphi}^2 f_{n_m}^*\|_p - 1 \geq n_m - 1,$$

and therefore

$$\omega_\varphi^2(f, t)_p = \sup_{h \leq t} \|\Delta_{h\varphi}^2 f^*\|_p = \infty. \qquad \square$$

5.3. Growth Orders of Type x^β with Arbitrary β

In Section 1.2, φ for $D = R_+$ was restricted in Condition II by $\varphi(x) \sim x^{\beta(0)}$ for $0 \leq x < 1$ where $\beta(0) \geq 0$ and $\varphi(x) \sim x^{\beta(\infty)}$ for $x > 1$ where $\beta(\infty) \leq 1$. With the concepts of $\omega_\varphi^r(f, t)_p$ as defined in (1) (and (2.1.2)) this was necessary as neither $\beta(0) < 0$ nor $\beta(\infty) > 1$ is compatible with the finite overlapping property, and therefore, as we saw in the last section, $\omega_\varphi^r(f, t)_p$ may be equal to infinity while $\|f\|_p < \infty$ and $K_{r,\varphi}(f, t^r)_p < \infty$. This means that for such functions φ we need a different measure of continuity to characterize the K-functional. We will denote this different modulus by $W_\varphi^r(f, t)_p$. We will define this new class of admissible step-weight functions φ on $D = R_+$ here. The interested reader can find the corresponding definitions for $D = R$ and $D = (0, 1)$ in Appendix A.

50   5. More General Step-Weight Functions φ

Definition 5.3.1. A measurable function φ on R_+ is admissible if for some fixed positive constants A_1, A, B_1, B, M, and t_0 it satisfies:

(i) $A_1 x^{\beta(0)} \leq \varphi(x) \leq A x^{\beta(0)}$, for $0 < x \leq 1$,
(ii) $B_1 x^{\beta(\infty)} \leq \varphi(x) \leq B x^{\beta(\infty)}$, for $1 < x < \infty$,
(iii) there exists a constant $C \geq 2$ such that for $0 < h < t_0$ and any interval $E \subset [Ch_0^*/2, 2h_1^*/C]$ we have

$$\text{meas}\{x : x \pm (rh/2)\varphi(x) \in E\} \leq M \text{ meas } E, \quad (5.3.1)$$

where

$$h_0^* \equiv \begin{cases} (Ar)^{1/1-\beta(0)} h^{1/1-\beta(0)} & \text{if } \beta(0) < 0 \\ 0 & \text{otherwise} \end{cases}$$

and

$$h_1^* \equiv \begin{cases} (Br)^{1/1-\beta(\infty)} h^{1/1-\beta(\infty)} & \text{if } \beta(\infty) > 1 \\ \infty & \text{otherwise} \end{cases}.$$

We note the following.

(a) $h_0^* \to 0$ and $h_1^* \to \infty$ as $h \to 0+$.
(b) In (5.3.1) any η satisfying $\eta \leq rh/2$ can replace $rh/2$ with the same M.
(c) Any φ satisfying Conditions I–III of Section 1.2 is admissible.
(d) The function $\varphi(x) = x^{\beta(0)}(1+x)^{\beta(\infty)}$ with $\beta(0) < 0$ or $\beta(\infty) > 1$ or both is admissible but does not satisfy the conditions of Section 1.2.

Furthermore, (5.3.1) is an analogue of the (stronger) finite overlapping condition, and it implies for every measurable f

$$\int_{Ch_0^*}^{h_1^*/C} |f(x \pm \eta\varphi(x))| \, dx \leq M \int_0^\infty |f(x)| \, dx \quad \text{for } |\eta| \leq rh/2. \quad (5.3.2)$$

For admissible φ on R_+ we define the modulus of smoothness W_φ^r by

$$W_\varphi^r(f,t)_p = \|f - P_{r,t}f\|_{L_p[0,t_0^*]} + \sup_{h \leq t} \|\Delta_{h\varphi}^r f\|_{L_p[Ch_0^*, h_1^*/C]} + \|f\|_{L_p[t_1^*, \infty)}, \quad (5.3.3)$$

where $P_{r,t}f$ is the $L_2[0, t_0^*]$ orthogonal projection of f onto the polynomials of degree at most $r - 1$. We observe that $P_{r,t}f$ exists for each r and t as the unique polynomial satisfying

$$\int_0^{t^*} f(x) x^i \, dx = \int_0^{t^*} P_{r,t} f(x) x^i \, dx \quad \text{for } 0 \leq i < r.$$

We note that for $\beta(0) \geq 0$ and $\beta(\infty) \leq 1$ $W_\varphi^r(f,h) \sim \omega_\varphi^r(f,h)$ as $h \to 0+$.

The equivalence of $W_\varphi^r(f,t)_p$ with the K-functional can now be stated and proved.

5.3. Growth Orders of Type x^β with Arbitrary β

Theorem 5.3.2. *Suppose φ is admissible (in the sense of Definition 5.3.1), $W_\varphi^r(f,t)_p$ is given in (5.33), and $K_{r,\varphi}(f,t^r)_p$ is given in (2.1.1). Then we have for $f \in L_p(R_+)$, $1 \le p \le \infty$*

$$M^{-1} W_\varphi^r(f,t)_p \le K_{r,\varphi}(f,t^r)_p \le M W_\varphi^r(f,t)_p, \qquad 0 < t \le t_0, \qquad (5.3.4)$$

where M and t_0 are independent of f.

For the proof of the lower estimate of $K_{r,\varphi}(f,t^r)_p$ we will need the following lemma.

Lemma 5.3.3. *For $f \in L_p[0,\theta]$, $1 \le p \le \infty$, $P_r f$ the $L_2[0,\theta]$ projection of f onto the polynomials of degree at most $r-1$, we have*

$$\|P_r f\|_{L_p[0,\theta]} \le M \|f\|_{L_p[0,\theta]},$$

where $M \equiv M_{r,p}$ does not depend on θ.

PROOF. The result is probably known, but independence of θ, that is crucial for us, was not emphasized. Let $p_k(x)$ $0 \le k < r$ be orthonormal polynomials of degree k in $L_2[0,1]$. We have

$$\|p_k\|_{L_p[0,1]} = A_k(p) \le A \quad \text{for } k < r \quad \text{and} \quad 1 \le p \le \infty.$$

In $[0,\theta]$ the orthonormal system is $q_k(x) = \theta^{-1/2} p_k(x/\theta)$, and

$$\|q_k\|_{L_p[0,\theta]} = A_k(p) \cdot \theta^{(1/p)-(1/2)}.$$

Thus, if

$$P_r f = \sum_{k=0}^{r-1} a_k q_k,$$

then with $q = p/(p-1)$

$$|a_k| = \left| \int_0^\theta f \cdot q_k \right| \le \|f\|_{L_p[0,\theta]} \|q_k\|_{L_q[0,\theta]} \le \|f\|_{L_p[0,\theta]} A_k(q) \theta^{(1/q)-(1/2)}$$

from which we obtain

$$\|P_r f\|_{L_p[0,\theta]} \le \|f\|_{L_p[0,\theta]} \sum_{k=0}^{r-1} A_k(p) \theta^{(1/p)-(1/2)} A_k(q) \theta^{(1/q)-(1/2)} \le M \|f\|_{L_p[0,\theta]}.$$

$\square$

PROOF OF THEOREM 5.3.2. In the proof we will separate the behavior near 0 and near ∞. That is, we deal first with $t_1^* = \infty$ and then with $t_0^* = 0$, which we call Case I and II respectively. When both $t_0^* > 0$ and $t_1^* < \infty$ (i.e., $\beta(0) < 0$ and $\beta(\infty) > 1$), both parts of the proof must be combined. First we establish the upper estimate, i.e.,

$$K_{r,\varphi}(f,t^r)_p \le M W_\varphi^r(f,t)_p \qquad (5.3.5)$$

5. More General Step-Weight Functions φ

in both cases. Following the proof in Section 2.2 and modifying the constant c in (2.2.7)–(2.2.9), we construct a function G_t such that

$$\|f - G_t\|_{L_p(t_0^*/4, 4t_1^*)} + t^r \|\varphi^r G_t^{(r)}\|_{L_p(t_0^*/4, 4t_1^*)}$$
$$\leq M \inf_{0 < h \leq t} \|\Delta_{h\varphi}^r f\|_{L_p[Ch_0^*, h_1^*/C]} \leq M W_\varphi^r(f, t)_p.$$

We also have

$$\|f - P_{r,t}f\|_{L_p(0, t_0^*)} + t^r \|\varphi^r (P_{r,t}f)^{(r)}\|_{L_p(0, t_0^*)} = \|f - P_{r,t}f\|_{L_p(0, t_0^*)} \leq W_\varphi^r(f, t)_p$$

and

$$\|f - 0\|_{L_p(t_1^*, \infty)} + t^r \|\varphi^r 0^{(r)}\|_{L_p(t_1^*, \infty)} = \|f\|_{L_p(t_1^*, \infty)} \leq W_\varphi^r(f, t)_p.$$

The function $g_t(x)$ given by

$$g_t(x) = (P_{r,t}f)(x)\psi(4x/t_0^*) + G_t(x)(1 - \psi(4x/t_0^*))\psi(x/t_1^*)$$

(for the definition of ψ see (2.2.3)) which we obtain by the "patching together" technique of Section 2.2 (cf. (2.2.12)–(2.2.15)) satisfies

$$\|f - g_t\|_p + t^r \|\varphi^r g_t^{(r)}\|_p \leq M W_\varphi^r(f, t)_p$$

and this proves (5.3.5).

To establish the lower estimate for Case I ($\beta(0) < 0, \beta(\infty) \leq 1$) we have to show only

$$\|f - P_{r,t}f\|_{L_p[0, t_0^*]} \leq M K_{r,\varphi}(f, t)_p \qquad (5.3.6)$$

as the inequality

$$\|\Delta_{h\varphi}^r f\|_{L_p[Ch_0^*, \infty)} \leq M K_{r,\varphi}(f, t^r)_p \quad \text{for } 0 < h < t \qquad (5.3.7)$$

follows from the consideration in Section 2.4. For proving (5.3.7) we have to use here (5.3.1) and (5.3.2) instead of (1.2.1). Observe also that for $0 < h < t$ and $x > Ch_0^*$ ($0 \leq h \leq t_0$) one has $x/2 \leq x \pm rh\varphi(x)/2 \leq 2x$ and therefore $\varphi(x \pm rh\varphi(x)/2) \sim \varphi(x)$ and hence the computations needed here are simpler than but similar to those of Section 2.4. To demonstrate (5.3.6) we choose g_t satisfying

$$\|f - g_t\|_p + t^r \|\varphi^r g_t^{(r)}\|_p \leq 2K_{r,\varphi}(f, t^r)_p, \qquad (5.3.8)$$

and using Lemma 5.3.3, we have to show only that

$$\|g_t - P_{r,t}g_t\|_{L_p[0, t_0^*]} \leq M_1 K_{r,\varphi}(f, t^r)_p.$$

Since $\varphi^r g_t^{(r)} \in L_p$ and $\varphi(x) \sim x^\beta$, $\beta = \beta(0) < 0$ as $x \to 0+$, it follows that $\lim_{x \to 0+} g_t^{(i)}(x)$ exist for $0 \leq i < r$, and we may assume that $g^{(i)}(0)$ is that limit. Therefore, for $Q_{r-1}(x) \equiv \sum_{i=0}^{r-1} g^{(i)}(0) x^i / i!$ we have,

$$g(x) = Q_{r-1}(x) + \int_0^x \int_0^{x_1} \cdots \int_0^{x_{r-1}} g^{(r)}(\tau) \, d\tau \, dx_{r-1} \cdots dx_1.$$

Using repeatedly Hardy's inequality (Hardy, Littlewood, and Pólya [1, p. 244, (991)]) given by

5.3. Growth Orders of Type x^β with Arbitrary β

$$\left\|\frac{1}{x}\int_0^x h(u)\,du\right\|_{L_p(R_+)} \leq \frac{p}{p-1}\|h\|_{L_p(R_+)}, \quad 1 < p \leq \infty, \qquad (5.3.9)$$

we obtain for $1 < p \leq \infty$

$$(\varphi(t_0^*)/t_0^*)^r \|g_t - Q_{r-1}\|_{L_p[0,t_0^*]}$$

$$\leq M \left\|\frac{1}{x}\int_0^x \frac{1}{x_1}\int_0^{x_1} \cdots \frac{1}{x_{r-1}}\int_0^{x_{r-1}} \varphi(\tau)^r g_t^{(r)}(\tau)\,d\tau\,dx_{r-1}\cdots dx_1\right\|_{L_p[0,t_0^*]}$$

$$\leq M \left(\frac{p}{p-1}\right)^r \|\varphi^r g_t^{(r)}\|_{L_p(R_+)}.$$

For $p = 1$ we have

$$(\varphi(t_0^*)/t_0^*)^r \|g_t - Q_{r-1}\|_{L_1[0,t_0^*]}$$

$$\leq M(t_0^*)^{-r} \int_0^{t_0^*}\int_0^x\int_0^{x_1}\cdots\int_0^{x_{r-1}} \varphi(\tau)^r |g_t^{(r)}(\tau)|\,d\tau\,dx_{r-1}\cdots dx_1\,dx$$

$$\leq M(t_0^*)^{-r} \int_0^{t_0^*}\int_0^{t_0^*}\cdots\int_0^{t_0^*} \varphi(\tau)^r |g_t^{(r)}(\tau)|\,d\tau\,dx_{r-1}\cdots dx_1\,dx$$

$$= M\|\varphi^r g_t^{(r)}\|_{L_1[0,t_0^*]}.$$

Recalling that $\varphi(t_0^*)/t_0^* \sim t^{-1}$, we have, using Lemma 5.3.3,

$$\|g_t - P_{r,t}g_t\|_{L_p[0,t_0^*]} \leq \|g_t - Q_{r-1}\|_{L_p[0,t_0^*]} + \|P_{r,t}(g_t - Q_{r-1})\|_{L_p[0,t_0^*]}$$

$$\leq M\|g_t - Q_{r-1}\|_{L_p[0,t_0^*]} \leq M_1 t^r \|\varphi^r g_t^{(r)}\|_{L_p(R_+)}$$

$$\leq 2M_1 K_{r,\varphi}(f,t^r)_p,$$

which completes the proof of Case I.

We have already discussed the estimate $K_{r,\varphi}(f,t^r)_p \leq M\omega_\varphi^r(f,t)_p$ in Case II. To complete the proof, we have to establish the lower estimate for the K-functional in Case II (i.e., $\beta(0) \geq 0$ and $\beta(\infty) > 1$). For that it is enough to show

$$\|f\|_{L_p[t_1^*,\infty)} \leq MK_{r,\varphi}(f,t^r)_p.$$

We choose g_t as in (5.3.8) and observe that $\varphi^r g_t^{(r)} \in L_p$ and $\varphi(x) \sim x^\beta$ as $x \to \infty$ with $\beta = \beta(\infty) > 1$ imply

$$g_t(x) = (-1)^r \int_x^\infty \int_{x_1}^\infty \cdots \int_{x_{r-1}}^\infty g_t^{(r)}(\tau)\,d\tau\,dx_{r-1}\cdots dx_1.$$

The convergence of the multiple integral follows from

$$\int_x^\infty \int_{x_1}^\infty \cdots \int_{x_{r-1}}^\infty |g_t^{(r)}(\tau)|\,d\tau\,dx_{r-1}\cdots dx_1$$

$$\leq M \int_x^\infty x_1^{-\beta} \int_{x_1}^\infty x_2^{-\beta} \cdots \int_{x_{r-1}}^\infty \tau^{-\beta}|\varphi(\tau)^r g_t^{(r)}(\tau)|\,d\tau\,dx_{r-1}\cdots dx_1$$

$$\leq M_1 \int_x^\infty x_1^{-\beta} \int_{x_1}^\infty \cdots \int_{x_{r-2}}^\infty x_{r-1}^{-\beta}\|\varphi^r g_t^{(r)}\|_p\,dx_{r-1}\cdots dx_1 \leq M_2\|\varphi^r g^{(r)}\|_{L_p}.$$

We now recall that

$$\left\|\int_x^\infty h(\tau)\,d\tau\right\|_{L_p(R_+)} \le p\|xh(x)\|_{L_p(R_+)}, \quad 1 \le p < \infty$$

(Stein [1, p. 272, A.4]). Since $\beta = \beta(\infty) > 1$, for $\tau > t_1^*$ we have $(\varphi(\tau)/\tau)^r \le M\varphi(t_1^*)/t_1^*$, and therefore,

$(\varphi(t_1^*)/t_1^*)^r \|g_t\|_{L_p[t_1^*,\infty)}$

$$\le M \left\|\int_x^\infty \cdots \int_{x_{r-1}}^\infty \frac{\varphi(\tau)^r}{\tau^r} |g_t^{(r)}(\tau)|\,d\tau\,dx_{r-1}\cdots dx\right\|_{L_p[t_1^*,\infty)}$$

$$\le M \left\|\int_x^\infty \frac{1}{x_1}\int_{x_1}^\infty \cdots \int_{x_{r-2}}^\infty \frac{1}{x_{r-1}}\int_{x_{r-1}}^\infty \frac{1}{\tau}\{\varphi(\tau)^r|g_t^{(r)}(\tau)|\}\,d\tau\,dx_{r-1}\cdots dx_1\right\|_{L_p(R_+)}$$

$$\le M p^r \|\varphi^r g_t^{(r)}\|_{L_p(R_+)}.$$

For $p = \infty$ and $x \ge t_1^*$

$$|g_t(x)| \le M \|\varphi^r g_t^{(r)}\|_\infty \int_x^\infty \int_{x_1}^\infty \cdots \int_{x_{r-1}}^\infty \tau^{-r\beta}\,d\tau\cdots dx_1 \le M_1 x^{-r\beta+r} \|\varphi^r g_t^{(r)}\|_\infty$$

and therefore

$$(\varphi(t_1^*)/t_1^*)^r \|g_t\|_{L_\infty[t_1^*,\infty)} \le M_2 \|\varphi^r g_t^{(r)}\|_{L_\infty(R_+)}.$$

Using again $\varphi(t_1^*)/t_1^* \sim t^{-1}$, we complete the proof of Case II for $1 \le p \le \infty$ as

$$\|f\|_{L_p(t_1^*,\infty)} \le \|f - g_t\|_{L_p(t_1^*,\infty)} + \|g_t\|_{L_p(t_1^*,\infty)}$$

$$\le M(\|f - g_t\|_p + t^r \|\varphi^r g_t^{(r)}\|_p) \le M_1 K_{r,\varphi}(f, t^r)_p. \quad \square$$

CHAPTER 6
WEIGHTED MODULI OF SMOOTHNESS

We shall need weighted moduli of smoothness in some significant situations of which the following are examples:

(a) Investigation of the K-functional of the pair of spaces: a weighted L_p space and a weighted Sobolev space.
(b) Relation between modulus of smoothness of a function and that of its derivative (which turns out to be a weighted modulus of smoothness).
(c) Applications to best weighted polynomial approximation.
(d) Applications to weighted approximation of some linear operators.

Though we will introduce a complete modulus equivalent to the corresponding K-functional, we will mostly deal with the main-part modulus, which is sufficient for application and much easier to compute.

6.1. Weighted Moduli of Smoothness and Weighted K-Functionals

For a weight function w and step-weight function φ satisfying the conditions of Section 1.2 (the standard conditions of this work), the weighted K-functional is given by

$$K_{r,\varphi}(f, t^r)_{w,p} = \inf\{\|w(f-g)\|_p + t^r\|w\varphi^r g^{(r)}\|_p; g^{(r-1)} \in \text{A.C.}_{\cdot\text{loc}}\}. \quad (6.1.1)$$

To introduce the weighted modulus, we have to assume some conditions on w in addition to the assumptions on φ given in Section 1.2. We require that the weight function w be measurable on D ($D = (a,b)$, $(a,b) = I$, R_+ or R), that $w \sim 1$ in every compact subinterval of D, and that

$$w(x) \sim \begin{cases} |x|^{\gamma(a)} & \text{as } x \to a+0 \\ x^{\gamma(\infty)} & \text{as } x \to \infty & \text{when } b = \infty \\ (1-x)^{\gamma(1)} & \text{as } x \to 1-0 & \text{when } b = 1. \end{cases} \quad (6.1.2)$$

In this section we assume that $\gamma(c) \geq 0$ if $0 \leq \beta(c) < 1$ (cf. Section 1.2) at a finite endpoint c, but beginning with the next section $\gamma(a)$ and $\gamma(b)$ will be arbitrary real numbers.

We will give the definition of the weighted modulus of smoothness only for $D = R_+$. The changes needed for $D = (0, 1)$ do not cause any difficulty and will be given in Appendix B. Thus, let $D = R_+$ and

$$A_1 x^{\beta(0)} \leq \varphi(x) \leq A x^{\beta(0)}, \qquad 0 < x < 1. \quad (6.1.3)$$

We set

$$t^* = \begin{cases} (Ar)^{1/1-\beta(0)} t^{1/1-\beta(0)} & \text{if } 0 \leq \beta(0) < 1 \\ 0 & \text{if } \beta(0) \geq 1 \end{cases} \quad (6.1.4)$$

and define

$$\omega_\varphi^r(f, t)_{w,p} = \sup_{0 < h \leq t} \|w \Delta_{h\varphi}^r f\|_p \quad \text{if either} \quad \gamma(0) = 0 \quad \text{or} \quad \beta(0) \geq 1 \quad (6.1.5)$$

and

$$\omega_\varphi^r(f, t)_{w,p} = \sup_{0 < h \leq t} \|w \Delta_{h\varphi}^r f\|_{L_p[t^*, \infty)} + \sup_{0 < h \leq t^*} \|w \vec{\Delta}_h^r f\|_{L_p[0, 12t^*]} \quad (6.1.6)$$

for $\gamma(0) > 0$ and $0 \leq \beta(0) < 1$. The definition of $\omega_\varphi^r(f, t)_{w,p}$ in (6.1.5) represents also the situation for $D = R$, and for $D = I$ if the pair $\gamma(i)$ and $\beta(i)$ satisfies $\gamma(i) = 0$ or $\beta(i) \geq 1$ ($i = 0$ or $i = 1$). For $D = I$ if $\gamma(1) > 0$ and $0 \leq \beta(1) < 1$ we will have to use backwards differences too in the analogue of (6.1.6) (see Appendix B).

With this definition of ω we can state and prove the theorem establishing its equivalence with the weighted K-functional.

Theorem 6.1.1. *Suppose $K_{r,\varphi}(f, t^r)_{w,p}$ is given by (6.1.1), $\omega_\varphi^r(f, t)_{w,p}$ is given by (6.1.5) or (6.1.6), φ satisfies the conditions of Section 1.2, and w satisfies (6.1.2) with $\gamma(c) \geq 0$ if c ($c = a$ or $c = b$) is finite and $0 < \beta(c) < 1$. Then for some M we have*

$$M^{-1} \omega_\varphi^r(f, t)_{w,p} \leq K_{r,\varphi}(f, t^r)_{w,p} \leq M \omega_\varphi^r(f, t)_{w,p} \quad \text{for } 0 < t \leq t_0. \quad (6.1.7)$$

Note that this constitutes an extension of Theorem 2.1.1 since for $w \equiv 1$ the weighted and "ordinary" K-functionals and moduli of smoothness coincide. The same result is valid for $D = R$ and $D = (0, 1)$. (See Appendix B for definitions of ω_φ^r on these domains.) We will prove here the result for $D = R_+$.

Remark 6.1.2. One should note that we cannot write $\sup_{0 < h \leq t} \|w \Delta_{h\varphi}^r f\|$ in (6.1.6) if $\gamma(0) > 0$ and $\beta(0) < 1$, as that expression may not be finite even if $wf \in L_p$. If $\gamma(0) < 0$, $\|w \vec{\Delta}_h^r f\|_p$ may not be finite even if $wf \in L_p$. These problems

6.1. Weighted Moduli of Smoothness and Weighted K-Functionals

are of the nonfinite overlapping nature (with the measure $w(x)\,dx$). These are the reasons for defining $\omega_\varphi^r(f,t)_{w,p}$ as we did and restricting $\gamma(0)$ by $\gamma(0) \geq 0$ (at least when $\beta(0) < 1$).

Remark 6.1.3. We could have chosen Ct^*, as long as $C > (1/2)^{1/1-\beta(0)}$, instead of t^* in (6.1.6). Also $12t^*$ could be replaced by any Rt^* with $R > 1$ (and if Ct^* is chosen instead of t^*, by RCt^*). The number 12 is convenient for using the construction of Chapter 2. Another expression which in fact is equivalent to (6.1.6) is

$$\tilde{\omega}_\varphi^r(f,t)_{w,p} = \sup_{0<h\leq t} \|w\Delta_{h\varphi}^r f\|_{L_p[h^*,\infty)} + \sup_{0<h\leq t^*} \|w\vec{\Delta}_h^r f\|_{L_p[0,t^*]}. \quad (6.1.8)$$

PROOF OF THEOREM 6.1.1. We may be brief in the proof, as most of the ideas were developed in Sections 2.2, 2.3, and 2.4. We will deal with $D = R_+$. We first notice that if $\beta(0) \geq 1$, then $u \in (x - rt\varphi(x)/2, x + rt\varphi(x)/2)$ implies $(x/2) < u < 2x$ (for $t \leq t_0$), and therefore $w(u) \sim w(x)$ and $\varphi(u) \sim \varphi(x)$, and the proof in Sections 2.2 and 2.4 can be followed word for word. The second possible situation is $\gamma(0) = 0$, but $\gamma(\infty)$ may be different from zero, as otherwise we are dealing with the original case of Theorem 2.1.1. In this case for $x < 1$ the treatment is that of Theorem 2.1.1, and for $x > 1$ and u as above $w(u) \sim w(x)$ and $\varphi(x) \sim \varphi(u)$, and we follow Sections 2.2 and 2.4 as remarked earlier. So we are left with the case $\gamma(0) > 0$ and $0 \leq \beta \equiv \beta(0) < 1$. To prove our result for $1 \leq p < \infty$ (the arguments below can be easily followed for the proof of the theorem for $p = \infty$) we define

$$\omega_\varphi^{*r}(f,t)_{w,p} = \left\{\frac{1}{t}\int_0^t\int_{t^*}^\infty |w(x)\Delta_{\tau\varphi(x)}^r f(x)|^p\,dx\,d\tau\right\}^{1/p} \\ + \left\{\frac{1}{t^*}\int_0^{t^*}\int_0^{12t^*} |w(x)\vec{\Delta}_u^r f(x)|^p\,dx\,du\right\}^{1/p}. \quad (6.1.9)$$

Minor changes in the construction given in Section 2.3, that is, using

$$G_{t,2}(x) = \psi_l(x)\frac{1}{t^*}\int_0^{t^*} \tilde{f}_v(x)\,dv$$

instead of that given in (2.3.6) with

$$\tilde{f}_v(x) = r^r \int_0^{1/r}\cdots\int_0^{1/r}\left[\sum_{k=1}^r (-1)^{k+1}\binom{r}{k}f(x+kv(u_1+\cdots+u_r))\right]du_1\cdots du_r$$

instead of $\tilde{f}_v(x)$ in (2.3.7), yield

$$K_{r,\varphi}(f,t^r)_{w,p} \leq M_1 \omega_\varphi^{*r}(f,t)_{w,p}.$$

Now the inequality

$$\omega_\varphi^{*r}(f,t)_{w,p} \leq M_2 \omega_\varphi^r(f,t)_{w,p}$$

with the expression in (6.1.6) follows in a simpler fashion and therefore

$$K_{r,\varphi}(f,t^r)_{w,p} \leq M\omega_\varphi^r(f,t)_{w,p}$$

has been established. To get the lower estimate for the K-functional, we recall that for $x > h^*$ we have $x - rh\varphi(x)/2 \geq x/2$ (for $x > Ch^*$, where $C > 2^{-1/1-\beta}$ we have $x - rh\varphi(x)/2 > \eta x$) and therefore for $u \in (x - rh\varphi(x)/2, x + rh\varphi(x)/2)$, $\varphi(u) \sim \varphi(x)$ and $w(u) \sim w(x)$, from which one can deduce, using Section 2.4, that for $0 < h \leq t$

$$\|w\Delta_{h\varphi}^r f\|_{L_p[h^*, \infty)} \leq MK_{r,\varphi}(f, t^r)_{w,p}$$

and this yields

$$\sup_{0 < h \leq t} \|w\Delta_{h\varphi}^r f\|_{L_p[h^*, \infty)} < MK_{r,\varphi}(f, t^r)_{w,p}.$$

To estimate $\|w\vec{\Delta}_{h\varphi}^r f\|_{L_p[0, 12t^*]}$ we follow the technique of Section 2.4. We choose g_t such that

$$\|w(f - g_t)\|_p \leq 2K_{r,\varphi}(f, t^r)_{w,p} \quad \text{and} \quad t^r \|w\varphi^r g_t^{(r)}\|_p \leq 2K_{r,\varphi}(f, t^r)_{w,p}.$$

Obviously

$$\|w\vec{\Delta}_h^r(f - g_t)\|_{L[0, 12t^*]} \leq M_1 2^r \|w(f - g_t)\|_{L_p(R_+)},$$

where $|w(x)/w(x + u)| \leq M_1$ for $0 < u \leq rh$. (Note that $t^* \neq 0$ implies $\gamma(0) \geq 0$ and hence such M_1 exists.) For $g_t^{(r-1)} \in \text{A.C.}_{\text{loc}}$ we use Taylor's formula with integral remainder, and write

$$|\vec{\Delta}_h^r g(x)| \leq M \int_0^{rh} \frac{((rh/2) - |rh/2 - u|)^{r-1}}{\varphi(x + u)^r} \varphi(x + u)^r |g^{(r)}(x + u)| \, du.$$

We observe that

$$\int_0^h \frac{((h/2) - |(h/2) - u|)^{r-1}}{\varphi(x + u)^r} du \leq M \int_0^h \frac{((h/2) - |(h/2) - u|)^{r-1}}{u^{r\beta}} du \leq M_1 h^{(1-\beta)r}$$

and

$$\int_{x:0 \leq x \leq v \leq x+h} ((h/2) - |(h/2) - (v - x)|)^{r-1} dx \leq M_2 \min(v^r, h^r)$$
$$\leq M_3(\varphi(v))^r h^{(1-\beta)r}.$$

Therefore, using Jensen's inequality (cf. Section 2.4) for $0 < h \leq t^* \equiv (Art)^{1/1-\beta}$, we obtain

$$\|w\vec{\Delta}_h^r g\|_{L_p(0, 12t^*)}$$
$$\leq M \left\{ \int_0^{12t^*} \left(w(x) \int_0^{rh} \frac{((rh/2) - |(rh/2) - u|)^{r-1}}{\varphi(x + u)^r} \varphi(x + u)^r |g^{(r)}(x + u)| \right)^p dx \right\}^{1/p}$$
$$\leq M_4 \left\{ \int_0^1 w(v)^p \varphi(v)^{r(p-1)} |g^{(r)}(v)|^p h^{(1-\beta)r(p-1)} \right.$$
$$\left. \times \int_{x:0 \leq x \leq v \leq x+rh} ((rh/2) - |(rh/2) - (v - x)|)^{r-1} dx \, dv \right\}^{1/p}$$
$$\leq M_5 h^{(1-\beta)r} \|w\varphi^r g^{(r)}\|_p \leq M_3 t^r \|w\varphi^r g^{(r)}\|_p.$$

This now implies
$$\sup_{0<h\leq t^*} \|w\vec{\Delta}_h^r f\|_{L_p[0,12t^*]} \leq M_6 K_{r,\varphi}(f,t^r)_{w,p}$$
and therefore the inequality on the left of (6.1.7), which completes the proof of Theorem 6.1.1. □

Following Theorem 4.1.3, one can also show the following result.

Theorem 6.1.4. *For $\omega_\varphi^j(f,t)_{w,p}$ satisfying (6.1.5) or (6.1.6) we have*
$$\omega_\varphi^{r+1}(f,t)_{w,p} \leq M\omega_\varphi^r(f,t)_{w,p}.$$

PROOF. Actually we estimate $\omega_\varphi^{r+1}(f,t)_{w,p}$ by $K_{r,\varphi}(f,t^r)_{w,p}$ and we may use for this the same type of calculation used for the lower estimate of the K-functional in Theorem 6.1.1. □

6.2. The Weighted Main-Part Modulus

In Section 6.1 the weighted modulus was discussed but the discussion did not, and could not, apply to the situation $\gamma(0) < 0$, $0 \leq \beta = \beta(0) < 1$. In applications, however, there are cases in which that situation occurs. We will introduce here the main-part modulus which will answer problems related to applications as well as having the other advantages and characteristics of the main-part modulus. In the rest of the chapter the weight function w will satisfy (6.1.2) *without any restriction* on $\gamma(c)$. We define the weighted main-part modulus for $D = R_+$ by

$$\Omega_\varphi^r(C,f,t)_{w,p} = \sup_{0<h\leq t} \|w\Delta_{h\varphi}^r f\|_{L_p[Ch^*,\infty)},$$
$$\Omega_\varphi^r(1,f,t)_{w,p} = \Omega_\varphi^r(f,t)_{w,p}, \tag{6.2.1}$$

where $C > 2^{1/\beta(0)-1}$ and h^* is given by (6.1.3) and (6.1.4). For $D = R$ the main-part modulus agrees with $\omega_\varphi^r(f,t)_{w,p}$ and for $D = (0,1)$ the main-part modulus will be given in Appendix B. We also define

$$\Omega_\varphi^{*r}(C,f,t)_{w,p} = \sup_{0<h\leq t}\left\{\frac{1}{h}\int_0^h\int_{Ch^*}^\infty |w(x)\Delta_{\tau\varphi(x)}^r f(x)|^p \, dx \, d\tau\right\}^{1/p} \tag{6.2.2}$$

$$\Omega_\varphi^{*r}(C,f,t)_{w,\infty} = \sup_{0<h\leq t}\sup_{x\geq Ch^*} \frac{1}{h}\int_0^h |w(x)\Delta_{\tau\varphi(x)}^r f(x)| \, d\tau.$$

The main-part K-functional, which will correspond to Ω, is given for $C > 0$ by

$$\mathcal{K}_{r,\varphi}(C,f,t^r)_{w,p} = \sup_{0<h\leq t}\inf_g \{\|w(f-g)\|_{L_p[Ch^*,\infty)}$$
$$+ h^r\|w\varphi^r g^{(r)}\|_{L_p[Ch^*,\infty)}; g^{(r-1)} \in \text{A.C.}_{\cdot\text{loc}}(Ch^*,\infty)\}, \tag{6.2.3}$$

$$\mathcal{K}_{r,\varphi}(1,f,t^r)_{w,p} \equiv \mathcal{K}_{r,\varphi}(f,t^r)_{w,p}.$$

60   6. Weighted Moduli of Smoothness

Notice that for $\beta(0) \geq 1$ the main-part K-functional agrees with the K-functional of (6.1.1).

The properties of $\Omega_\varphi^r(C, f, t)_{w,p}$, their relations for different C satisfying $C > 2^{1/\beta - 1}$ and with $\mathscr{K}_{r,\varphi}(C, f, t^r)_{w,p}$ are given in the following theorem.

Theorem 6.2.1. *For* $\Omega_\varphi^r(C, f, t)_{w,p}$ *and* $\mathscr{K}_{r,\varphi}(C, f, t^r)_{w,p}$ *given in this section (where φ satisfies the conditions of Section 1.2 and w satisfies (6.1.2)) any C, $C_1 > 2^{1/\beta(0)-1}$ and any $C_2, C_3 > 0$, we have:*

$$\Omega_\varphi^r(C, f, t)_{w,p} \sim \Omega_\varphi^r(C_1, f, t)_{w,p}. \tag{6.2.4}$$

$$\mathscr{K}_{r,\varphi}(C_2, f, t^r)_{w,p} \sim \mathscr{K}_{r,\varphi}(C_3, f, t^r)_{w,p}. \tag{6.2.5}$$

$$\Omega_\varphi^r(C, f, t)_{w,p} \sim \mathscr{K}_{r,\varphi}(C_2, f, t^r)_{w,p} \sim \Omega_\varphi^{*r}(C, f, t)_{w,p}. \tag{6.2.6}$$

$$\mathscr{K}_{r,\varphi}(C_2, f, (\lambda t)^r)_{w,p} \leq M\lambda^r \mathscr{K}_{r,\varphi}(C_2, f, t^r)_{w,p}, \quad \lambda > 1. \tag{6.2.7}$$

$$\Omega_\varphi^r(C, f, (\lambda t)^r)_{w,p} \leq M\lambda^r \Omega_\varphi^r(C, f, t^r)_{w,p}, \quad \lambda > 1, t, \lambda t \in [0, t_0]. \tag{6.2.8}$$

Moreover, the equivalence relations (6.2.4)–(6.2.6) hold uniformly in f and small t for fixed C, C_1, C_2, and C_3.

As a result of this theorem we may deal with $\Omega_\varphi^r(f, t)_{w,p}$ and $\mathscr{K}_{r,\varphi}(f, t^r)_{w,p}$; that is, with the case $C = 1$.

PROOF. In view of Section 6.1, we can assume $0 \leq \beta(0) \equiv \beta < 1$, i.e., $t^* > 0$ for $t > 0$. For $C > C_1$ and $C_2 > C_3$ we have

$$\Omega_\varphi^r(C, f, t)_{w,p} \leq \Omega_\varphi^r(C_1, f, t)_{w,p} \quad \text{and} \quad \mathscr{K}_{r,\varphi}(C_2, f, t^r)_{w,p} \leq \mathscr{K}_{r,\varphi}(C_3, f, t^r)_{w,p}.$$

We follow the proof of the upper estimate for the K-functional in Section 2.3 with a minor change in the definition of $G_{t,1}$ (which now will be all that counts, as G_2 is not needed), namely replace $\psi_l(x)$ by 0 in (2.3.6). We now obtain (as in (2.3.8) and (2.3.9))

$$\inf_g \{\|w(f - g)\|_{L_p[12Ch^*, \infty)} + h^r \|w\varphi^r g^{(r)}\|_{L_p[12Ch^*, \infty)}; g^{(r-1)} \in \text{A.C.}_{\text{loc}}\}$$

$$\leq M\Omega_\varphi^{*r}(C, f, h)_{w,p} \leq M\Omega_\varphi^r(C, f, h)_{w,p} \leq M\Omega_\varphi^r(C, f, t)_{w,p},$$

when we recall that for $\tau \leq h \leq t$, $x \geq Ch^*$ and $u \in [x - (r/2)\tau\varphi(x), x + (r/2)\tau\varphi(x)]$ one has $R^{-1}w(x) \leq w(u) \leq Rw(x)$, where R depends on C but not on h or t. Therefore,

$$\mathscr{K}_{r,\varphi}(12C, f, t^r)_{w,p} \leq \Omega_\varphi^r(C, f, t)_{w,p}.$$

Following the technique of Section 2.4 and recalling that with some $\eta > 0$ we have for $x > Ch^*$, $\eta x < x - (r/2)h\varphi(x) < x + (r/2)h\varphi(x) < 2x$, we can write with $\eta_1 \equiv \eta^{1-\beta}$,

$$\Omega_\varphi^r(C, f, t) = \sup_{0 < h \leq t} \|w\Delta_{h\varphi}^r f\|_{L_p[Ch^*, \infty)}$$

$$\leq M \sup_{0 < h \leq t} \left(\inf_g (\|w(f - g)\|_{L_p[\eta Ch^*, \infty]} + h^r \|w\varphi^r g^{(r)}\|_{L_p[\eta Ch^*, \infty)} \right)$$

6.2. The Weighted Main-part Modulus

$$\leq M_1 \eta_1^{-r} \operatorname*{Sup}_{h \leq t} \left(\inf_g (\|w(f-g)\|_{L_p[C(\eta_1 h)^*, \infty]} \right.$$

$$\left. + (\eta_1 h)^r \|w\varphi^r g^{(r)}\|_{L_p[C(\eta_1 h)^*, \infty]} \right)$$

$$\leq M_2 \eta_1^{-r} \mathscr{K}_{r,\varphi}(C, f, t^r)_{w,p}.$$

This completes the proof of (6.2.6) pending the proof of (6.2.5).
We have for $C_3 < C_2$ and $h_1 = h(C_3/C_2)^{1-\beta}$

$$\mathscr{K}_{r,\varphi}(C_3, f, t^r)_{w,p} = \operatorname*{Sup}_{0 < h \leq t} \inf_g \{\|w(f-g)\|_{L_p[C_3 h^*, \infty)} + h^r \|w\varphi^r g^{(r)}\|_{L_p[C_3 h^*, \infty)}\}$$

$$\leq \operatorname*{Sup}_{0 < h \leq t} \inf_g \{\|w(f-g)\|_{L_p[C_2 h_1^*, \infty]}$$

$$+ (C_3/C_2)^{r(\beta-1)} h_1^r \|w\varphi^r g^{(r)}\|_{L_p[C_2 h_1^*, \infty)}\}$$

$$\leq (C_2/C_3)^{r(1-\beta)} \mathscr{K}_{r,\varphi}(C_2, f, (t(C_3/C_2)^{1-\beta})^r)_{w,p}$$

$$\leq (C_2/C_3)^{r(1-\beta)} \mathscr{K}_{r,\varphi}(C_2, f, t^r)_{w,p},$$

which, combined with the above proof, implies the assertions (6.2.4) and (6.2.5). To show (6.2.7), from which (6.2.8) also follows, we set

$$C_* = C_2 \lambda^{1/1-\beta} \quad (\lambda^{1/1-\beta} > 1) \quad \text{and} \quad h_1 = h/\lambda,$$

and write

$$\mathscr{K}_{r,\varphi}(C_2, f, (\lambda t)^r)_{w,p} = \operatorname*{Sup}_{0 < h \leq \lambda t} \inf_g \{\|w(f-g)\|_{L_p[C_2 h^*, \infty)} + h^r \|w\varphi^r g^{(r)}\|_{L_p[C_2 h^*, \infty)}\}$$

$$= \operatorname*{Sup}_{0 < h_1 \leq t} \inf_g (\|w(f-g)\|_{L_p[C_* h_1^*, \infty)}$$

$$+ \lambda^r h_1^r \|w\varphi^r g^{(r)}\|_{L_p[C_* h_1^*, \infty)})$$

$$\leq \lambda^r \mathscr{K}_{r,\varphi}(C_*, f, t^r)_{w,p} \leq \lambda^r \mathscr{K}_{r,\varphi}(C_2, f, t^r)_{w,p}. \quad \square$$

Theorem 6.2.2. *Suppose w satisfies (6.1.2) with $\gamma(0) \geq 0$, ω is given by (6.1.5) and (6.1.6), and Ω is given by (6.2.1). Then*

$$M^{-1} \Omega_\varphi^r(f, t)_{w,p} \leq \omega_\varphi^r(f, t)_{w,p} \leq M \int_0^t (\Omega_\varphi^r(f, \tau)_{w,p}/\tau) \, d\tau. \tag{6.2.9}$$

We can also state the following equivalent result.

Corollary 6.2.3. *Suppose w satisfies (6.1.2) with $\gamma(0) \geq 0$, K is given by (6.1.1), and $\mathscr{K}$ is given by (6.2.3). Then*

$$M^{-1} \mathscr{K}_{r,\varphi}(f, t^r)_{w,p} \leq K_{r,\varphi}(f, t^r)_{w,p}$$

$$\leq M \int_0^t (\mathscr{K}_{r,\varphi}(f, \tau^r)_{w,p}/\tau) \, d\tau. \tag{6.2.10}$$

PROOF. The proof of Theorem 6.2.2 follows word for word that of Theorem 3.3.2. In the proof we can replace $V(t)$ of (3.3.5) (in case of $D = R_+$) by

$$V(t) = \left\{ \frac{1}{t^*} \int_0^{t^*} \int_0^{(12+r)t^*} w(x) |\vec{\Delta}_u^r f(x)|^p \, dx \, du \right\}^{1/p}$$

and split $V(t)$ into $V(t/2)$ and a remainder that depends on $\Omega^*(C, f, tB)$ in exactly the same way as was done in the proof of Theorem 3.3.2. Corollary 6.2.3 follows from Theorem 6.2.2, Theorem 6.2.1 (6.2.6), and Theorem 6.1.1. □

Remark 6.2.4. While in Theorem 6.2.2 the restriction $\gamma(0) \geq 0$ cannot be dropped without changing the definition of ω in (6.1.6), Corollary 6.2.3 can be stated (but does not follow from the present proof) for other $\gamma(0)$ (not only $\gamma \geq 0$).

We can also show that the following result is valid.

Theorem 6.2.5. For $C > 2^{1/\beta(0)-1}$, an integer r and $\Omega_\varphi^j(C, f, t)_{w,p}$ given by (6.2.1), we have $\Omega_\varphi^{r+1}(C, f, t)_{w,p} \leq M\Omega_\varphi^r(C, f, t)_{w,p}$.

PROOF. We estimate $\Omega_\varphi^{r+1}(C, f, t)_{w,p}$ by $\mathcal{K}_{r,\varphi}(C, f, t^r)_{w,p}$ following the proof of Theorem 4.1.3 and then we use Theorem 6.2.1. □

Remark 6.2.6. For $\varphi_1 \sim \varphi_2$ it follows from Theorem 6.2.1 that

$$\Omega_{\varphi_1}^r(f, t)_{w,p} \sim \Omega_{\varphi_2}^r(f, t)_{w,p}.$$

Using this, it is possible to extend many properties of ω_φ^r given in Chapter 4 to $\Omega_\varphi^r(f, t)_{w,p}$. This, however is not applicable to the Marchaud-type inequality. (See also Section 6.4.)

6.3. Smoothness Properties of Derivatives

For the classical moduli of smoothness ($\varphi \equiv 1$) it is well-known that for $l < \alpha \leq r$, $\omega^r(f, t)_p = O(t^\alpha)$ and $\omega^{r-l}(f^{(l)}, t)_p = O(t^{\alpha-l})$, ($f^{(l-1)} \in$ A.C. and $f^{(l)} \in L_p$) are equivalent. The extension of this result to our moduli involves weighted moduli of smoothness even for the nonweighted moduli of (2.1.2).

Theorem 6.3.1. Suppose φ satisfies the conditions of Section 1.2, w satisfies (6.1.2), $1 \leq p \leq \infty$, $r \geq 1$, $0 < l < r$, and Ω is given by (6.2.1) or if $D = [0, 1]$ by Appendix B. Then

(a) $\int_0^1 (\Omega_\varphi^r(f, \tau)_{w,p}/\tau^{l+1}) d\tau < \infty$ implies that $f^{(l-1)}$ is locally absolutely continuous, $f^{(l)} \in L_p$ locally (i.e., $f^{(l)} \in L_p[c, d]$ for all c, d satisfying $a < c < d < b$) and

6.3. Smoothness Properties of Derivatives

$$\Omega_\varphi^{r-l}(f^{(l)}, t)_{w\varphi^l, p} \leq L \int_0^t (\Omega_\varphi^r(f, \tau)_{w, p}/\tau^{l+1}) \, d\tau, \qquad 0 < t < 1, \quad (6.3.1)$$

where L is independent of t and f;

(b) if $f^{(l-1)}$ is locally absolutely continuous and $f^{(l)} \in L_p$ locally, we have

$$\Omega_\varphi^r(f, t)_{w, p} \leq L t^l \Omega_\varphi^{r-l}(f^{(l)}, t)_{w\varphi^l, p}, \qquad 0 < t < 1, \quad (6.3.2)$$

where L is independent of t and f; and in particular

(c) for $l < \alpha$

$$\Omega_\varphi^r(f, t)_{w, p} = O(t^\alpha) \quad \text{and} \quad \Omega_\varphi^{r-l}(f^{(l)}, t)_{w\varphi^l, p} = O(t^{\alpha-l})$$

are equivalent.

Combining (c) with Theorem 6.2.2, we obtain

Corollary 6.3.2. *For $1 \leq l \leq \alpha$*

$$\omega_\varphi^r(f, t)_p = O(t^\alpha) \quad \text{and} \quad \omega_\varphi^{r-l}(f^{(l)}, t)_{\varphi^l, r} = O(t^{\alpha-l})$$

are equivalent provided f is locally in L_p.

The following special case of Theorem 6.3.1 is worth mentioning separately.

Corollary 6.3.3. *Suppose $\beta(c) \geq 1$ (cf. Section 1.2) if c is finite. Then*

(a) $\int_0^1 (\omega_\varphi^r(f, \tau)_{w, p}/\tau^{l+1}) \, d\tau < \infty$ *implies that $f^{(l-1)}$ is locally absolutely continuous, $w\varphi^l f^{(l)} \in L_p$ locally, and*

$$\omega_\varphi^{r-l}(f^{(l)}, t)_{w\varphi^l, p} \leq L \int_0^t (\omega_\varphi^r(f, \tau)_{w, p}/\tau^{l+1}) \, d\tau, \quad \text{for } 0 < t < 1;$$

(b) *if $f^{(l-1)}$ is locally absolutely continuous and $w\varphi^l f^{(l)} \in L_p$ locally, we have*

$$\omega_\varphi^r(f, t)_{w, p} \leq L t^l \omega_\varphi^{r-l}(f^{(l)}, t)_{w\varphi^l, p},$$

where L is independent of t ($0 < t < 1$) and f; and in particular

(c) *for $l < \alpha$*

$$\omega_\varphi^r(f, t)_{w, p} = O(t^\alpha) \quad \text{and} \quad \omega_\varphi^{r-l}(f^{(l)}, t)_{w\varphi^l, p} = O(t^{\alpha-l})$$

are equivalent.

Remark 6.3.4. (a) Even if $w = 1$, we have weighted moduli in our estimates in case $\varphi \neq 1$.

(b) Actually the main result is for $l = 1$, as the result for other l can be achieved by simple iteration from it.

PROOF OF THEOREM 6.3.1. We shall prove the theorem only for $D = R_+$, as for other domains the proof is similar. We first consider $0 \leq \beta(0) < 1$ for which

$t^* > 0$. Using Theorem 6.2.1, we can replace $\Omega_\varphi^r(f,t)_{w,p}$ by $\mathscr{K}_{r,\varphi}(f,t^r)_{w,p}$. For given t we can choose g_t satisfying

$$\|w(f - g_t)\|_{L_p[t^*,\infty)} + t^r\|w\varphi^r g_t^{(r)}\|_{L_p[t^*,\infty)} \le 2\mathscr{K}_{r,\varphi}(f,t^r)_{w,p}. \qquad (6.3.3)$$

We now observe that for $0 < h < d - c$, $0 < j < r$, and $g^{(r-1)}$ absolutely continuous in $[c,d]$ we have, following Ditzian [3, p. 311, Lemma 2.1],

$$\|g^{(j)}\|_{L_p[c,d]} \le M(h^{-j}\|g\|_{L_p[c,d]} + h^{r-j}\|g^{(r)}\|_{L_p[c,d]}), \qquad (6.3.4)$$

where M does not depend on c, d, p, or g. Actually in the statement of Lemma 2.1 in the paper quoted $d - c$ replaces h. In that proof the expression (6.3.4) appears and then $h = (d - c)/2r$ is substituted, but any $h \le (d - c)/2r$ could be substituted, and changing to $h \le d - c$ only alters the constant (in a way that depends only on r).

For a fixed $\eta > 0$ and $\eta t \le t_1 < t$ we will show

$$\|w\varphi^j(g_t - g_{t_1})^{(j)}\|_{L_p[t^*,\infty)} \le Bt^{-j}\mathscr{K}_{r,\varphi}(f,t^r)_{w,p}, \qquad (6.3.5)$$

where B does not depend on t, $0 \le j \le r$, or f. For $j = 0$ and $j = r$ this follows from (6.3.3) when we recall that $t_1 < t$ implies $t_1^* < t^*$ and $\eta < t_1/t < 1$. For other j we use (6.3.4) with $g = g_t - g_{t_1}$, $[c,d] = [2^k t^*, 2^{k+1} t^*]$, and $h = (2^k t^*)^{\beta(0)} t$ if $2^k t^* < 1$ and $h = (2^k t^*)^{\beta(\infty)} t$ if $2^k t^* \ge 1$ ($k = 0, 1, 2, \ldots$) to obtain

$$\|(g_t - g_{t_1})^{(j)}\|^p_{L_p[2^k t^*, 2^{k+1} t^*]} \le 2^p M((2^k t^*)^{-\beta j p} t^{-jp}\|g_t - g_{t_1}\|^p_{L_p[2^k t^*, 2^{k+1} t^*]}$$
$$+ (2^k t^*)^{\beta(r-j)p} t^{(r-j)p}\|g_t^{(r)} - g_{t_1}^{(r)}\|^p_{L_p[2^k t^*, 2^{k+1} t^*]}),$$

where β is $\beta(0)$ or $\beta(\infty)$ for the appropriate k. Multiplying the above expression by $w(2^k t^*)^p (2^k t^*)^{\beta j p} \sim [w(x)\varphi(x)^j]^p$ in $x \in (2^k t^*, 2^{k+1} t^*)$, and summing on k, we obtain, using (6.3.5) for $j = 0$ and $j = r$,

$$\|w\varphi^j(g_t - g_{t_1})^{(j)}\|^p_{L_p[t^*,\infty)} \le M_1(t^{-jp}\|w(g_t - g_{t_1})\|^p_{L_p[t^*,\infty)}$$
$$+ t^{(r-j)p}\|w\varphi^r(g_t - g_{t_1})^{(r)}\|^p_{L_p[t^*,\infty)})$$
$$\le M_2 t^{-jp}\mathscr{K}_{r,\varphi}(f,t^r)^p_{w,p}.$$

For $p = \infty$ we do not take the pth power but the supremum in the last few expressions and still obtain (6.3.5).

We now write

$$f = g_t + \sum_{n=0}^{\infty}(g_{t/2^{n+1}} - g_{t/2^n}).$$

Proving our theorem for $l = 1$, we use $j = 1$ in the above which implies that

$$g_t' + \sum_{n=0}^{\infty}(g'_{t/2^{n+1}} - g'_{t/2^n})$$

is locally convergent in L_p and therefore f is locally absolutely continuous. Moreover,

6.3. Smoothness Properties of Derivatives

$$\|w\varphi(f' - g'_t)\|_{L_p[t^*,\infty)} \leq \sum_{n=0}^{\infty} \|w\varphi(g'_{t/2^{n+1}} - g'_{t/2^n})\|_{L_p[t^*,\infty)}$$

$$\leq \sum_{n=0}^{\infty} \|w\varphi(g'_{t/2^{n+1}} - g'_{t/2^n})\|_{L_p[(t/2^n)^*,\infty)}$$

$$\leq B \sum_{n=0}^{\infty} \left(\frac{2^n}{t}\right) \mathcal{K}_{r,\varphi}(f, (t/2^n)^r)_{w,p}$$

$$\leq B_1 \int_0^t \mathcal{K}_{r,\varphi}(f, \tau^r)/\tau^2 \, d\tau$$

and

$$t^{r-1}\|w\varphi\varphi^{r-1}(g'_t)^{(r-1)}\|_{L_p[t^*,\infty)} \leq 2\frac{1}{t}\mathcal{K}_{r,\varphi}(f, t^r)_{w,p}$$

$$\leq M \int_0^t \mathcal{K}_{r,\varphi}(f, \tau^r)/\tau^2 \, d\tau,$$

which completes the proof of (6.3.1) for $l = 1$. For $l > 1$ we use induction on l. We have actually shown that the $L_p[t^*, \infty)$ norm of $w\varphi f'$ is bounded by

$$\|w\varphi g'_t\|_{L_p[t^*,\infty)} + M \int_0^t (\mathcal{K}_{r,\varphi}(f, \tau^r)/\tau^2) \, d\tau.$$

To prove (6.3.2) (again we treat just $l = 1$ and $D = R_+$) we will show that for $1 \leq p < \infty$ and some $L \geq 1$ to be chosen later

$$\left\{\int_{h^*}^{\infty} w(x)^p \frac{1}{h} \int_0^{h/L} |\Delta^r_{\tau\varphi(x)} f(x)|^p \, d\tau \, dx\right\}^{1/p} \leq Mt\Omega^{r-1}_\varphi(f', t)_{w\varphi, p}, \quad h \leq t \quad (6.3.6)$$

and

$$\sup_{\tau \leq (t/L)} \sup_{x \geq t^*} w(x)|\Delta^r_{\tau\varphi(x)} f(x)| \leq Mt\Omega^{r-1}_\varphi(f', t)_{w\varphi, \infty}. \quad (6.3.7)$$

We recall first that

$$\Delta^r_{\tau\varphi(x)} f(x) = \int_{-\tau\varphi(x)/2}^{\tau\varphi(x)/2} \Delta^{r-1}_{\tau\varphi(x)} f'(x + u) \, du \quad \text{a.e.}$$

We observe that it is enough to demonstrate (6.3.6) for $h = t$ and write

$$\int_{t^*}^{\infty} w(x)^p \frac{1}{t} \int_0^{t/L} |\Delta^r_{\tau\varphi(x)} f(x)|^p \, d\tau \, dx$$

$$\leq \int_{t^*}^{\infty} w(x)^p \frac{1}{t} \int_0^{t/L} (t\varphi(x)/L)^{p-1} \int_{-t\varphi(x)/2L}^{t\varphi(x)/2L} |\Delta^{r-1}_{\tau\varphi(x)} f'(x+u)|^p \, du \, d\tau \, dx$$

$$\leq M_1 \int_{t^*}^{\infty} w(x)^p \varphi(x)^{p-1} t^{p-2} \int_{-t\varphi(x)/2L}^{t\varphi(x)/2L} \int_0^{t/L} |\Delta^{r-1}_{\tau\varphi(x)} f'(x+u)|^p \, d\tau \, du \, dx \equiv I.$$

We recall that for $L \geq 1$, $|u| \leq t\varphi(x)/2$, $t^* < x < 1$, and $\beta = \beta(0)$ we have

$$x + u \geq x - t\varphi(x)/2 \geq x - Atx^\beta/2 \geq (t^*)^\beta((t^*)^{1-\beta} - At/2) = t^*\left(1 - \frac{1}{2r}\right)$$

which implies $x/2 \leq x + u \leq 2x$ and therefore $L^{-1}\varphi(x+u) \leq \varphi(x) \leq L\varphi(x+u)$ (and this is our choice of L). We now write following Lemma 2.2.1 with $\varphi_1(x+u)$ chosen as our $\varphi(x)$

$$\int_0^{t/L} |\Delta_{\tau\varphi(x)}^{r-1} f'(x+u)|^p \, d\tau = \frac{1}{\varphi(x)} \int_0^{t\varphi(x)/L} |\Delta_v^{r-1} f'(x+u)|^p \, dv$$

$$\leq \frac{L}{\varphi(x+u)} \int_0^{t\varphi(x+u)} |\Delta_v^{r-1} f'(x+u)| \, dv$$

$$\leq L \int_0^t |\Delta_{\tau\varphi(x+u)}^{r-1} f'(x+u)|^p \, d\tau.$$

Therefore

$$I \leq M_2 \int_0^t \int_{t^*}^\infty w(x)^p \varphi(x)^{p-1} t^{p-2} \int_{-t\varphi(x)/2L}^{t\varphi(x)/2L} |\Delta_{\tau\varphi(x+u)}^{r-1} f(x+u)|^p \, du \, dx \, d\tau$$

$$\leq M_3 t^{p-1} \sup_{\tau \leq t} \int_{t^*}^\infty \int_{-t\varphi(x)/2L}^{t\varphi(x)/2L} w(x+u)^p \varphi(x+u)^{p-1} |\Delta_{\tau\varphi(x+u)}^{r-1} f'(x+u)|^p \, du \, dx$$

$$\leq M_3 t^{p-1} \sup_{\tau \leq t} \int_{t^*(1-(1/2r))}^\infty w(v)^p \varphi(v)^{p-1} |\Delta_{\tau\varphi(v)}^{r-1} f'(v)|^p \left\{ \int_{|x-v| \leq t\varphi(x)/2L} dx \right\} dv$$

$$\leq M_4 t^p \Omega_\varphi^{r-1}(f', t)_{w\varphi, p}^p$$

as

$$\int_{|x-v| \leq t\varphi(x)/2L} dx \leq Ct\varphi(v)$$

and

$$t^*\left(1 - \frac{1}{2r}\right) \geq (A(r-1))^{1/1-\beta(0)} t^{1/1-\beta(0)}.$$

The proof of (6.3.7) is similar but simpler, as the left side of (6.3.7) is bounded by

$$\sup_{h \leq t/L} \sup_{x \leq t^*} w(x) \int_{-h\varphi(x)/2}^{h\varphi(x)/2} |\Delta_{h\varphi(x)}^{r-1} f'(x+u)| \, du$$

$$\leq \sup_{h \leq t/L} \sup_{x \geq t^*} \sup_{|u| \leq h\varphi(x)/2} hw(x)\varphi(x) |\Delta_{h\varphi(x)}^{r-1} f'(x+u)|$$

$$\leq M_5 t \sup_{h \leq t} \sup_{v \geq t^*(1-(1/2r))} hw(v)\varphi(v) |\Delta_{h\varphi(v)}^{r-1} f'(v)|.$$

To complete the proof we recall (see Theorem 6.2.1) that $\mathcal{K}_{r,\varphi}(L^{1/1-\beta},f,(t/L)^r)_{w,p}$ is bounded from above by the supremum of the left hand side of (6.3.6) for $h \leq t$, or by (6.3.7), and use Theorem 6.2.1 (6.2.6).

The proof for β satisfying $\beta(0) \geq 1$, that is, of Corollary 6.3.3, does not change much from that of the above (Theorem 6.3.1). For part (a) we divide R_+ into $\{[2^{-k}, 2^{-k+1}]\}$, $-\infty < k < \infty$. To prove (6.3.5) we now use $\beta(0)$ for positive k and $\beta(\infty)$ for negative k. The rest follows word for word but on R_+ rather than on $[t^*, \infty)$. For part (b) we just use

$$\frac{x}{2} \leq x - rt\varphi(x)/2 < x + rt\varphi(x)/2 \leq 2x$$

for t sufficiently small and therefore the inequality follows on R_+ instead of $[t^*, \infty)$. □

6.4. Marchaud Inequality for Weighted Main-Part Moduli

Inequality of the Marchaud type was already proved in Section 4.3 in relation to $\omega_\varphi^r(f,t)_p$, and it is valid for the main-part modulus (nonweighted), as will be an easy consequence of the present section for $p < \infty$. As a result of Sections 6.1 and 6.2 the reader might have the feeling that every result proved for ω_φ^r can be extended to the weighted main-part modulus, and what is more, the proofs are usually simpler because we "chopped off" some neighborhoods of finite endpoints that caused so much difficulty in Sections 2.3 and 2.4. The major exception to that situation is the Machaud inequality. The difficulty with extending this inequality is that we want to estimate $\Omega(f,t)$ by an expression of the type $\int_{u \geq t} \Omega(f,u)/u^2 \, du$, but for $u > t$ the support of the integral defining Ω (see (6.2.1)) is smaller than the corresponding support for t, i.e., (Ct^*, ∞) shrinks considerably as t increases. In fact, for the weighted main-part modulus the Marchaud inequality does not always hold, as we see from the following example, and therefore some additional conditions on the weights will be necessary.

EXAMPLE 6.4.1. Suppose $D = R_+$, $r = 1$, $1 \leq p < \infty$, $w(x) = x^{-\gamma}$ with $(1/2) + (1/p) \leq \gamma < 1 + (1/p)$, $\varphi(x) = \sqrt{x}$ and $f \in C^2(R_+)$, $f(x) = x \log x - x$ for $x < 1/2$ and $f(x) = 0$ for $x > 1$. The behavior of $\Omega_\varphi^2(f,t)_{w,p}$ is given by

$$\Omega_\varphi^2(f,t)_{w,p} \sim \left\{ t^{2p} \int_{4t^2} |w(x)\varphi(x)^2 f''(\xi(x))|^p \, dx \right\}^{1/p}$$

$$\sim \left\{ t^{2p} \int_{4t^2} (x^{-\gamma} x x^{-1})^p \, dx \right\}^{1/p}$$

$$\sim t^{2-2\gamma+2/p},$$

which is $o(1)$ when $2 - 2\gamma + (2/p) > 0$, i.e., when $\gamma < 1 + (1/p)$. Similarly, we calculate $\Omega_\varphi^1(f, t)_{w, p}$ by

$$\Omega_\varphi^1(f, t)_{w, p} \sim \left\{ t^p \int_{t^2} |w(x)\varphi(x) f'(\xi(x))|^p \, dx \right\}^{1/p}$$

$$\sim t \left\{ \int_{t^2} (x^{-\gamma} x^{1/2} |\log x|)^p \, dx \right\}^{1/p}$$

$$\sim t \left(\log \frac{1}{t} \right)^{1 + 1/p}$$

for $\gamma = (1/2) + (1/p)$ ($-\gamma p + (p/2) = -1$) and

$$\Omega_\varphi^1(f, t)_{w, p} \sim t^{2 - 2\gamma + 2/p} \log \frac{1}{t}$$

for $\gamma > (1/2) + (1/p)$ and again $\Omega_\varphi^1(f, t)_{w, p} = o(1)$ if $\gamma < 1 + (1/p)$. Therefore,

$$\Omega_\varphi^1(f, t)_{w, p} \geq C(\log 1/t)^{1/p} \left(t \int_t^1 (\Omega_\varphi^2(f, \tau)_{w, p}/\tau^2) \, d\tau + t \|wf\|_p \right).$$

From the above example it is clear that the Marchaud-type inequality, that is,

$$\Omega_\varphi^1(f, t)_{w, p} \leq Ct \left(\int_t (\Omega_\varphi^2(f, \tau)_{w, p}/\tau^2) \, d\tau + \|wf\|_p \right),$$

is not valid in general.

In Example 6.4.1 $\gamma(0) = -\gamma$, $\beta(0) = 1/2$, and $r = 1$, and therefore, the condition $(1/2) + (1/p) \leq \gamma$ for which the Marchaud inequality was shown to be invalid corresponds to $\gamma(0) + r\beta(0) \leq -1/p$. In the following theorem Marchaud's inequality is proved under the opposite hypothesis, i.e., $\gamma(0) + r\beta(0) > -1/p$.

Theorem 6.4.2. *Suppose $\gamma(c) + r\beta(c) > -1/p$ if c is a finite endpoint with $0 \leq \beta(c) < 1$. Then*

$$\Omega_\varphi^r(C, f, t)_{w, p} \leq Mt^r \left[\int_t^1 (\Omega_\varphi^{r+1}(C, f, \tau)_{w, p}/\tau^{r+1}) \, d\tau + \|wf\|_p \right]. \quad (6.4.1)$$

For the proof we will need the following lemma.

Lemma 6.4.3. *Suppose $\delta > -1/p$, $\xi > 0$, $a > 1$, $H(x) \in$ A.C. $[\xi, \xi a^2]$, and $H'(x) \in L_p[\xi, \xi a^2]$, where $1 \leq p \leq \infty$. Then*

$$\|x^\delta H(x)\|_{L_p[\xi, \xi a]} \leq \eta \|x^\delta H(x)\|_{L_p[\xi a, \xi a^2]} + M\xi \|x^\delta H'(x)\|_{L_p[\xi, \xi a^2]}, \quad (6.4.2)$$

where $\eta = \eta(a, \delta, p) < 1$ and $M = M(a, \delta, p) > 0$, both independent of ξ and $H(x)$.

6.4. Marchaud Inequality for Weighted Main-Part Moduli

PROOF. We write $H(x) = H(ax) - \int_x^{ax} H'(u)\,du$. Therefore for $p < \infty$ and $\delta_1 = \min(\delta, 0)$

$$\left\{\int_\xi^{\xi a} |x^\delta H(x)|^p\,dx\right\}^{1/p} \leq \left\{\int_\xi^{\xi a} |x^\delta H(ax)|^p\,dx\right\}^{1/p}$$

$$+ \left\{\int_\xi^{\xi a} \left|x^\delta \int_x^{ax} H'(u)\,du\right|^p dx\right\}^{1/p}$$

$$\leq a^{-\delta-(1/p)}\left\{\int_{\xi a}^{\xi a^2} |u^\delta H(u)|^p\,du\right\}^{1/p}$$

$$+ a^{-\delta_1}\left\{\int_\xi^{\xi a} (x(a-1))^{p-1}\left(\int_x^{xa} |u^\delta H'(u)|^p\,du\right)dx\right\}^{1/p}$$

$$\leq \eta \|x^\delta H(x)\|_{L_p[\xi a, \xi a^2]}$$

$$+ a^{-\delta_1}(a-1)a^{1/p}\xi\left\{\int_\xi^{\xi a^2} |u^\delta H'(u)|^p\,du\right\}^{1/p}.$$

For $p = \infty$ for which $\delta > 0$, we have

$$\|x^\delta H(x)\|_{L_\infty[\xi,\xi a]} \leq a^{-\delta}\|x^\delta H(x)\|_{L_\infty[\xi a,\xi a^2]} + \xi a(a-1)\|x^\delta H'(x)\|_{L_\infty[\xi,\xi a^2]}. \quad \square$$

PROOF OF THEOREM 6.4.2. We shall carry out the proof only for $D = R_+$. If $\beta(0) \geq 1$, then $t^* = 0$ (see (6.1.4)) and $x \pm rh\varphi(x)/2 \in (x/2, 2x)$ for small h and every x; hence the considerations of Section 4.3 can be repeated with minor changes. Therefore, in what follows let $0 \leq \beta(0) < 1$, i.e., $t^* > 0$ for $t > 0$. Using Theorem 6.2.1, we can replace the inequality (6.4.1) by the corresponding inequality on $\mathscr{K}$

$$\mathscr{K}_{r,\varphi}(C,f,t^r)_{w,p} \leq Mt^r\left[\int_t^1 (\mathscr{K}_{r+1,\varphi}(C,f,\tau^{r+1})_{w,p}/\tau^{r+1})\,d\tau + \|wf\|_p\right] \quad (6.4.3)$$

for any $C > 0$, and in fact it is sufficient to prove (6.4.3) with different C on the right and left side (any fixed pair of them). As we are dealing with $\mathscr{K}$, we can replace $\varphi(x)$ by $x^{\beta(0)}$ and $x^{\beta(\infty)}$ in $[0,1]$ and $[1,\infty)$ respectively, and replace $w(x)$ by $x^{\gamma(0)}$ and $x^{\gamma(\infty)}$ in $[0,1]$ and $[1,\infty)$ respectively (cf. (6.1.2)). We set $t_n = 2^n t$ for $n = 0, 1, \ldots$ and set $t_n^* = (Ar)^{1/1-\beta}t_n^{1/1-\beta}$ with $\beta = \beta(0)$ (that is t_n^* refers to r). We can now choose g_{t_n} for $n \geq 0$ such that

$$\|w(f - g_{t_n})\|_{L_p[t_n^*, \infty)} + t_n^{r+1}\|w\varphi^{r+1}g_{t_n}^{(r+1)}\|_{L_p[t_n^*, \infty)}$$
$$\leq 2\mathscr{K}_{r+1,\varphi}(2^{1/\beta-1}, f, t_n^{r+1})_{w,p}, \quad (6.4.4)$$

which is possible recalling the definition of $\mathscr{K}$ from (6.2.3), $2 \geq (r+1)/r$ and $2^{1/\beta-1} \leq ((r+1)/r)^{1/\beta-1}$. In what follows we write $B = 2^{1/\beta-1}$. We will construct G_t satisfying

$$\|w(f - G_t)\|_{L_p[t^*,\infty)} + t^r \|w\varphi^r G_t^{(r)}\|_{L_p[t^*,\infty)} \qquad (6.4.5)$$
$$\le Mt^r \left[\int_t^1 (\mathcal{K}_{r+1,\varphi}(B,f,u^{r+1}))_{w,p}/u^{r+1}) \, du + \|wf\|_p \right],$$

where M is independent of t. Inequality (6.4.5) for every t will imply

$$\mathcal{K}_{r,\varphi}(f,t)_{w,p} \le \sup_{h \le t} \{ \|w(f - G_h)\|_{L_p[h^*,\infty)} + h^r \|w\varphi G_h^{(r)}\|_{L_p[h^*,\infty)} \}$$

$$\le \sup_{h \le t} Mh^r \left[\int_h^1 (\mathcal{K}_{r+1,\varphi}(B,f,u^{r+1}))_{w,p}/u^{r+1}) \, du + \|wf\|_p \right]$$

$$\le Mt^r \sup_{h \le t} \left[\int_t^{t/h} \left(\mathcal{K}_{r+1,\varphi}\left(B,f,\left(\frac{\tau h}{t}\right)^{r+1}\right) \right)_{w,p} / \tau^{r+1} \right) d\tau + \|wf\|_p \right]$$

$$\le Mt^r \sup_{h \le t} \left[\int_t^1 (\mathcal{K}_{r+1,\varphi}(B,f,\tau^{r+1}))_{w,p}/\tau^{r+1}) \, d\tau \right.$$
$$\left. + \int_1^{t/h} (\|wf\|_p/\tau^{r+1}) \, d\tau + \|wf\|_p \right].$$

This is (6.4.3) with $C = 1$ on the left and $C = B$ on the right hand side. We will show (with some effort) that the function g_{t_0} of (6.4.4) will satisfy (6.4.5), i.e., we can set $G_t = g_{t_0} = g_t$, and so complete the proof.

To estimate the first term of (6.4.5) we write

$$\|w(f - G_t)\|_{L_p[t^*,\infty)} \equiv \|w(f - g_{t_0})\|_{L_p[t^*,\infty)} \le 2\mathcal{K}_{r+1,\varphi}(B,f,t)_{w,p}.$$

To estimate

$$\|w\varphi^r g_{t_0}^{(r)}\|_{L_p[t_0^*,\infty)} = \|w\varphi^r G_t^{(r)}\|_{L_p[t_0^*,\infty)} \equiv S_0$$

we write

$$S_n \equiv \|w\varphi^r g_{t_n}^{(r)}\|_{L_p[t_n^*,\infty)}$$
$$\le \|w\varphi^r g_{t_n}^{(r)}\|_{L_p[t_n^*,t_{n+1}^*]} + \|w\varphi^r (g_{t_n}^{(r)} - g_{t_{n+1}}^{(r)})\|_{L_p[t_{n+1}^*,\infty)} + \|w\varphi^r g_{t_{n+1}}^{(r)}\|_{L_p[t_{n+1}^*,\infty)}$$
$$\equiv I_n + J_n + S_{n+1}.$$

Iterating the above, $S_0 \le \sum_{n=0}^m I_n + \sum_{n=0}^m J_n + S_{m+1}$ follows. Choosing m such that $2^m t \ge 1 \ge 2^{m-1} t$, we will show the following:

(a) $J_n \le L t_{n+1}^{-r} \mathcal{K}_{r+1,\varphi}(B,f,t_{n+1}^{r+1})_{w,p}$,
(b) $I_n \le \eta I_{n+1} + R t_{n+1}^{-r} \mathcal{K}_{r+1,\varphi}(B,f,t_{n+1}^{r+1})_{w,p}$, and
(c) $S_{m+1} \le N \|wf\|_p$ and $I_{m+1} \le N \|wf\|_p$;

where L, R, N, and $\eta < 1$ are independent of n.

Using (b) and (c), we have

$$\sum_{n=0}^m I_n \le \eta \sum_{n=1}^m I_n + \sum_{n=1}^{m+1} t_{n+1}^{-r} \mathcal{K}_{r+1,\varphi}(B,f,t_{n+1}^{r+1})_{w,p} + \eta I_{m+1}$$

and therefore

6.5. Connection with Ordinary Weighted Moduli

$$\sum_{n=0}^{m} I_n \le M_1 \left[\int_t^1 (\mathcal{K}_{r+1,\varphi}(B,f,u^{r+1})_{w,p}/u^{r+1})\,du + \|wf\|_p \right].$$

Combining this with (a) and (c), we have

$$S_0 \le M_2 \left[\int_t^1 (\mathcal{K}_{r+1,\varphi}(B,f,u^{r+1})_{w,p}/u^{r+1})\,du + \|wf\|_p \right]$$

and therefore we need to show only (a), (b), and (c) to complete the proof.

To show (a) we observe that if we choose in the proof of Theorem 6.3.1 in (6.3.3) g_{t_n} and $g_{t_{n+1}}$ as in (6.4.4), we get for $j = r$ from (6.3.5) (with $r + 1$ replacing r),

$$\|w\varphi^r(g_{t_n} - g_{t_{n+1}})^{(r)}\|_{L_p[t_{n+1}^*,\infty)} \le L t_{n+1}^{-r} \mathcal{K}_{r+1,\varphi}(B,f,t_{n+1}^{r+1})_{w,p}.$$

To prove (b), which is the main step, we use Lemma 6.4.3 with $w(x)(\varphi(x))^r = x^\delta$, $\delta = \gamma(0) + r\beta(0) > -1/p$, $a = 2^{1/1-\beta}$, $\xi = t_n^*$, and $H(x) = g_{t_n}^{(r)}(x)$ to obtain

$$\|x^\delta g_{t_n}^{(r)}(x)\|_{L_p[t_n^*,t_{n+1}^*]} \le \eta \|x^\delta g_{t_n}^{(r)}(x)\|_{L_p[t_{n+1}^*,t_{n+2}^*]}$$

$$+ M_1 t_n^* \|x^\delta g_{t_n}^{(r+1)}(x)\|_{L_p[t_n^*,t_{n+2}^*]}$$

$$\le \eta \|x^\delta(g_{t_n}^{(r)}(x) - g_{t_{n+1}}^{(r)}(x))\|_{L_p[t_{n+1}^*,t_{n+2}^*]}$$

$$+ \eta \|x^\delta g_{t_{n+1}}^{(r)}(x)\|_{L_p[t_{n+1}^*,t_{n+2}^*]}$$

$$+ M_2 t_n \|x^\delta \varphi(x) g_{t_n}^{(r+1)}(x)\|_{L_p[t_n^*,t_{n+2}^*]}$$

$$\le \eta J_n + \eta I_{n+1} + M_2 t_n^{-r} \mathcal{K}_{r+1,\varphi}(B,f,t_n^{r+1})_{w,p}$$

$$\le \eta I_{n+1} + M_3 t_{n+1}^{-r} \mathcal{K}_{r+1,\varphi}(B,f,t_{n+1}^{r+1})_{w,p},$$

where in the last inequality we used (a). Finally to prove (c) we recall that $\mathcal{K}_{r+1,\varphi}(B,f,t^{r+1})_{w,p} \le \|wf\|_p$. Using the consideration by which we proved (6.3.5) for $g_{t_{m+1}}$ instead of $g_t - g_{t_1}$ and for $r + 1$ instead of r, we obtain

$$\|w\varphi^j g_{t_{m+1}}^{(j)}\|_{L_p(t_{m+1}^*,\infty)} \le M_4(t_{m+1}^{-j}\|wg_{t_{m+1}}\|_{L_p(t_{m+1}^*,\infty)}$$

$$+ t_{m+1}^{r+1-j}\|w\varphi^{r+1}g_{t_{m+1}}^{(r+1)}\|_{L_p(t_{m+1}^*,\infty)}$$

$$\le M_5(\|wf\|_{L_p(t_{m+1}^*,\infty)} + \mathcal{K}_{r+1,\varphi}(B,f,t_{m+1}^{r+1})_{w,p})$$

$$\le M_6\|wf\|_p$$

(note that $t_{m+1} \sim 1$) and for $j = r$ this yields (c) and therefore the proof of our theorem is complete. □

6.5. Connection with Ordinary Weighted Moduli

In this section we will discuss further the natural question of varying increment $h\varphi(x)$ in

$$\omega_\varphi^r(f,t)_p = \sup_{0 < h \le t} \|\Delta_{h\varphi}^r f\|_p. \tag{6.5.1}$$

In particular, is (6.5.1) equivalent to a usual (maybe weighted) modulus of smoothness (with $\varphi \equiv 1$)? The answer is negative if we compare our new moduli $\omega_\varphi^r(f,t)_p$ with the classical weighted moduli $\omega^r(f,t)_{w,p}$ of the same function f (see Totik [8]). However, we will indicate here (and prove elsewhere) that equivalence between $\omega_\varphi^r(f,t)_p$ and $\omega^r(g,t)_{w,p}$ for some g related to f can be established. We will concentrate only on characterizing

$$\omega_\varphi^r(F,t)_p = O(t^\beta) \qquad (t \to 0).$$

Using Theorems 4.1.3, 4.3.1, Corollary 3.3.4, and Theorem 6.3.1 successively, first we can assume that $r \geq 2$, $0 < r - \beta \leq 1$, then it is enough to consider

$$\Omega_\varphi^r(F,t)_p = O(t^\beta)$$

and this is equivalent to

$$\Omega_\varphi^2(f,t)_{w,p} = O(t^\alpha) \tag{6.5.2}$$

with $0 < \alpha = \beta - (r-2)$, $w = \varphi^{r-2}$, and $f = F^{(r-2)}$. Hence it is sufficient to deal with (6.5.2).

For the sake of simplicity, we will assume that φ is locally absolutely continuous and that besides the conditions of Section 1.2 it also satisfies

$$\varphi'(x) \sim \begin{cases} |x|^{\beta(a)-1} \operatorname{sign} x & \text{as } x \to a + 0 \\ x^{\beta(\infty)-1} & \text{as } x \to \infty \quad (D = R_+ \text{ or } R) \\ -(1-x)^{\beta(1)-1} & \text{as } x \to 1 - 0 \quad (D = (0,1)). \end{cases} \tag{6.5.3}$$

Remark 6.2.6 shows that this is not a severe restriction. Let

$$\Gamma(x) = \int_{1/2}^x \frac{1}{\varphi(\tau)} d\tau \quad \text{and} \quad \theta(y) = \Gamma^{-1}(y) \tag{6.5.4}$$

the inverse of Γ. Then Γ maps $D = (a,b)$ onto some interval (A,B) and θ maps (A,B) onto (a,b). With these mappings the following theorem, which we state without proof, answers the equivalence problem mentioned at the beginning of this section.

Theorem 6.5.1. *Suppose w and φ satisfy (6.2.1) and the condition of Section 1.2 and (6.5.3), respectively. Furthermore, $\beta(c) + \gamma(c) > 1 - 1/p$ if c is a finite endpoint with $0 \leq \beta(c) < 1$. Then with $g = f \circ \theta$ we have*

$$\Omega_1^2(g,t)_{w(\theta)\varphi(\theta)^{1/p},p} \leq M(\Omega_\varphi^2(f,t)_{w,p} + t^2 \|wf\|_p)$$

and

$$\Omega_\varphi^2(f,t)_{w,p} \leq M(\Omega_1^2(g,t)_{w(\theta)\varphi(\theta)^{1/p},p} + t^2 \|w(\theta)\varphi(\theta)^{1/p}g\|_p)$$

with an M independent of f and small t. In particular, if $wf \in L_p(D)$, then for $\alpha > 0$

$$\Omega_\varphi^2(f,t)_{w,p} = O(t^\alpha)$$

6.5. Connection with Ordinary Weighted Moduli

and
$$\Omega_1^2(g,t)_{w(\theta)(\varphi(\theta))^{1/p},\,p} = O(t^\alpha) \tag{6.5.5}$$

are equivalent. The same result holds for every θ such that θ' is locally absolutely continuous and

$$\theta' \sim \varphi(\theta), \qquad \theta'' \sim \varphi'(\theta)\varphi(\theta)$$

are uniformly satisfied on (A, B) (the domain of θ).

Note that the modulus in (6.5.5) is taken on (A, B).

We immediately observe that the result is false if, say, $0 \leq \beta(0) < 1$ and $\gamma(0) + \beta(0) \leq 1 - 1/p$. In fact, setting $D = (0, \infty)$, $\varphi(x) = x^\beta$, $0 \leq \beta < 1$, $w \equiv 1$ and $p \geq 1/1 - \beta$, we have

$$\gamma(0) + \beta(0) = \beta \leq 1 - 1/p.$$

Simple calculations show that for $\theta(x) = x^{1/1-\beta}$ and $f \in C^2(0, \infty)$ with

$$f(x) = \begin{cases} (1-\beta)x^{1-\beta}\log x - x^{1-\beta} & \text{if } 0 < x < 1/2 \\ 0 & \text{if } x > 1 \end{cases}$$

we have

$$\Omega_\varphi^2(f,t)_{1,p} \sim \begin{cases} t^2(\log 1/t)^{1+1/p} & \text{if } p = 1/1 - \beta \\ t^{1+(1/(1-\beta)p)}\log 1/t & \text{if } p > 1/1 - \beta \end{cases}$$

and (for $g = f \circ \theta$)

$$\Omega_1^2(g,t)_{\varphi(\theta)^{1/p},\,p} \sim \begin{cases} t^2 \log 1/t & \text{if } p = 1/1 - \beta \\ t^{1+(\beta/(1-\beta)p)+(1/p)} & \text{if } p > 1/1 - \beta. \end{cases}$$

So these two moduli have completely different orders.

Theorem 6.5.1 also answers the question "When does $\omega_\varphi^r(f,t)_\infty$ tend to zero as $t \to 0+?$" (see Section 4.1). In fact, one can easily see (use Theorems 4.1.3, 4.3.1, 6.2.5, and the proof of Theorem 4.3.1) that for bounded f and $g = f \circ \theta$ each of the relations

$$\omega_\varphi^r(f,t)_\infty = o(1), \qquad \Omega_\varphi^2(f,t)_\infty = o(1), \qquad \Omega_1^2(g,t)_{1,\infty} = o(1),$$

$$\Omega^1(g,t)_\infty = o(1), \qquad \omega^1(g,t)_\infty = o(1)$$

is equivalent to the next one under the assumption that

(*) f and g are continuous at c if c is a finite endpoint of their domain with $0 \leq \beta(c) < 1$;

furthermore, this (*) statement follows both from the first and the last relation. Thus we get:

Corollary 6.5.2. *If f is bounded, then for any $r > 0$*

$$\omega_\varphi^r(f,t)_\infty = o(1) \qquad (t \to 0+)$$

holds if and only if $f \circ \theta$ is uniformly continuous on (A, B).

It is worth illustrating Theorem 6.5.1 through some examples in which we take a different θ and not the one given concretely in (6.5.4).

EXAMPLE 6.5.3. The relations below characterize weighted polynomial approximations on $[-1,1]$. For $\alpha > 0$ set $\alpha = s + \alpha_1$, where s is an integer and $0 < \alpha_1 \le 1$ and let $D = (-1,1)$, $\varphi(x) = \sqrt{1-x^2}$, $1 \le p \le \infty$, and $w(x) = (1+x)^{\gamma_1}(1-x)^{\gamma_2}$ with $\min(\gamma_1, \gamma_2) > 1/2 - 1/p - s/2$. The introduction to this section and Theorems 6.3.1, 6.4.2, and 6.5.1 imply that for $0 < \alpha < r$ and $wf \in L_p(-1,1)$

$$\|w\Delta_{h\varphi}^r\|_{L_p(-1+(2rh)^2, 1-(2rh)^2)} = O(h^\alpha)$$

is equivalent to

$$\left\|\left(\cos\frac{x}{2}\right)^{2\gamma_1+s+1/p}\left(\sin\frac{x}{2}\right)^{2\gamma_2+s+1/p}\Delta_h^2 f^{(s)}(\cos x)\right\|_{L_p(2h, \pi-2h)} = O(h^{\alpha_1}).$$

EXAMPLE 6.5.4. Let $D = (0, \infty)$, $\varphi(x) = x$, $w(x) = x^{\gamma_1}(1+x)^{\gamma_2}$ with arbitrary γ_1 and γ_2 and $1 \le p \le \infty$. Then for $0 < \alpha \le 2$

$$\|w\Delta_{h\varphi}^2 f\|_{L_p[0,\infty)} = O(h^\alpha) \tag{6.5.6}$$

is equivalent to

$$\|w(e^x)e^{x/p}\Delta_h^2 f(e^x)\|_{L_p(-\infty,\infty)} = O(h^\alpha).$$

Since (6.5.6) is closely related to approximation by the Post–Widder and Gamma operators (see Chapter 10), the above mentioned equivalence enables one to characterize in a constructive way the weighted Lipschitz property

$$\|w\Delta_h^2 g\| = O(h^\alpha)$$

for any exponential weight $w(x) = e^{m_1 x} + e^{m_2 x}$.

EXAMPLE 6.5.5. If $D = (0, \infty)$ and $\varphi(x) = x$, then $\omega_\varphi^r(f, t)_\infty = o(1)$ if and only if $f(e^x)$ is uniformly continuous on $(-\infty, \infty)$. For $\varphi(x) = \sqrt{x}$ the function $f(x^2)$ must be uniformly continuous on $(0, \infty)$.

PART II
APPLICATIONS

CHAPTER 7
ALGEBRAIC POLYNOMIAL APPROXIMATION

In this chapter we relate the rate of convergence of best polynomial approximation to our new modulus of smoothness. Asymptotic behavior of the derivatives of the optimal algebraic polynomials will also be related to that modulus of smoothness. The results are of both direct and converse type and yield necessary and sufficient conditions whenever the rate of best approximation is $n^{-\alpha}$ for instance.

7.1. Background

For trigonometric polynomials investigation of the relation between the rate of convergence and the classical modulus of continuity leads to a beautifully simple theory due to S.N. Bernstein, D. Jackson, and S.B. Stechkin (see Lorentz [2] for example) which was extended to any Banach space of functions on $(-\pi, \pi)$ for which translation is a continuous isometry (see Shapiro [1] and Ditzian [5]). We will denote the class of trigonometric polynomials of degree at most n by τ_n. It was shown for any Banach space B of periodic functions for which translation is a continuous isometry that

$$E_n^*(f)_B \equiv \inf_{T_n \in \tau_n} \|f - T_n\|_B \le M\omega^r(f, n^{-1})_B \qquad (7.1.1)$$

and

$$\omega^r(f, h)_B \le M_r h^r \sum_{0 \le n < 1/h} (n+1)^{r-1} E_n^*(f)_B \qquad (7.1.2)$$

which yield in particular

$$\omega^r(f, h)_B = O(h^\alpha) \quad \text{for } \alpha < r \quad \text{if and only if} \quad E_n^*(f)_B = O(n^{-\alpha}). \qquad (7.1.3)$$

7. Algebraic Polynomial Approximation

We denote the set of algebraic polynomials of degree at most n by Π_n and the best approximation on $[-1, 1]$ by

$$E_n(f)_B = \inf_{P_n \in \Pi_n} \|f - P_n\|_B.$$

It was shown early by D. Jackson [1] that for $r = 1$

$$E_n(f)_C \leq M\omega^r\left(f, \frac{1}{n}\right)_C.$$

This inequality however, has no converse, and in fact one cannot characterize $f \in \text{Lip } \alpha$ for example, by $E_n(f)_C$ as was explained by S.M. Nikol'skii [1]. Later investigation by A.F. Timan [1], G. Freud [1], V.K. Dzjadyk [1], and Yu. Brudny [1] established:

(a) There exists a polynomial P_n, $P_n \in \Pi_n$ such that

$$|f(x) - P_n(x)| \leq M_r \omega^r(f, \Delta_n(x))_C,$$

for all $x \in (-1, 1)$, where $\Delta_n(x) \equiv n^{-1}\sqrt{1 - x^2} + n^{-2}$.

(b) If there exists for every n a polynomial P_n, $P_n \in \Pi_n$ such that

$$|f(x) - P_n(x)| \leq \omega(\Delta_n(x))$$

for some increasing function $\omega(x)$, then

$$\omega^r(f, h)_C \leq Mh^r \sum_{0 \leq n < 1/h} (n+1)^{r-1} \omega\left(\frac{1}{n}\right).$$

Therefore the classical modulus of continuity is related to approximation by polynomials which are not necessarily the best approximating polynomials, and it is not the norm that is measuring the rate of approximation. It was shown by V.P. Motornii [1] (see also DeVore [2] and Oswald [1]) that the natural analogue for L_p ($p < \infty$), that is $\|\delta_n(x)^{-\alpha}(f - P_n)(x)\|_{L_p} = O(n^{-\alpha})$, where $\delta_n(x) = \sqrt{1 - x^2} + n^{-1}$, is not equivalent to $f \in \text{Lip } \alpha$ in L_p. Moreover, it is the investigation of the class of functions for which $E_n(f)_p$ tends to zero at a certain rate which is the main question. The algebraic analogue of (7.1.1) and (7.1.2) requires a new kind of modulus of smoothness. Attempts in this direction were made in many recent investigations (see Ivanov [1–7], Butzer, Stens, and Wehrens [1–3] and Potapov [2–4]). However, our present modulus is the simplest and most easily computable expression that solves the problem. (See for more detailed comparison Chapter 13.) We will also investigate in the next chapter the problem of weighted polynomial approximation and relate it to the weighted main-part modulus. Crucial to the present discussion and that of the next chapter are some inequalities proved by P. Nevai [1] and M.K. Potapov [1]. These inequalities are sufficient for Jacobi weights but will be generalized to include other weights in Chapter 8, and a complete proof will be included there.

Another problem with which we will deal is that of investigating the smoothness using the asymptotic behavior of the derivatives of the polynomials of best approximation. For trigonometric polynomials, $B = L_p[-\pi, \pi]$ and the class Lip α this was done by M. Zamanski [1] and G.I. Sunouchi [1], and from this it follows for other Banach spaces (see Ditzian [5]). It should be noted that sometimes we can acquire from the asymptotic behavior of the derivatives of the best-polynomial (algebraic or trigonometric) approximation information which we cannot obtain from the rate of approximation. The simplest such situation for trigonometric polynomials is when $E_n^*(f) = O(n^{-r})$ which implies $\|\Delta_h^{r+1} f\| = O(h^r)$, and we cannot tell if f is Lip r, that is, $\|\Delta_h^r f\| = O(h^r)$ or not, but $\|T_n^{(r)}\| = O(1)$ will imply it and from $\|T_n^{(r)}\| = O(\psi(n))$ we learn the behavior of $\|\Delta_h^r f\|$. For some other situations $E_n^*(f)$ yields more information. For this reason we find the investigation of the asymptotic behavior of the derivatives of polynomials of best approximation interesting, and we do not try to force the expressions to be exactly like those for $E_n(f)$, as both $E_n(f)$ and $P_n^{(r)}$ are sensitive and informative on the measure of smoothness of f in slightly different ranges.

7.2. Best Polynomial Approximation

In this section we will relate

$$\omega_\varphi^r(f,t)_{L_p[-1,1]} \equiv \omega_\varphi^r(f,t)_p = \sup_{0 < h \leq t} \|\Delta_{h\sqrt{1-x^2}}^r f(x)\|_{L_p[-1,1]},$$

where $\varphi(x) = \sqrt{1-x^2}$ to $E_n(f)_p$ given by

$$E_n(f)_p = \inf_{P_n \in \Pi_n} \|f - P_n\|_{L_p[-1,1]}. \tag{7.2.1}$$

The direct result is given in the following theorem.

Theorem 7.2.1. *For $E_n(f)_p$ given in (7.2.1) and $\varphi(x) = \sqrt{1-x^2}$*

$$E_n(f)_p \leq M\omega_\varphi^r(f, n^{-1})_p, \qquad n > r, \tag{7.2.2}$$

where $M = M(r)$ is independent of $n > r$ and $f \in L_p[-1,1]$.

Since $\varphi(x) \leq 1$ on $[-1,1]$, Theorem 4.1.1 shows that (7.2.2) is stronger than Jackson's classical estimate (which is (7.2.2) with $r = 1$ and $\varphi \equiv 1$). In fact, examining $f(x) = |1-x|^\alpha$ (see Section 8.5), one observes that (7.2.2) is actually much stronger. From Theorem 7.2.4 below it follows that (7.2.2) is the best possible refinement of Jackson's estimate that has held for over 70 years.

We will need for the proof of Theorem 7.2.1 the following well-known lemma (Lorentz [2, p. 57]).

Lemma 7.2.2. *For any integer l there exists a sequence of even trigonometric polynomials $T_n(t)$ of degree n satisfying $T_n(t) \geq 0$, $\int_{-\pi}^{\pi} T_n(t)\,dt = 1$ and $\int_{-\pi}^{\pi} |t|^l T_n(t)\,dt \sim n^{-l}$ and therefore $\int_{-\pi}^{\pi} |t|^\gamma T_n(t)\,dt \leq L(\gamma) n^{-\gamma}$, for $0 < \gamma < l$.*

We define
$$\Delta_{1/t}(x) = |t|(|t| + \sqrt{1-x^2}) \quad \text{and} \quad \delta_n(x) = n\Delta_n(x). \tag{7.2.3}$$

With this notation we state and prove the following lemma which is crucial in the proof of Theorem 7.2.1.

Lemma 7.2.3. *For $T_n(t)$ of Lemma 7.2.2 with $l = 2m + 3$, $\delta_n(x)$ given by (7.2.3), $1 \le p \le \infty$, $f \in A.C.[-1,1]$ and $f' \in L_p[-1,1]$ we have*

$$\left\| \delta_n(x)^m \int_{-\pi}^{\pi} \left\{ \int_{\cos(\arccos x - t)}^{x} f'(u)\, du \right\} T_n(t)\, dt \right\|_{L_p[-1,1]} \tag{7.2.4}$$
$$\le \frac{M}{n} \|\delta_n(x)^{m+1} f'(x)\|_{L_p[-1,1]}.$$

PROOF OF LEMMA 7.2.3. The proof depends on some simple computations that will be detailed because the lemma is crucial. We first observe that

$$|x - \cos(\arccos x \mp t)| = |x(1 - \cos t) \mp \sqrt{1-x^2} \sin t| \le \Delta_{1/t}(x).$$

We can now deduce for u between x and $\cos(\arccos x - t)$ that

$$|\delta_n(x) - \delta_n(u)| = \frac{|x-u||x+u|}{\sqrt{1-u^2} + \sqrt{1-x^2}} \le \frac{2\Delta_{1/t}(x)}{\sqrt{1-u^2} + \sqrt{1-x^2}}.$$

We will need that on the right hand side of the above $\Delta_{1/t}(x)$ be replaced by $\Delta_{1/t}(u)$. For $|u| \le |x|$ we have $\Delta_{1/t}(x) \le \Delta_{1/t}(u)$. On the other hand, if $|u| \ge |x|$ and u is between x and $\cos(\arccos x - t)$, then x is between u and $\cos(\arccos u + t)$, and hence $|x-u| \le \Delta_{1/t}(u)$. Therefore, for u between x and $\cos(\arccos x - t)$, we have also $|x-u| \le \min(\Delta_{1/t}(u), \Delta_{1/t}(x))$. We now have

$$|\delta_n(x) - \delta_n(u)| \le 2\Delta_{1/t}(u)/(\sqrt{1-u^2} + \sqrt{1-x^2}).$$

If $\sqrt{1-x^2} + \sqrt{1-u^2} \ge 1/n$, then $|\delta_n(x) - \delta_n(u)| \le 2n\Delta_{1/t}(u)$; otherwise $1/n \le \delta_n(x), \delta_n(u) \le 2/n$, which implies $|\delta_n(x) - \delta_n(u)| \le 4/n \le 4\delta_n(u)$. Using

$$n\Delta_{1/t}(u) = nt^2 + n|t|\sqrt{1-u^2} \le n^2 t^2 \frac{1}{n} + n|t|\sqrt{1-u^2} \le (n^2 t^2 + n|t|)\delta_n(u),$$

we have

$$\delta_n(x) \le \delta_n(u) + \max(2n\Delta_{1/t}(u), 4\delta_n(u))$$
$$\le \delta_n(u) + \max(2n^2 t^2 + 2n|t|, 4)\delta_n(u)$$
$$\le (2n^2 t^2 + 2n|t| + 5)\delta_n(u).$$

From what we have proved it also follows that

$$|\Delta_{1/t}(x) - \Delta_{1/t}(u)| = |t||\delta_n(x) - \delta_n(u)| \le |t|\max(2n\Delta_{1/t}(u), 4\delta_n(u))$$
$$\le |t|(2n^2 t^2 + 2n|t| + 4)\delta_n(u),$$

and therefore

7.2. Best Polynomial Approximation 81

$$\Delta_{1/t}(x) \leq (2n^2|t|^3 + 2nt^2 + 4|t| + nt^2 + |t|)\delta_n(u)$$
$$\leq |t|(2n^2t^2 + 3n|t| + 5)\delta_n(u).$$

Since $n|t|$ is between $(nt)^2$ and 1, we can summarize the above estimates by

$$\delta_n(x) \leq 9\max(n^2t^2, 1)\delta_n(u) \quad \text{and}$$
$$\Delta_{1/t}(x) \leq 10|t|\max(n^2t^2, 1)\delta_n(u), \tag{7.25}$$

for any u between x and $\cos(\arccos x - t)$. In Chapter 8 we will need (7.2.5) to be valid for any x and u satisfying $|x - u| \leq \min(\Delta_{1/t}(u), \Delta_{1/t}(x))$ (which is obvious from the proof).

For $1 \leq p < \infty$ we now write, using (7.2.5),

$$I^p \equiv \int_{-1}^{1} \delta_n(x)^{mp} \left| \int_{-\pi}^{\pi} \left\{ \int_{\cos(\arccos x - t)}^{x} f'(u)\,du \right\} T_n(t)\,dt \right|^p dx$$

$$\leq 9^{mp} \int_{-1}^{1} \left| \int_{-\pi}^{\pi} \left| \int_{\cos(\arccos x - t)}^{x} \max(1, (nt)^{2m})\delta_n(u)^m |f'(u)|\,du \right| T_n(t)\,dt \right|^p dx.$$

For $1 < p < \infty$ we write, using (7.2.5) again and earlier estimates,

$$I^p \leq 10^{(m+1)p} \int_{-1}^{1} \left| \int_{-\pi}^{\pi} \frac{1}{x - \cos(\arccos x - t)} \int_{\cos(\arccos x - t)}^{x} \delta_n(u)^{m+1} \right.$$
$$\left. \times |f'(u)|\,du \left| \max((nt)^{2m+2}, 1)|t|\,T_n(t)\,dt \right|^p dx \right.$$

$$\leq 10^{(m+1)p} \int_{-1}^{1} M(F, x)^p \left| \int_{-\pi}^{\pi} |t|(1 + (nt)^{2m+2}) T_n(t)\,dt \right|^p dx,$$

where $M(F, x)$ is the maximal function of $F(u) = \delta_n(u)^{m+1} f'(u)$. Using Lemma 7.2.2 and the inequality $\|M(F)\|_p \leq A_p \|F\|_p$, we have

$$I^p \leq L n^{-p} \|M(F)\|_p^p \leq L_1 n^{-p} \|F\|_p^p = L_1 n^{-p} \|\delta_n(u)^{m+1} f'(u)\|_{L_p}^p.$$

For $p = \infty$ we estimate

$$I(\infty) = \sup_x \delta_n(x)^m \left| \int_{-\pi}^{\pi} \left\{ \int_{\cos(\arccos x - t)}^{x} f'(u)\,du \right\} T_n(t)\,dt \right|$$

$$\leq L \sup_x \left| \int_{-\pi}^{\pi} \frac{1}{x - \cos(\arccos x - t)} \int_{\cos(\arccos x - t)}^{x} \delta_n(u)^{n+1} \right.$$
$$\left. \times |f'(u)|\,du \left| ((nt)^{2m+2} + 1)|t|\,T_n(t)\,dt \right| \right.$$

$$\leq L_1 n^{-1} \|M(F)\|_\infty \leq L_1 n^{-1} \|F\|_\infty$$
$$= L_1 n^{-1} \|\delta_n(u)^{m+1} f'(u)\|_{L_\infty[-1,1]}.$$

At the beginning of the proof we verified

$$|x - u| \leq \Delta_{1/t}(u) \leq |t|(1 + n|t|)\delta_n(u)$$

7. Algebraic Polynomial Approximation

provided u is between x and $\cos(\arccos x - t)$. Hence for $p = 1$ we can use Fubini's theorem and write

$$I \leq 9^m \int_{-1}^{1} \int_{-\pi}^{\pi} \max(1,(nt)^{2m}) T_n(t) \left| \int_{\cos(\arccos x - t)}^{x} \delta_n(u)^m |f'(u)| \, du \right| dt \, dx$$

$$\leq 9^m \int_{-1}^{1} |f'(u)| \delta_n(u)^m \int_{-\pi}^{\pi} \max(1,(nt)^{2m}) T_n(t) \int_{\{x;\, u \in (x,\, \cos(\arccos x + t)\}} dx \, dt \, du$$

$$\leq 9^m \int_{-1}^{1} |f'(u)| \delta_n(u)^m \int_{-\pi}^{\pi} \max(1,(nt)^{2m}) T_n(t) 2\Delta_{1/t}(u) \, dt \, du$$

$$\leq L \int_{-1}^{1} |f'(u)| \delta_n(u)^{m+1} \int_{-\pi}^{\pi} |t| (1 + (nt)^{2m+2}) T_n(t) \, dt \, du$$

$$\leq L_1 n^{-1} \| f'(u) \delta_n(u)^{m+1} \|_{L_1[-1,1]}. \qquad \square$$

PROOF OF THEOREM 7.2.1. Using Theorem 3.1.1, we obtain the equivalence of

$$\bar{K}_{r,\varphi}(f, t^r)_p \equiv \inf_g \{ \|f - g\|_p + t^r \|\varphi^r g^{(r)}\|_p + t^{2r} \|g^{(r)}\|_p \}$$

with $\omega_\varphi^r(f, t)_p$. We choose g_n such that

$$\|f - g_n\|_p \leq 2\bar{K}_{r,\varphi}(f, n^{-r})_p, \qquad n^{-r} \|\varphi^r g_n^{(r)}\|_p \leq 2\bar{K}_{r,\varphi}(f, n^{-r})_p$$

and

$$n^{-2r} \|g_n^{(r)}\|_p \leq 2\bar{K}_{r,\varphi}(f, n^{-r})_p$$

are satisfied. For this choice of g_n we have

$$n^{-r} \left\| \left(\varphi + \frac{1}{n} \right)^r g_n^{(r)} \right\|_p \equiv n^{-r} \|\delta_n^r g_n^{(r)}\|_p \leq 2^{r+1} \bar{K}_{r,\varphi}(f, n^{-r})_p.$$

We recall that Lemma 7.2.3 actually implies for a function g satisfying $\delta_n^{m+1} g' \in L_p$ that there exists $P_n(x)$ given by

$$P_n(x) = \int_{-\pi}^{\pi} g(\cos(\arccos x - t)) T_n(t) \, dt$$

such that

$$\|\delta_n^m (P_n - g)\|_p \leq \frac{M}{n} \|\delta_n^{m+1} g'\|_p \qquad (0 \leq m \leq r).$$

We choose first $g(x) = g_n^{(r-1)}(x)$ and $m = r - 1$, and we obtain by the above a polynomial $P_{n,1}(x) \in \Pi_n$ satisfying

$$\|\delta_n^{r-1}(P_{n,1} - g_n^{(r-1)})\|_p \leq M_1 n^{-1} \bar{K}_{r,\varphi}(f, n^{-r})_p.$$

Using the same lemma with $g(u) = \int_0^u (P_{n,1}(x) - g^{(r-1)}(x)) \, dx$, we obtain $P_{n,2} \in \Pi_{n+1}$ satisfying

$$\|\delta_n^{r-2}(P_{n,2} - g_n^{(r-2)})\|_p \leq M_2 n^{r-2} \bar{K}_{r,\varphi}(f, n^{-r})_p.$$

7.2. Best Polynomial Approximation

Repeating the above process enough times, we get for some $P_{n,r} \in \Pi_{n+r-1}$

$$\|P_{n,r} - g_n\|_p \le M_r \bar{K}_{r,\varphi}(f, n^{-r})_p,$$

and therefore

$$\|P_{n,r} - f\|_p \le (M_r + 2)\bar{K}_{r,\varphi}(f, n^{-r})_p \le M\omega_\varphi^r(f, n^{-1})_p,$$

which completes the proof of our theorem since $\omega_\varphi^r(f, n^{-1})_p \sim \omega_\varphi^r(f, (n+r-1)^{-1})_p$. □

The converse result is given in the following theorem.

Theorem 7.2.4. *For an integer $r > 0$ there exists $M(r)$ independent of $0 < t < 1$ and $f \in L_p[-1, 1]$ for which*

$$\omega_\varphi^r(f, t)_p \le M(r) t^r \sum_{0 \le n \le 1/t} (n+1)^{r-1} E_n(f)_p, \qquad (7.2.6)$$

where $\omega_\varphi^r(f, t)$ and $E_n(f)_p$ are given in (2.1.2) and (7.2.1).

PROOF. Crucial for the proof is the Markov–Bernstein-type inequality given by

$$\|(1-x^2)^{r/2} P_n^{(r)}(x)\|_{L_p[-1,1]} \le Mn^r \|P_n(x)\|_{L_p[-1,1]}, \qquad P_n \in \Pi_n. \quad (7.2.7)$$

This result was proved by M.K. Potapov [1]. We will need later a more general result than (7.2.7), which we will prove in detail in Theorem 8.4.7 of Chapter 8. In the following let P_n be the best approximation of f by Π_n in L_p. With $l = \max\{k: 2^k \le 1/t\}$ we get from Theorem 2.1.1

$$\omega_\varphi^r(f, t)_p \le M_1 K_{r,\varphi}(f, t^r)_p \le M_1(\|f - P_{2^l}\|_p + t^r \|\varphi^r P_{2^l}^{(r)}\|_p), \quad (7.2.8)$$

where $\varphi(x) = \sqrt{1-x^2}$. We write $P_{2^l}^{(r)} = \sum_{k=0}^{l-1}(P_{2^{k+1}} - P_{2^k})^{(r)}$ and use (7.2.7) to obtain

$$\|\varphi^r P_{2^l}^{(r)}\|_p \le \sum_{k=0}^{l-1} \|\varphi^r (P_{2^{k+1}} - P_{2^k})^{(r)}\|_p$$

$$\le M \sum_{k=0}^{l-1} 2^{(k+1)r} E_{2^k}(f)_p \qquad (7.2.9)$$

$$\le M \sum_{0 \le n \le 1/t} (n+1)^{r-1} E_n(f)$$

which, together with (7.2.8), completes the proof. □

From Theorem 7.2.1 and Theorem 7.2.4 we can derive the following corollary.

Corollary 7.2.5. *For $0 < \alpha < r$ and $f \in L_p[-1, 1]$,*

$$E_n(f)_p \sim n^{-\alpha} \quad \text{and} \quad \|\Delta_{h\sqrt{1-x^2}}^r f(x)\|_{L_p[-1,1]} \sim h^\alpha$$

are equivalent.

7.3. Asymptotic Behavior of Derivatives of Best Approximating Polynomials

We will relate the behavior of $P_n^{(r)}(x)$ to the modulus of continuity of the function.

Theorem 7.3.1. *For P_n the best nth degree polynomial approximation to f in $L_p[-1, 1]$ and an integer r, $r > 0$ we have*

$$\|\varphi^r P_n^{(r)}\|_{L_p[-1,1]} \leq Mn^r \omega_\varphi^r(f, n^{-1})_p, \tag{7.3.1}$$

where $\varphi(x) = \sqrt{1 - x^2}$ and M is independent of n and f.

In this theorem the polynomial P_n described above can be exchanged with any P_n satisfying $\|f - P_n\| \leq M_1 E_n(f)_p$ with some fixed M_1.

PROOF. We first prove the estimate

$$\|\varphi^{r+1} P_n^{(r+1)}\|_{L_p[-1,1]} \leq Ln^{r+1} \omega_\varphi^r(f, n^{-1})_p. \tag{7.3.2}$$

For l given by $l = \max(k: 2^k < n)$ we expand $P_n(x)$ by

$$P_n(x) - P_0(x) = P_n(x) - P_{2^l}(x) + (P_{2^l}(x) - P_{2^{l-1}}(x)) + \cdots + (P_1(x) - P_0(x)).$$

We recall that for $m < n$ $\|P_n - P_m\|_p \leq 2E_m(f)_p$ and use (7.2.7) with $r + 1$ instead of r to obtain

$$\|\varphi^{r+1} P_n^{(r+1)}\|_{L_p[-1,1]} \leq M_1 \sum_{k=0}^{l} 2^{k(r+1)} E_{2^k}(f)_p.$$

Theorems 7.2.1 and 2.1.1 imply $E_n(f)_p \leq M_2 \omega_\varphi^r(f, n^{-1})_p$ and

$$E_{2^k}(f)_p \leq M_2 \omega_\varphi^r(f, 2^{-k})_p \leq M_3 K_{r,\varphi}(f, 2^{-kr})_p \leq M_3 2^r 2^{(l-k)r} K_{r,\varphi}(f, 2^{-(l+1)r})_p$$
$$\leq M_4 2^{(l-k)r} \omega_\varphi^r(f, 2^{-(l+1)})_p \leq M_4 2^{(l-k)r} \omega_\varphi^r(f, n^{-1})_p,$$

where $M_2, M_3,$ and M_4 do not depend on k or l. Combining the above, we write

$$\|\varphi^{r+1} P_n^{(r+1)}\|_{L_p[-1,1]} \leq M_5 \omega_\varphi^r(f, n^{-1})_p \sum_{k=0}^{l} 2^{k(r+1)} 2^{(l-k)r} \leq Ln^{r+1} \omega_\varphi^r(f, n^{-1})_p.$$

For $\varphi(x) = \sqrt{1 - x^2}$ it follows from a well-known result on differences that

$$n^{-r}(1 - x^2)^{r/2} P_n^{(r)}(\xi(x)) = \Delta_{\varphi(x)/n}^r P_n(x),$$

for some $\xi(x)$ satisfying $x - r\varphi(x)/2n \leq \xi(x) \leq x + r\varphi(x)/2n$. To estimate $\Delta_{\varphi(x)/n}^r P_n(x)$ in $L_p(D_n)$, where $D_n = [-1 + 2r^2 n^{-2}, 1 - 2r^2 n^{-2}]$, which guarantees $x \pm (r/2n)\varphi(x) \in [-1 + r^2 n^{-2}, 1 - r^2 n^{-2}]$ ($x \in D_n$), we write

$$\|\Delta_{\varphi/n}^r P_n\|_{L_p[D_n]} \leq \|\Delta_{\varphi/n}^r (P_n - f)\|_{L_p[D_n]} + \|\Delta_{\varphi/n}^r f\|_{L_p[-1,1]}$$
$$\leq CE_n(f)_p + \omega_\varphi^r(f, n^{-1})_p$$
$$\leq C_1 \omega_\varphi^r(f, n^{-1})_p.$$

7.3. Asymptotic Behavior of Derivatives of Best Approximating Polynomials

To obtain (7.3.1) we write (note that $P_n^{(r)}(\xi(x))$ is a measurable function of x)

$$\|n^{-r}\varphi^r P_n^{(r)}\|_{L_p(D_n)} \leq \|n^{-r}\varphi^r P_n^{(r)} - \Delta_{\varphi/n}^r P_n\|_{L_p(D_n)} + \|\Delta_{\varphi/n}^r P_n\|_{L_p[-1,1]}$$

$$\leq n^{-r}\|(1-x^2)^{r/2}\{P_n^{(r)}(x) - P_n^{(r)}(\xi(x))\}\|_{L_p(D_n)}$$
$$+ C_1 \omega_\varphi^r(f, n^{-1})_p$$

$$\leq n^{-r}\left\|(1-x^2)^{r/2}\int_x^{\xi(x)} P_n^{(r+1)}(u)\,du\right\|_{L_p(D_n)} + C_1\omega_\varphi^r(f,n^{-1})_p.$$

We have for $L_p(D_n)$

$$\left\|(1-x^2)^{r/2}\int_x^{\xi(x)} P_n^{(r+1)}(u)\,du\right\|_{L_p(D_n)}$$

$$\leq \left\|(1-x^2)^{r/2}\int_{x-r\varphi(x)/2n}^{x+r\varphi(x)/2n}|P_n^{(r+1)}(u)|\,du\right\|_{L_p(D_n)} \equiv I_p.$$

For $1 < p \leq \infty$ we can deduce from (7.3.2), recalling that the maximal function $M(F)$ satisfies $\|M(F)\|_p \leq A_p\|F\|$,

$$I_p \leq \frac{r}{n}\left\|\varphi(x)^{r+1}\frac{1}{r\varphi(x)/n}\int_{x-r\varphi(x)/2n}^{x+r\varphi(x)/2n}|P_n^{(r+1)}(u)|\,du\right\|_{L_p(D_n)}$$

$$\leq Bn^{-1}\left\|\frac{1}{r\varphi(x)/n}\int_{x-r\varphi(x)/2n}^{x+r\varphi(x)/2n}\varphi(u)^{r+1}P_n^{(r+1)}(u)\,du\right\|_{L_p(D_n)}$$

$$\leq B_1 n^{-1}\|M(\varphi^{r+1}P_n^{(r+1)},x)\|_{L_p(D_n)}$$

$$\leq B_2 n^{-1}\|\varphi^{r+1}P_n^{(r+1)}\|_p$$

$$\leq B_3 n^r \omega_\varphi^r(f,n^{-1})_p.$$

For $p = 1$ we have

$$I_1 = \int_{D_n}(1-x^2)^{r/2}\int_{x-r\varphi(x)/2n}^{x+r\varphi(x)/2n}|P_n^{(r+1)}(u)|\,du\,dx$$

$$\leq B\int_{D_n}\int_{x-r\varphi(x)/2n}^{x+r\varphi(x)/2n}\varphi(u)^r|P_n^{(r+1)}(u)|\,du\,dx$$

$$\leq B\int \varphi(u)^r|P_n^{(r+1)}(u)|\left\{\int_{|x-u|<r\varphi(x)/2n}dx\right\}du$$

$$\leq B_1 n^{-1}\|\varphi^{r+1}P_n^{(r+1)}\|_{L_1[-1,1]}.$$

We now have

$$\|(1-x^2)^{r/2}P_n^{(r)}(x)\|_{L_p(D_n)} \leq M_1 n^r \omega_\varphi^r(f,n^{-1})_p.$$

To exchange D_n on the left with $[-1,1]$ it is enough to show

$$\|\varphi^r Q_n\|_{L_p(D_n)} \leq C\|\varphi^r Q_n\|_{L_p[-1,1]}, \qquad Q_n \in \Pi_n. \tag{7.3.3}$$

For $1 \le p < \infty$, (7.3.3) is essentially given in Nevai [1, Th. 14, p. 113] with somewhat different formulation. For $p = \infty$ the inequality

$$|R_n(\lambda)| \le (\lambda + \sqrt{\lambda^2 - 1})^n \max_{x \in [-1,1]} |R_n(x)|, \qquad |\lambda| > 1, \qquad R_n \in \Pi_n$$

by S.N. Bernstein (see Timan [2])) applied to D_n instead of $[-1, 1]$ implies (7.3.3). In any case, we will prove a much more general result in Section 8.4. □

The converse result for derivatives is given in the following theorem.

Theorem 7.3.2. *Suppose* $\|\varphi^r P_n^{(r)}\|_{L_p[-1,1]} \le M n^r \psi(1/n)$, *where* $\varphi(x) = \sqrt{1-x^2}$, P_n *is the best nth degree polynomial approximation to f in L_p and $\psi(\tau) \searrow 0$ as $\tau \to 0+$. Then*

$$E_n(f)_p \le M \int_0^{1/n} \psi(\tau)/\tau \, d\tau$$

and

$$\omega_\varphi^r(f, t)_p \le M \int_0^t \psi(\tau)/\tau \, d\tau.$$

In particular, if

$$\int_0^t \psi(\tau)/\tau \, d\tau = O(\psi(t)) \qquad (0 < t < 1),$$

the above implies

$$E_n(f)_p \le M_1 \psi\left(\frac{1}{n}\right) \quad \text{and} \quad \omega_\varphi^r(f,t) \le M_1 \psi(t).$$

PROOF. We follow an argument by Sunouchi [1] and write

$$I_n = \|P_{2n} - P_n(P_{2n})\|_p \ge \|f - P_n(P_{2n})\|_p - \|f - P_{2n}\|_p \ge E_n(f)_p - E_{2n}(f)_p,$$

where $P_n(P_{2n})$ is the best nth degree polynomial approximation to P_{2n} in L_p. Using Theorem 7.2.1 for $f = P_{2n}$, and Theorem 2.1.1 we have

$$I_n = \|P_{2n} - P_n(P_{2n})\|_p = E_n(P_{2n})_p \le M\omega_\varphi^r(P_{2n}, n^{-1})_p \le M_1 K_{r,\varphi}(P_{2n}, n^{-r})_p$$

$$\le M_1 n^{-r} \|\varphi^r P_{2n}^{(r)}\|_{L_p[-1,1]} \le M_2 \psi(1/2n)$$

with M_2 independent of n. We can now write

$$E_n(f)_p = \sum_{k=0}^{\infty} (E_{2^k n}(f)_p - E_{2^{k+1} n}(f)_p) \le \sum_{k=0}^{\infty} I_{2^k n}$$

$$\le M_2 \sum_{k=1}^{\infty} \psi(2^{-k} n^{-1}) \le M_3 \int_0^{1/n} (\psi(\tau)/\tau) \, d\tau.$$

7.4. Error Bounds for Gaussian Quadrature

To obtain the estimate on $\omega_\varphi^r(f, n^{-1})_p$ (Theorem 7.2.4 is not applicable) we write for $(n+1)^{-1} \le h \le n^{-1}$

$$\|\Delta_{h\varphi}^r f\|_p \le \|\Delta_{h\varphi}^r(f - P_{2n})\|_p + \|\Delta_{h\varphi}^r P_{2n}\|_p \le M_4 E_{2n}(f)_p + M_5(hn)^r \psi(1/2n)$$

$$\le M_6 \sum_{k=0}^{\infty} \psi(2^{-k-1}n).$$

The rest of the conclusions are immediate. □

We can also state the following consequence to Theorems 7.3.1 and 7.3.2.

Corollary 7.3.3. *For* $0 < \alpha \le r$ $\|\varphi^r P_n^{(r)}\|_p \le Mn^{r-\alpha}$ *and* $\omega_\varphi^r(f, n^{-1})_p \le Mn^{-\alpha}$ *are equivalent.*

Remark 7.3.4. (a) It is well known that $E_n(f)_p = O(n^{-r})$ does not imply $\omega_\varphi^r(f, n^{-1})_p = O(n^{-r})$ but as we have seen above $\|\varphi^r P_n^{(r)}\|_p = O(1)$ does. Moreover, $E_n(f)_p = O(n^{-r}(\log n)^\sigma)$ does not imply $\omega_\varphi^r(f, n^{-1})_p = O(n^{-r}(\log n)^\sigma)$ (it implies $\omega_\varphi^{r+1}(f, n^{-1}) = O(n^{-r}(\log n)^\sigma))$ but $\|\varphi^r P_n^{(r)}\|_p = O((\log n)^\sigma)$ implies $\omega_\varphi^r(f, n^{-1}) = O(n^{-r}(\log n)^\sigma)$. That is, in the range for which $\omega_\varphi^r(f, n^{-1})_p$ and $\omega_\varphi^{r+1}(f, n^{-1})_p$ may have different behavior $E_n(f)$ yields information on $\omega_\varphi^{r+1}(f, n^{-1})$ while for $j = r$ and $j = r+1$ $\|\varphi^j P_n^{(j)}\|_p$ yield information on $\omega_\varphi^j(f, n^{-1})_p$.

(b) If $\psi(\tau)$ tends to zero slowly like $(\log 1/\tau)^{-\sigma}$, $\sigma > 0$ for example, the assumption $\|\varphi^r P_n^{(r)}\|_p = O(n^r \psi(n^{-1}))$ yields essentially no information on the smoothness of the function while $E_n(f)_p = O(\psi(n))$ implies $\omega_\varphi^r(f, n^{-1}) = O(\psi(n))$.

These remarks illustrate that the theorems in Sections 7.2 and 7.3 are applicable to different ranges and they supplement each other.

7.4. Error Bounds for Gaussian Quadrature

Consider the Gaussian quadrature process

$$\int_{-1}^{1} f(x)\,dx \sim \sum_{j=1}^{n} \omega_j f(x_j) = I_n(f) \tag{7.4.1}$$

based on the roots $-1 < x_1 < \cdots < x_n < 1$ of the nth Legendre polynomial. Since this is exact for polynomials of degree less than $2n$, we get for the error

$$e_n(f) = \int_{-1}^{1} f(x)\,dx - I_n(f)$$

in (7.4.1) the trivial bound

$$e_n(f) \le 2E_{2n-1}(f)_C \tag{7.4.2}$$

(note that $\omega_j \geq 0$ and $\sum_{j=1}^{n} \omega_j = 1$). The crude method (which is used in the literature) of estimating $e_n(f)$ consists of applying Jackson's estimate on the right of (7.4.2). From Section 7.2 we get the sharper inequality

$$e_n(f) \leq M\omega_\varphi^r(f, 1/n)_C \qquad (\varphi(x) = \sqrt{1-x^2}) \tag{7.4.3}$$

which already takes into account the possibly less smooth behavior of f at ± 1. However, the supremum norm in (7.4.3) is still too rough, and the natural question is whether for smooth functions one can get upper bounds for $e_n(f)$ using certain L_1 norms. R.A. DeVore and L.R. Scott [1] found such estimates; they proved

$$e_n(f) \leq M_s n^{-s} \int_{-1}^{1} |f^{(s)}(x)| (1-x^2)^{s/2} \, dx \tag{7.4.4}$$

first for $s = 1$ which obviously implies

$$e_n(f) \leq M n^{-1} E_{2n-2}(f')_{\varphi,1}, \tag{7.4.5}$$

where $E_n(f')_{\varphi,1}$ means the best weighted approximation (with weight $\varphi(x) = \sqrt{1-x^2}$) of f' in the L_1 norm. They then proceeded to estimate $E_n(f')_{\varphi,1}$ using higher order derivatives of f which finally yielded (7.4.4) for any $s \geq 1$. Combining our results with (7.4.5) we obtain the following theorem.

Theorem 7.4.1. *With $\varphi(x) = \sqrt{1-x^2}$ we have*

$$e_n(f) \leq M_r n^{-1} \int_0^{1/n} \omega_\varphi^r(f,\tau)_1/\tau^2 \, d\tau, \qquad n > r, \tag{7.4.6}$$

where M_r is independent of $n > r$ and $f \in L_1[-1,1]$.

Of course the convergence of the integral on the right implies that f is L_1 equivalent of a locally absolutely continuous function. We use this equivalent representative of f in the quadrature formula. (Otherwise, we do not have even $e_n(f) = o(1)$.)

PROOF. Let $P_n \in \Pi_n$ be the best approximating polynomial for f in $L_1[-1,1]$. Then

$$f = P_n + \sum_{k=0}^{\infty} (P_{2^{k+1}n} - P_{2^k n})$$

in L_1 (i.e., the expression on the right is the L_1 equivalent of f which we need). From (7.4.5), (7.2.7), and Theorem 7.2.1 we obtain

$$e_n(f) \leq M n^{-1} \|\varphi(f' - P_n')\|_1$$

$$\leq M n^{-1} \sum_{k=0}^{\infty} \|\varphi(P_{2^{k+1}n} - P_{2^k n})'\|_1$$

$$\leq M_1 n^{-1} \sum_{k=0}^{\infty} 2^{k+1} n \|P_{2^{k+1}n} - P_{2^k n}\|_1$$

7.4. Error Bounds for Gaussian Quadrature

$$\leq 2M_1 n^{-1} \sum_{k=0}^{\infty} 2^{k+1} n E_{2^k n}(f)_1$$

$$\leq M_2 n^{-1} \sum_{k=0}^{\infty} 2^k n \omega_\varphi^r(f, 2^{-k} n^{-1})_1$$

$$\leq M_3 n^{-1} \int_0^{1/n} \omega_\varphi^r(f, \tau)_1 / \tau^2 \, d\tau,$$

provided the last integral converges. □

As a final remark, we mention that similar bounds hold for many other systems of nodes and in (7.4.6) the right hand side has the order

$$\int_{-1}^{x_1} |f| + \int_{x_n}^{1} |f|$$

for any f constructed from analytic functions, $|x \pm 1|^s$ and iterated logarithms of these, which means that (7.4.6) is the best possible estimate for such functions.

CHAPTER 8
WEIGHTED BEST POLYNOMIAL APPROXIMATION

For some weight functions, including Jacobi weights, we relate the best weighted polynomial approximation and asymptotic behavior of the derivatives of the optimal polynomials to weighted main-part moduli of smoothness.

8.1. Some Concepts and Description of the Weight

For a weight function $w(x)$ on $[-1, 1]$ the best nth degree weighted polynomial approximation is given by

$$E_n(f)_{w,p} = \inf_{P_n \in \Pi_n} \|w(f - P_n)\|_{L_p[-1,1]}. \quad (8.1.1)$$

We will investigate the relation between $E_n(f)_{w,p}$ or the asymptotic behavior of the derivative of P_n (the optimal polynomial in (8.1.1)) and the weighted main-part moduli given by

$$\Omega_\varphi^r(f,t)_{w,p} = \sup_{0 < h \leq t} \|w\Delta_{h\varphi}^r f\|_{L_p[-1+2r^2h^2, 1-2r^2h^2]}, \quad \varphi(x) = \sqrt{1-x^2}. \quad (8.1.2)$$

The definition in (8.1.2) coincides with definitions in Chapters 3 and 6 as $h^* = 2r^2 h^2$ in the present situation ($\sqrt{1 \pm x} \leq \sqrt{1-x^2} \leq \sqrt{2}\sqrt{1 \pm x}$, for $\mp x \geq 0$). We will allow, however, in this chapter somewhat more general weights than in Chapter 6.

The most important example of weight $w(x)$ treated in this chapter is that of the Jacobi weights $w(x) = (1+x)^{\gamma_1}(1-x)^{\gamma_2}$ with the restriction $\gamma_i > -1/p$ for $p < \infty$ and $\gamma_i \geq 0$ for $p = \infty$. M.K. Potapov [2–4] and P.L. Butzer et al. [1, 2] investigated the rate of best polynomial approximation with such weights, relating them to moduli based on generalized translations. K.G. Ivanov [7]

8.1. Some Concepts and Description of the Weight

investigated it for some of the above weights, relating them to a sequence of conditions about his moduli. Direct estimates were also proved by N.X. Ky [1] and for $p = 1$ by R. DeVore and L.R. Scott [1]. These results are more restrictive than those presented in this chapter with regard to $E_n(f)_{w,p}$ and γ_i. Although the main achievement here is the simpler and computable modulus, our results will also be proved for more general weights that can include logarithmic behavior near ± 1. Comparison of our results with those mentioned above will be given in Chapter 13.

We define the class of weights J_p^* as follows.

Definition 8.1.1. The positive weight function $w(x)$ on $(-1, 1)$ is of class J_p^* if

(a) $w(x) = w_-(\sqrt{1+x})w_+(\sqrt{1-x})$,
(b) $w_+(y) = y^{\gamma_1}v_+(y)$, $w_-(y) = y^{\gamma_2}v_-(y)$, where $\gamma_i > -2/p$ and $v_\pm(y) \sim 1$ on every interval $[\delta, \sqrt{2}]$, $\delta > 0$,
(c) for every $\varepsilon > 0$ $y^\varepsilon v_\pm(y)$ are increasing and $y^{-\varepsilon} v_\pm(y)$ are decreasing in $(0, \delta(\varepsilon))$ for some $\delta(\varepsilon) > 0$, and
(d) for $p = \infty$ we may have $\gamma_1 = 0$ or $\gamma_2 = 0$ in which case $v_-(y)$ or $v_+(y)$ have to be nondecreasing for small y.

The Jacobi weights $w(x) = (1+x)^{\gamma_1}(1-x)^{\gamma_2}$ with $\gamma_i > -1/p$ and for $p = \infty$ $\gamma_i \geq 0$ are a special but very important class of weights belonging to J_p^*.

For the theorems about best weighted polynomial approximation in Sections 8.2 and 8.3 one needs the following properties of the weights of class J_p^*. Properties (C), (D), and (E) will be proved below, while the deeper properties (A) and (B) will be proved in Section 8.4.

(A) For any integer $r \geq 0$ $w(x) \in J_p^*$ implies $(1-x^2)^{r/2} w(x) \in J_p^*$ and

$$\|w(x)(1-x^2)^{r/2} P_n^{(r)}(x)\|_{L_p[-1,1]} \leq Mn^r \|w(x) P_n(x)\|_{L_p[-1,1]}, \quad P_n \in \Pi_n. \tag{8.1.3}$$

(B) For all $c > 0$ there exists a constant $M(c)$ independent of $n > \sqrt{2c}$ such that

$$\|wP_n\|_{L_p[-1,1]} \leq M(c) \|wP_n\|_{L_p[-1+cn^{-2}, 1-cn^{-2}]}, \quad P_n \in \Pi_n. \tag{8.1.4}$$

(C) The weight function can be written as $w(x) = w_-(\sqrt{1+x})w_+(\sqrt{1-x})$, where $w_\pm$ satisfies $A_\delta^{-1} \leq w_\pm(x) \leq A_\delta$ in $[\delta, \sqrt{2}]$ $\delta > 0$.

(D) For $w_n(x) \equiv w_-(\sqrt{1+x} + (1/n))w_+(\sqrt{1-x} + (1/n))$ and $c > 0$ we have

$$A^{-1} w_n(x) \leq w(x) \leq A w_n(x) \quad \text{for } x \in [-1+cn^{-2}, 1-cn^{-2}], \tag{8.1.5}$$

where $A \equiv A(c)$ is independent of n, and for some $s \geq 0$

$$w_n(x) \leq B(1 + (nt)^2)^s w_n(u) \quad \text{for } |x-u| \leq \min(\Delta_{1/t}(u), \Delta_{1/t}(x)) \tag{8.1.6}$$

$(\Delta_{1/t}(u) = |t|(|t| + \sqrt{1-u^2}))$, where B is independent of n and t.

(E) For some $C > 0$ and $\varphi(x) = \sqrt{1-x^2}$ we have

$$M^{-1} \Omega_\varphi^r(f,t)_{w,p} \leq \mathcal{K}_{\varphi,r}(C/2r^2, f, t^r)_{w,p} \leq M \Omega_\varphi^r(f,t)_{w,p}.$$

We recall (see (6.2.1) and (6.2.3)) that $\Omega_\varphi^r(f,t)_{w,p}$ and $\mathscr{K}_{\varphi,r}(C/2r^2, f, t^r)_{w,p}$ are given by

$$\Omega_\varphi^r(f,t)_{w,p} = \sup_{0<h\le t} \|w\Delta_{h\varphi}^r f\|_{L_p[-1+2r^2h^2, 1-2r^2h^2]}$$

and

$$\mathscr{K}_{\varphi,r}(C/2r^2, f, t^r)_{w,p} \equiv \sup_{0<h\le t} \inf_g \{\|w(f-g)\|_{L_p[-1+Ch^2, 1-Ch^2]}$$
$$+ h^r \|w\varphi^r g^{(r)}\|_{L_p[-1+Ch^2, 1-Ch^2]};$$
$$g^{(r-1)} \in \text{A.C.}[-1+Ch^2, 1-Ch^2]\},$$

respectively.

The two inequalities (8.1.3) and (8.1.4) are of fundamental importance in polynomial approximation. For any Jacobi weight w (8.1.3) was proved by M.K. Potapov [1] and B. Khalilova [1] (rediscovered by S.V. Konjagin [1]) and (for $p < \infty$) (8.1.4) (in a slightly different form) by P. Nevai [1]. In these proofs, especially in that of (8.1.4), the fact that w was a Jacobi weight was used extensively. For example, the proof of (8.1.4) depends on the behavior of the Christoffel functions for Jacobi polynomials and on the fine distribution law of the roots of these polynomials. If the weight w contains a logarithmic factor, for example, then very little is known about the corresponding orthogonal polynomials. (There is no generating function, differential equation, or explicit recurrence formula to help.) Therefore, to prove (8.1.4) for such weights a totally new approach has to be used. In Section 8.4 we will utilize some ideas of M.K. Potapov and prove both (8.1.3) and (8.1.4) at one stroke, which will enable us to extend the theory presented in Chapter 7 to weighted polynomial approximation with general weights, that is, with weights w satisfying $w \in J_p^*$.

Lemma 8.1.2. *For $w \in J_p^*$ conditions* (C), (D), *and* (E) *on the weights w given above are satisfied.*

PROOF. Condition (C) is obvious for $w \in J_p^*$. To demonstrate condition (D) for $w \in J_p^*$ (which, together with condition (C), does not depend on p), we observe that we can prove it separately for $w_+(\sqrt{1-x})$ and $w_-(\sqrt{1+x})$, and because of symmetry the proof for one of them will be sufficient. To show (8.1.5) we observe that $y^{-\gamma_2+\varepsilon} w_-(y)$ ($\varepsilon > 0$) is increasing for small y ((c) of Definition 8.1.1), and therefore

$$(\sqrt{1+x} + n^{-1})^{-\gamma_2+\varepsilon} w_-(\sqrt{1+x} + n^{-1}) \ge C_1 (1+x)^{-\gamma_2+\varepsilon} w_-(\sqrt{1+x}).$$

As $x > -1 + cn^{-2}$, we have $1 \le (\sqrt{1+x} + n^{-1})/\sqrt{1+x} \le c^{-1/2} + 1$ which implies

$$w_-(\sqrt{1+x}) \le M(1 + c^{-1/2})^{|\gamma_2|+\varepsilon} w_-(\sqrt{1+x} + n^{-1}).$$

Similarly we can prove the other inequality of (8.1.5) using the fact that $y^{-\gamma_2-\varepsilon} w_-(y)$ is decreasing for small y. To show (8.1.6), again just for w_-, we

8.1. Some Concepts and Description of the Weight

recall that for $u > x$ one has $\sqrt{1+u} + n^{-1} > \sqrt{1+x} + n^{-1}$ and therefore

$$(\sqrt{1+x} + n^{-1})^{-\gamma_2+\varepsilon} w_-(\sqrt{1+x} + n^{-1})$$
$$\leq M(\sqrt{1+u} + n^{-1})^{-\gamma_2+\varepsilon} w_-(\sqrt{1+u} + n^{-1}),$$

and for $u < x$

$$(\sqrt{1+x} + n^{-1})^{-\gamma_2-\varepsilon} w_-(\sqrt{1+x} + n^{-1})$$
$$\leq M(\sqrt{1+u} + n^{-1})^{-\gamma_2-\varepsilon} w_-(\sqrt{1+u} + n^{-1}).$$

It is sufficient to show now for $|x - u| < \min(\Delta_{1/t}(u), \Delta_{1/t}(x))$, where $x, u \in (-1, 1)$, that

$$\sqrt{1+u} + n^{-1} \leq B_1(1 + (nt)^2)(\sqrt{1+x} + n^{-1})$$
$$\sqrt{1+x} + n^{-1} \leq B_2(1 + (nt)^2)(\sqrt{1+u} + n^{-1}).$$
(8.1.7)

We divide the proof of (8.1.7) into 4 cases,

(i) $x \leq 0$ and $u \leq 0$,
(ii) $x \geq -1/2$ and $u \geq -1/2$,
(iii) $x < -1/2$ and $u > 0$, and
(iv) $u < -1/2$ and $x > 0$.

(Note that (i), (ii), (iii), and (iv) cover all possible cases.) Cases (ii), (iii), and (iv) are almost trivial. To show case (i) we use (7.2.5) (see the remark made after it) to obtain

$$\sqrt{1+x} + n^{-1} \leq \sqrt{1-x^2} + n^{-1} \equiv \delta_n(x) \leq 9(1 + n^2 t^2)\delta_n(u)$$
$$\leq 18(1 + n^2 t^2)(\sqrt{1+u} + n^{-1}),$$

and (7.2.5) with the roles of x and u reversed to obtain

$$\sqrt{1+u} + n^{-1} \leq 18(1 + n^2 t^2)(\sqrt{1+x} + n^{-1}).$$

Condition (E) was proved in Chapter 6 in Theorem 6.2.1 for Jacobi-type weights, and all we have to show in order to imitate the proof of Theorem 6.2.1 is

$$A^{-1} w(x) \leq w(u) \leq A w(x) \quad \text{for } x \in [-1 + 2r^2 h^2, 1 - 2r^2 h^2] \text{ and}$$
$$u \in \left(x - \frac{r}{2} h\varphi(x), x + \frac{r}{2} h\varphi(x)\right).$$
(8.1.8)

But $(-1 + x)/2 < x - (r/2)h\varphi(x) < x + (r/2)h\varphi(x) < (1 + x)/2$ (because of the range of x) or $(1 + x)/2 < 1 + u \leq 2(1 + x)$ and $2(1 - x) \leq 1 - u \leq (1 - x)/2$ which imply (8.1.8) using simple properties of $w \in J_p^*$. The result of (E) is achieved now with $C = r^2$. □

Remark 8.1.3. Ivanov [5–6] introduced the following condition on a weight function $w(x)$: $|x - u| \leq \lambda \Delta_n(u)$ implies $w(x) \leq B\lambda^s w(u)$ for $\lambda \geq 1$ and $s \geq 0$.

Weight functions satisfying that condition can have neither a zero nor a singularity near ± 1 (or elsewhere).

We will prove the theorems of Sections 8.2 and 8.3 using only conditions (A), (B), (C), (D), and (E) of w. We know that Jacobi weights satisfy them but we still need to show for $w \in J_p^*$ that (A) and (B) are satisfied and that will be done in Section 8.4.

8.2. Best Weighted Algebraic Polynomial Approximation

The relation between the best polynomial approximation and the smoothness is given in the following theorem.

Theorem 8.2.1. *For* $w \in J_p^*$, $E_n(f)_{w,p}$ *given in* (8.1.1), $\Omega_\varphi^r(f,t)_{w,p}$ *given in* (8.1.2), *and* $\varphi(x) = \sqrt{1-x^2}$, *we have*

$$E_n(f)_{w,p} \leq M \sum_{k=0}^{\infty} \Omega_\varphi^r(f, n^{-1}2^{-k})_{w,p} \sim \int_0^{1/n} (\Omega_\varphi^r(f,\tau)_{w,p}/\tau) \, d\tau \quad (8.2.1)$$

and

$$\Omega_\varphi^r(f,h)_{w,p} \leq Mh^r \sum_{0 \leq n < 1/h} (n+1)^{r-1} E_n(f)_{w,p}. \quad (8.2.2)$$

From Theorem 8.2.1, one can deduce the following immediate but important corollary.

Corollary 8.2.2. *For* $w \in J_p^*$ *and* $\alpha < r$, $E_n(f)_{w,p} = O(n^{-\alpha})$ *is equivalent to* $\Omega_\varphi^r(f,h)_{w,p} = O(h^\alpha)$.

Paraphrasing Corollary 8.2.2, one can write

$$\inf_{P_n \in \Pi_n} \|w(f - P_n)\|_p = O(n^{-\alpha}) \Leftrightarrow \|w \Delta_{h\varphi}^r f\|_{L_p[-1+2r^2h^2, 1-2r^2h^2]} = O(h^\alpha). \quad (8.2.3)$$

PROOF OF THEOREM 8.2.1. We assume conditions (A)–(E) of Section 8.1. (Conditions (C), (D), and (E) were proved in Lemma 8.1.2 and conditions (A) and (B) will be proved in Section 8.4.) The two parts of our theorem are actually independent results. We first prove (8.2.1). The first crucial step is to show that for some fixed $C > 0$ there exists an algebraic polynomial $P_n^* \in \Pi_n$ (for every n) such that

$$\|w(f - P_n^*)\|_{L_p[-1+Cn^{-2}, 1-Cn^{-2}]} \leq B\Omega_\varphi^r\left(f, \frac{1}{n}\right)_{w,p}, \quad (8.2.4)$$

where $B = B(C)$ does not depend on $n \geq r$. Using (8.2.4) and (8.1.4), we have

$$\|w(P_{2^{k+1}n}^* - P_{2^k n}^*)\|_{L_p[-1,1]} \leq B_1 \|w(P_{2^{k+1}n}^* - P_{2^k n}^*)\|_{L_p[-1+C2^{-2k}n^{-2}, 1-C2^{-2k}n^{-2}]}$$

$$\leq B_2 \Omega_\varphi^r(f, 2^{-k}n^{-1})_{w,p}.$$

8.2. Best Weighted Algebraic Polynomial Approximation

From the convergence of $\sum_{k=0}^{\infty} \Omega_\varphi^r(f, n^{-1}2^{-k})_{w,p}$ (otherwise (8.2.1) is trivial), we deduce that the series $P_n^*(x) + \sum_{k=1}^{\infty} (P_{2^k n}^*(x) - P_{2^{k-1} n}^*(x))$ converges a.e. in every subinterval of $[-1, 1]$, and therefore a.e. in $[-1, 1]$. Moreover,

$$f(x) = P_n^*(x) + \sum_{k=1}^{\infty} (P_{2^k n}^*(x) - P_{2^{k-1} n}^*(x))$$

a.e. in $[-1 + C2^{-2k}n^{-2}, 1 - C2^{-2k}n^{-2}]$

for every k and therefore a.e. in $[-1, 1]$. Therefore,

$$\|w(f - P_n^*)\|_{L_p[-1,1]} \leq \sum_{k=1}^{\infty} \|w(P_{2^{k+1}n}^* - P_{2^k n}^*)\|_{L_p[-1,1]}$$

$$\leq B_2 \sum_{k=0}^{\infty} \Omega_\varphi^r(f, 2^{-k}n^{-1})_{w,p},$$

which implies (8.2.1) provided that we prove (8.2.4), which is our next goal.

By Lemma 8.1.2 (E) there exists a sequence of functions g_n^* such that for some $C > 0$ and M the inequalities

$$\|w(f - g_n^*)\|_{L_p[-1+Cn^{-2}, 1-Cn^{-2}]} \leq M\Omega_\varphi^r(f, n^{-1})_{w,p} \quad (\varphi(x) = \sqrt{1 - x^2})$$

and

$$\|w\varphi^r(g_n^*)^{(r)}\|_{L_p[-1+Cn^{-2}, 1-Cn^{-2}]} \leq Mn^r \Omega_\varphi^r(f, n^{-1})_{w,p}$$

are satisfied. We set

$$g_n(x) = \begin{cases} g_n^*(x) & \text{if } x \in (-1 + Cn^{-2}, 1 - Cn^{-2}) \\ Q_{n,1}(x) & \text{if } x \in (-1, -1 + Cn^{-2}] \\ Q_{n,2}(x) & \text{if } x \in [1 - Cn^{-2}, 1), \end{cases}$$

where $Q_{n,1}$ and $Q_{n,2}$ are the polynomials of degree at most $(r-1)$ with ith derivatives equal to the ith derivative of g_n^* at $-1 + Cn^{-2}$ and $1 - Cn^{-2}$ respectively, for every $0 \leq i < r$. Then $g_n^{(r-1)} \in A.C._{\text{loc}}[-1, 1]$,

$$\|w(f - g_n)\|_{L_p[-1+Cn^{-2}, 1-Cn^{-2}]} \leq M\Omega_\varphi^r(f, n^{-1})_{w,p}$$

and

$$\|w\varphi^r g_n^{(r)}\|_{L_p[-1,1]} \leq Mn^r \Omega_\varphi^r(f, n^{-1})_{w,p}. \tag{8.2.5}$$

Since

$$\inf_{P_n \in \Pi_n} \|w(f - P_n)\|_{L_p[-1+Cn^{-2}, 1-Cn^{-2}]}$$

$$\leq \|w(f - g_n)\|_{L_p[-1+Cn^{-2}, 1-Cn^{-2}]} + \inf_{P_n \in \Pi_n} \|w(g_n - P_n)\|_{L_p[-1+Cn^{-2}, 1-Cn^{-2}]},$$

it is enough to show that (8.2.5) implies

$$\inf_{P_n \in \Pi_n} \|w(g_n - P_n)\|_{L_p[-1+Cn^{-2}, 1-Cn^{-2}]} \leq M_1 \Omega_\varphi^r(f, n^{-1})_{w,p}.$$

On $[-1 + Cn^{-2}, 1 - Cn^{-2}]$ we have (see Lemma 8.1.2(D)) $w \sim w_n$ and
$$\delta_n(x) = \sqrt{1-x^2} + 1/n \sim \sqrt{1-x^2} = \varphi(x).$$
Therefore, it suffices to prove (note that $g_n^{(r)}(x) = 0$ if $x \notin [-1 + Cn^{-2}, 1 - Cn^{-2}]$)

$$\inf_{P_n \in \Pi_n} \|w_n(g_n - P_n)\|_{L_p[-1,1]} \le M_2 n^{-r} \|w_n \delta_n^r g_n^{(r)}\|_{L_p[-1,1]} \tag{8.2.6}$$

$(g^{(r-1)} \in \text{A.C.})$ which follows from

$$\inf_{P_n \in \Pi_n} \|w_n(G - P_n)\|_{L_p[-1,1]} \le M_3 n^{-1} \|w_n \delta_n G'\|_{L_p[-1,1]} \tag{8.2.7}$$

$(G \in \text{A.C.}[-1, 1])$. In fact, applying (8.2.7) to

$$G'(x) - Q_{n-1}(x) = \left(G(x) - \int_0^x Q_{n-1}(u)\,du\right)'$$

with optimal $Q_{n-1} \in \Pi_{n-1}$, we get

$$E_n(G)_{w_n, p} \le M_3 n^{-1} E_{n-1}(G')_{w_n \delta_n, p}, \tag{8.2.8}$$

and iteration of this leads to (8.2.6). (Note that $w \in J_p^*$ implies $w\varphi \in J_p^*$; hence in the iteration of (8.2.8) $w_n \delta_n^s$ can play the role of w_n.)

That (8.2.7) is true with

$$P_n(x) = \int_{-\pi}^{\pi} G(\cos(\arccos x - t)) T_n(t)\,dt$$

(see the proof of Theorem 7.2.1) is the content of the following lemma.

Lemma 8.2.3. *For $T_n(t)$ of Lemma 7.2.2 with l sufficiently large, $G \in \text{A.C.}[-1, 1]$, $G' \in L_p[-1, 1]$, and $w \in J_p^*$, we have*

$$\left\| w_n(x) \int_{-\pi}^{\pi} \left\{ \int_{\cos(\arccos x - t)}^x G'(u)\,du \right\} T_n(t)\,dt \right\|_{L_p[-1,1]} \le \frac{M}{n} \|w_n \delta_n G'\|_{L_p[-1,1]}. \tag{8.2.9}$$

Using properties (C) and (D) of Lemma 8.1.2 about the weight $w \in J_p^*$, the proof is almost identical with that of Lemma 7.2.3.

The proof of (8.2.2) follows the same lines as that of Theorem 7.2.4. For $g = P_{2^l}$ Lemma 8.1.2(E) implies

$$\|w \Delta_{h\varphi}^r f\|_{L_p[-1+2hr^2h^2, 1-2r^2h^2]} \le M \sup_{\tau \le h} \inf_l (\|w(f - P_{2^l})\|_p + \tau^r \|w\varphi^r P_{2^l}^{(r)}\|_p).$$

We choose P_n such that $P_n \in \Pi_n$ and $\|w(f - P_n)\|_p \le E_n(f)_{w,p}$, and l such that $2^l < 1/\tau \le 2^{l+1}$. The estimate of $\|w\varphi^r P_{2^l}^{(r)}\|$ is now similar to (7.2.9) if we use (8.1.3) instead of (7.2.7). □

Following Section 6.1 for Jacobi weights $w(x) = (1+x)^{\gamma_1}(1-x)^{\gamma_2}$ with $\gamma_i \ge 0$ one can define the weighted modulus $\omega_\varphi^r(f, t)_{w,p}$ by

8.2. Best Weighted Algebraic Polynomial Approximation

$$\omega_\varphi^r(f,t)_{w,p} = \sup_{0<h\leq t} \|w\Delta_{h\sqrt{1-x^2}}^r f(x)\|_{L_p[-1+2r^2h^2,\,1-2r^2h^2]}$$

$$+ \sup_{0<h\leq 2r^2t^2} \|w\vec{\Delta}_h^r f\|_{L_p[-1,\,-1+2r^2t^2]} \qquad (8.2.10)$$

$$+ \sup_{0<h\leq 2r^2t^2} \|w\overleftarrow{\Delta}_h^r f\|_{L_p[1-2r^2t^2,\,1]}.$$

We will prove that the following result is valid.

Theorem 8.2.4. *Suppose* $\omega_\varphi^r(f,t)_{w,p}$ *is given in* (8.2.10) *for Jacobi weights* $w(x) = (1+x)^{\gamma_1}(1-x)^{\gamma_2}$ *with* $\gamma_i \geq 0$. *Then*

$$\omega_\varphi^r(f,h)_{w,p} \leq Mh^r \sum_{0\leq n<1/h} (n+1)^{r-1} E_n(f)_{w,p}. \qquad (8.2.11)$$

This result, though it relates to a much more restricted weight than that of Theorem 8.2.1, is somewhat more general than (8.2.2) for those weights. Nevertheless, it will be proved as a minor addition to the proof of Theorem 8.2.1. We observe that $\omega_\varphi^r(f,t)_{w,p} \geq \Omega_\varphi^r(f,t)_{w,p}$ and therefore ω_φ can replace Ω_φ in (8.2.1).

PROOF OF THEOREM 8.2.4. Essentially we have to show only that $\sup_{0<h\leq 2r^2n^{-2}} \|w\vec{\Delta}_h^r f\|_{L_p[-1,\,1-2r^2n^{-2}]}$ and the corresponding expression near 1 are bounded by $Mn^{-r}\sum_{0\leq k\leq n}(k+1)^{r-1}E_k(f)_{w,p}$. Repeating the procedure of the last theorem in the proof of (8.2.2), we write $f = (f - P_n) + P_n$ and observe that

$$\sup_{0<h\leq 2r^2n^{-2}} \|w\vec{\Delta}_h^r(f-P_n)\|_{L_p[-1,\,-1+2r^2n^{-2}]} \leq M_r E_n(f)_{w,p}.$$

Using Taylor's formula and the assumption $\gamma_1 \geq 0$ we have

$$\sup_{0<h\leq 2r^2n^{-2}} \|w\vec{\Delta}_h^r P_n(x)\|_{L_p[-1,\,-1+2r^2n^{-2}]}$$

$$\leq M \sup_{0<j\leq r,\,0<h\leq 2r^2n^{-2}} \left\| w(x)\int_0^{jh} (jh-u)^{r-1}|P_n^{(r)}(x+u)|\,du \right\|_{L_p[-1,\,-1+2r^2n^{-2}]}$$

$$\leq M \sup_{0<j\leq r,\,0<h\leq 2r^2n^{-2}} \left\| \int_0^{jh} (jh)^{r-1} w(x+u)|P_n^{(r)}(x+u)|\,du \right\|_{L_p[-1,\,-1+2r^2n^{-2}]}$$

$$\leq M_1 n^{-2r} \|wP_n^{(r)}\|_{L_p[-1,\,-1+Ar^2n^{-2}]} \leq M_2 n^{-2r}\|wP_n^{(r)}\|_{L_p[-1+Ar^2n^{-2},\,1-Ar^2n^{-2}]}$$

$$\leq M_2 n^{-r}\|w\varphi^r P_n^{(r)}\|_{L_p[-1,1]}.$$

The rest now follows the proof of (8.2.2). □

Corollary 8.2.5. *For* $w \in J_p^*$ *and* $\varphi(x) = \sqrt{1-x^2}$, $\mathscr{K}_{r,\varphi}(f,\tau^r)_{w,p} = O(\tau^\alpha)$ *for* $\alpha < r$ *implies* $K_{r,\varphi}(f,t^r)_{w,p} = O(t^\alpha)$.

PROOF. This result follows as $\mathscr{K}_{r,\varphi}(f,\tau^r)_{w,p} = O(\tau^\alpha)$ implies $\Omega_\varphi^r(f,t)_{w,p} = O(t^\alpha)$, and hence $E_n(f)_{w,p} = O(n^{-\alpha})$ which implies (cf. Theorem 8.3.1 below)

$\|w\varphi^r P_n^{(r)}\| \leq O(n^{r-\alpha})$. For $t \sim 1/n$ we have
$$K_{r,\varphi}(f,t^r)_{w,p} \leq \|w(f - P_n)\|_p + n^{-r}\|w\varphi^r P_n^{(r)}\|_p = O(n^{-\alpha}) = O(t^\alpha). \quad \square$$

This corollary illustrates a nice interplay between the general theory presented in the first part of this book and its applications. On the one hand, Part I was used in the solution of the problem of this section, and on the other hand, the results we achieved enabled us to derive Corollary 8.2.5, which we could not prove in Sections 6.1–6.2 (cf. the introduction to Section 6.2). Recall now that in Corollary 6.2.3 we showed for Jacobi weights with $\gamma_i \geq 0$ the inequality

$$K_{r,\varphi}(f,t^r)_{w,p} \leq A \int_0^t (\mathcal{K}_{r,\varphi}(f,\tau^r)_{w,p}/\tau)\,d\tau$$

from which the implication "if $\mathcal{K}_{r,\varphi}(f,\tau^r)_{w,p} = O(\tau^\alpha)$ for some $\alpha < r$, then $K_{r,\varphi}(f,t^r)_{w,p} = O(t^\alpha)$" followed for a more restricted weight w and less restricted step-weight φ.

8.3. Derivatives of the Optimal Polynomials

We shall investigate the relation between the asymptotic behavior of derivatives of polynomials of best weighted approximation in a way analogous to that of Section 7.2 and for similar reasons. In this section when we write $P_n(x)$ we will mean the optimal weighted polynomial approximation, that is $\|w(f - P_n)\|_p = E_n(f)_{w,p}$.

Theorem 8.3.1. *For P_n satisfying $\|w(f - P_n)\|_p = E_n(f)_{w,p}$ and $w \in J_p^*$ we have*

$$\|w\varphi^r P_n^{(r)}\|_p \leq M n^r \int_0^{1/n} (\Omega_\varphi^r(f,\tau)_{w,p}/\tau)\,d\tau, \tag{8.3.1}$$

$$\Omega_\varphi^r(f,t)_{w,p} \leq M \sum_{k=1}^{\infty} 2^{-kr} n^{-r} \|w\varphi^r P_{2^k n}^{(r)}\|_p \quad \text{for } n = \left[\frac{1}{t}\right], \tag{8.3.2}$$

$$\|w\varphi^r P_n^{(r)}\|_p \leq M \sum_{k=0}^{n} (k+1)^{r-1} E_k(f)_{w,p}, \tag{8.3.3}$$

and

$$E_n(f)_{w,p} \leq M \sum_{k=1}^{\infty} 2^{-kr} n^{-r} \|w\varphi^r P_{2^k n}^{(r)}\|_p. \tag{8.3.4}$$

From this theorem we can deduce the following useful corollary.

Corollary 8.3.2. *The conditions $\|w\varphi^r P_n^{(r)}\|_p = O(n^{r-\alpha})$ and $\Omega_\varphi^r(f,t)_{w,p} = O(t^\alpha)$ are equivalent for $\alpha \leq r$.*

8.3. Derivatives of the Optimal Polynomials

PROOF OF THEOREM 8.3.1. As a corollary of the later stages of the proof of Theorem 8.2.1 we obtain (8.3.3) (see also (7.2.9)). We will now prove (8.3.4) and (8.3.2). We define $P_n(P_{2n})$ by $\|w(P_{2n} - P_n(P_{2n}))\|_p = E_n(P_{2n})_{w,p} \equiv I_n$ and write

$$I_n \geq \|w(f - P_n(P_{2n}))\|_p - \|w(f - P_{2n})\|_p \geq E_n(f)_{w,p} - E_{2n}(f)_{w,p}.$$

Using (8.1.4) and (8.2.4), and recalling

$$\Omega_\varphi^r(P_{2n}, t)_{w,p} \leq M \mathcal{K}_{r,\varphi}(1, P_n, t^r)_{w,p} \leq M \sup_{h \leq t} h^r \|w\varphi^r P_{2n}^{(r)}\|_{L_p[-1+r^2h^2, 1-r^2h^2]}$$

$$\leq M t^r \|w\varphi^r P_{2n}^{(r)}\|_{L_p[-1,1]},$$

we have

$$I_n \leq M n^{-r} \|w\varphi^r P_{2n}^{(r)}\|_{L_p[-1,1]}.$$

This now implies

$$E_n(f)_{w,p} = \sum_{k=0}^{\infty} (E_{2^k n}(f)_{w,p} - E_{2^{k+1}n}(f)_{w,p}) \leq \sum_{k=0}^{\infty} I_{2^k n}$$

$$\leq M \sum_{k=0}^{\infty} 2^{-kr} n^{-r} \|w\varphi^r P_{2^{k+1}n}^{(r)}\|_p$$

which is (8.3.4). To show (8.3.2) we write

$$\Omega_\varphi^r(f, t)_{w,p} \leq \Omega_\varphi^r(f - P_n, t)_{w,p} + \Omega_\varphi^r(P_n, t)_{w,p},$$

and, as $\Omega_\varphi^r(f - P_n, t)_{w,p} \leq M E_n(f)_{w,p}$ and $\Omega_\varphi^r(P_n, t)_{w,p} \leq M t^r \|w\varphi^r P_n^{(r)}\|_p$, we obtain (8.3.2) from (8.3.4).

Now we turn to (8.3.1). We first prove

$$\|w\varphi^{r+1} P_n^{(r+1)}\|_p \leq M_1 n^{r+1} \int_0^{1/n} (\Omega_\varphi^r(f, \tau)_{w,p}/\tau) \, d\tau. \tag{8.3.5}$$

Using (8.3.3) and Theorem 8.2.1, we obtain for $l = \max(k: 2^k < n)$

$$\|w\varphi^{r+1} P_n^{(r+1)}\|_p \leq M_2 \left\{ \sum_{k=1}^{l} 2^{k(r+1)} \int_0^{2^{-k}} (\Omega_\varphi^r(f, \tau)_{w,p}/\tau) \, d\tau \right\}.$$

From this estimate (8.3.5) will follow when we show for $k < l$

$$\int_0^{2^{-k}} (\Omega_\varphi^r(f, \tau)_{w,p}/\tau) \, d\tau \leq M_3 2^{(l-k)r} \int_0^{2^{-l-1}} (\Omega_\varphi^r(f, \tau)_{w,p}/\tau) \, d\tau, \tag{8.3.6}$$

where M_3 is independent of k, l, and f, as this will imply

$$\|w\varphi^{r+1} P_n^{(r+1)}\|_p \leq M_4 \left(2^{l(r+1)} \sum_{k=1}^{l} 2^{(k-l)} \int_0^{2^{-l-1}} (\Omega_\varphi^r(f, \tau)_{w,p}/\tau) \, d\tau \right)$$

$$\leq M_5 n^{r+1} \int_0^{1/n} (\Omega_\varphi^r(f, \tau)_{w,p}/\tau) \, d\tau.$$

To prove (8.3.6) we observe that it is a consequence of

$$\Omega^r_\varphi(f,\tau)_{w,p} \leq M 2^{vr}\Omega^r_\varphi(f,\tau/2^v)_{w,p}.$$

This was proved in Section 6.2 (see (6.2.8)) for Jacobi weights and because of Lemma 8.1.2(E) the proof of (6.2.8) can be repeated word for word when $w \in J_p^*$.

We now follow the proof of Theorem 7.3.1; that is, we use $n^{-r}\varphi(x)^r P_n^{(r)}(\xi(x)) = \Delta^r_{\varphi(x)/n} P_n(x)$ for $x \in [-1 + 2r^2 n^{-2}, 1 - 2r^2 n^{-2}] \equiv D_n$ and $\xi(x) \in [x - r\varphi(x)/2n, x + r\varphi(x)/2n] = I(x) \subset [-1 + r^2 n^{-2}, 1 - r^2 n^{-2}]$ and recall from (8.1.8) that if $u \in I(x)$ and $w(x) \in J_p^*$, then $w(x) \sim w(u)$. All of this implies

$$n^{-r}\|w\varphi^r P_n^{(r)}\|_{L_p(D_n)} = \|n^{-r}w\varphi^r P_n^{(r)} - w\Delta^r_{\varphi/n} P_n\|_{L_p(D_n)}$$
$$+ \|w\Delta^r_{\varphi/n}(P_n - f)\|_{L_p(D_n)} + \|w\Delta^r_{\varphi/n} f\|_{L_p(D_n)}$$
$$\leq n^{-r}\left\| w(x)\varphi(x)^r \int_x^{\xi(x)} P_n^{(r+1)}(u)\,du \right\|_{L_p(D_n)}$$
$$+ L_1\left(E_n(f)_{w,p} + \Omega^r_\varphi\left(f, \frac{1}{n}\right)_{w,p} \right).$$

To complete the proof we write

$$\left\| w(x)\varphi(x)^r \int_x^{\xi(x)} P_n^{(r+1)}(u)\,du \right\|_{L_p(D_n)}$$
$$\leq L_2 \left\| \int_{x-r\varphi(x)/2n}^{x+r\varphi(x)/2n} w(u)\varphi(u)^r |P_n^{(r+1)}(u)|\,du \right\|_{L_p(D_n)}$$

and this expression is smaller than $Cn^{-1}\|w\varphi^{r+1} P_n^{(r+1)}\|_{L_p[-1,1]}$ (see Section 2.4). Combining the above estimates, we now have

$$\|w\varphi^r P_n^{(r)}\|_{L_p(D_n)} \leq Cn^r \int_0^{1/n} (\Omega^r_\varphi(f,\tau)_{w,p}/\tau)\,d\tau,$$

which implies (8.3.1) by using (8.1.4), which is applicable since $w \in J_p^*$ and $P_n^{(r)} \in \Pi_n$. □

8.4. Proof of Some Crucial Inequalities for $w \in J_p^*$

In this section we will prove for $w \in J_p^*$ the inequalities (8.1.3) and (8.1.4) crucial for the proof of the theorems in Sections 8.2 and 8.3. These inequalities were already known for Jacobi weights and some of the methods of the present proof appeared in earlier papers of M.K. Potapov [1] and B. Khalilova [1] (see also Nevai [1] and [2]). However, the class of weights treated here is larger and a complete proof will be given in one place (as opposed to a proof spread over several papers and two languages). Moreover, (8.1.4) will be proved without referring to orthogonal polynomials as done by P. Nevai [1–2].

8.4. Proof of Some Crucial Inequalities for $w \in J_p^*$

First we recall that $w(x)$ on $[-1, 1]$ is of class J_p^* if

(a) $w(x) = w_-(\sqrt{1+x}) w_+(\sqrt{1-x})$,
(b) $w_-(y) = y^{\gamma_1} v_-(y)$, $w_+(y) = y^{\gamma_2} v_+(y)$ for $\gamma_i > -2/p$, and $v_\pm(y) \sim 1$ on every interval $[\delta, \sqrt{2}]$, $\delta > 0$,
(c) for every $\varepsilon > 0$ $y^\varepsilon v_\pm(y)$ and $y^{-\varepsilon} v_\pm(y)$ are increasing and decreasing, respectively, in $(0, \delta(\varepsilon))$ for some $\delta(\varepsilon) > 0$, and
(d) for $p = \infty$ we may have $\gamma_1 = 0$ or $\gamma_2 = 0$ in which case $v_-(y)$ or $v_+(y)$ respectively has to be nondecreasing for small y.

We now assume V_i essentially satisfies the above conditions on $v_\pm$; that is, $y^\varepsilon V_i(y)$ and $y^{-\varepsilon} V_i(y)$ are increasing and decreasing for small y, and define

$$W_\gamma(t) = |\sin t|^\gamma V_1(|\sin t|), \quad W_{\gamma,n}(t) = \left(|\sin t| + \frac{1}{n}\right)^\gamma V_1\left(|\sin t| + \frac{1}{n}\right) \quad (8.4.1)$$

and

$$\overline{W}_\gamma(t) = |\cos t|^\gamma V_2(|\cos t|), \quad \overline{W}_{\gamma,n}(t) = \left(|\cos t| + \frac{1}{n}\right)^\gamma V_2\left(|\cos t| + \frac{1}{n}\right). \quad (8.4.2)$$

We are now ready to state and prove a sequence of results that will lead to the proof of (8.1.3) and (8.1.4).

Lemma 8.4.1. *For $W_{\gamma,n}$ given in (8.4.1) $1 \leq p \leq \infty$ and $\gamma > -1$ we have*

$$\|T_n'(t) W_{\gamma,n}(t)\|_{L_p(T)} \leq Mn \|T_n(t) W_{\gamma,n}(t)\|_{L_p(T)}, \quad T_n \in \tau_n, \quad (8.4.3)$$

where M is independent of n. (T is the "circle" $[-\pi, \pi]$ and τ_n is the set of trigonometric polynomials of degree at most n.)

PROOF. We first show (8.4.3) for $|\gamma| < 1$. Using the M. Riesz formula (see Timan [2, p. 202]) given by

$$T_n'(t) = \frac{1}{4n} \sum_{k=1}^{2n} T_n(t + t_k)(-1)^{k+1}/\sin^2(t_k/2) \quad \text{where } t_k = \frac{2k-1}{2n}\pi, \quad (8.4.4)$$

we have

$$I(t) = T_n'(t) W_{\gamma,n}(t) = \frac{1}{4n} \sum_{k=1}^{2n} T_n(t + t_k) W_{\gamma,n}(t + t_k) \frac{(-1)^{k+1}}{\sin^2(t_k/2)} \frac{W_{\gamma,n}(t)}{W_{\gamma,n}(t + t_k)}.$$

We will show first that

$$A_n(t, t_k) \equiv \frac{1 + n|\sin t|}{1 + n|\sin(t + t_k)|} \leq 2(1 + n|\sin t_k|)$$

for all t, and this implies for $t = -u - t_k$

$$\frac{1 + n|\sin(u + t_k)|}{1 + n|\sin u|} \leq 2(1 + n|\sin t_k|).$$

8. Weighted Best Polynomial Approximation

It is enough to estimate $A_n(t, t_k)$ for $|t| \leq \pi/2$ as $A_n(t + \pi, t_k) = A_n(t, t_k)$. We assume $|t_k - r\pi| < \pi/4$ for some r, as otherwise $(1 + n|\sin t_k|) \geq 1 + (n/2)\sqrt{2}$, and since $A_n(t, t_k) \leq 1 + n$ for any t and t_k, the estimate is clear. Using symmetry, we may assume $0 < t_k \leq \pi/4$. If $0 \leq t \leq \pi/2$, we have $\sin t \leq 2|\sin(t + t_k)|$ and $A_n(t, t_k) \leq 2$. If $-2t_k \leq t \leq 0$, then

$$\frac{1 + n|\sin t|}{1 + n|\sin(t + t_k)|} \leq 1 + n|\sin 2t_k| \leq 1 + 2n \sin t_k.$$

Finally, if $-\pi/2 \leq t \leq -2t_k$, then $|\sin(t + t_k)| \geq |\sin t/2|$ and $A_n(t, t_k) \leq 2$ and this completes the estimate of $A_n(t, t_k)$. We now have for fixed $\varepsilon > 0$

$$\frac{W_{\gamma,n}(t)}{W_{\gamma,n}(t + t_k)} = A_n(t, t_k)^\gamma \frac{V_1\left(|\sin t| + \dfrac{1}{n}\right)}{V_1\left(|\sin(t + t_k)| + \dfrac{1}{n}\right)}$$

$$\leq M_\varepsilon \max(A_n(t, t_k)^{\gamma + \varepsilon}, A_n(t, t_k)^{\gamma - \varepsilon})$$

as, if $|\sin t| < |\sin(t + t_k)|$, then

$$A_n(t, t_k)^\varepsilon \frac{V_1\left(|\sin t| + \dfrac{1}{n}\right)}{V_1\left(|\sin(t + t_k)| + \dfrac{1}{n}\right)} \leq M,$$

and if $|\sin t| > |\sin(t + t_k)|$, then

$$A_n(t, t_k)^{-\varepsilon} \frac{V_1\left(|\sin t| + \dfrac{1}{n}\right)}{V_1\left(|\sin(t + t_k)| + \dfrac{1}{n}\right)} \leq M.$$

Choosing $\varepsilon > 0$ such that $|\gamma \pm \varepsilon| < 1$, we have $(W_{\gamma,n}(t)/W_{\gamma,n}(t + t_k)) \leq M(1 + n|\sin t_k|)^{|\gamma|+\varepsilon}$. We now use the triangle inequality to obtain

$$\|I\|_p \leq \frac{C}{4n} \sum_{k=1}^{2n} \frac{(1 + n|\sin t_k|)^{|\gamma|+\varepsilon}}{\sin^2(t_k/2)} \|T_n(t + t_k) W_{\gamma,n}(t + t_k)\|_p$$

$$\leq C_1 \|T_n(t) W_{\gamma,n}(t)\|_p \frac{1}{n} \sum_{k=1}^{2n} (1 + n|\sin t_k|)^{|\gamma|+\varepsilon}/\sin^2(t_k/2)$$

$$\leq C_2 n \|T_n(t) W_{\gamma,n}(t)\|_p.$$

We now prove the lemma for $\gamma = 1$. Let $T_{n+1}(t) = T_n(t) \sin t$ and therefore $T'_{n+1}(t) = T'_n(t) \sin t + T_n(t) \cos t$. This now implies

8.4. Proof of Some Crucial Inequalities for $w \in J_p^*$

$\|T_n'(t)W_{1,n}(t)\|_p$

$\leq \|T_n'(t)W_{0,n}(t)\sin t\|_p + \|T_n'(t)n^{-1}W_{0,n}(t)\|_p$

$\leq \|T_{n+1}'(t)W_{0,n}(t)\|_p + \|T_n(t)W_{0,n}(t)\cos t\|_p + n^{-1}\|T_n'(t)W_{0,n}(t)\|_p$

$\leq C_3 n\|T_{n+1}(t)W_{0,n}(t)\|_p + \|T_n(t)W_{0,n}(t)\|_p + C_3\|T_n(t)W_{0,n}(t)\|_p$

$\leq C_4 n\|T_n(t)W_{1,n}(t)\|_p$

as both $|\sin t|$ and $1/n$ are smaller than $(|\sin t| + 1/n)$.

We now prove the result for $\gamma > 1$. Define $\mu = k$ for $\gamma = 2k + 1$ and $\mu = [(\gamma + 1)/2]$ otherwise. Obviously, $-1 < \gamma - 2u \leq 1$, and writing $T_{n+2\mu}(t) = T_n(t)(\sin t)^{2\mu}$, we have

$$T_{n+2\mu}'(t) = T_n'(t)(\sin t)^{2\mu} + 2\mu T_n(t)(\sin t)^{2\mu-1}\cos t,$$

and therefore

$\|T_n'(t)W_{\gamma,n}(t)\|_p$

$\leq C\{\|T_n'(t)W_{\gamma-2\mu,n}(t)(\sin t)^{2\mu}\|_p + n^{-2\mu}\|T_n'(t)W_{\gamma-2\mu,n}(t)\|_p\}$

$\leq C\{\|T_{n+2\mu}'(t)W_{\gamma-2\mu,n}(t)\|_p + 2\mu\|T_n(t)W_{\gamma-2\mu,n}(t)\sin^{2\mu-1}t\cos t\|_p$

$\quad + C_1 n^{-2\mu+1}\|T_n(t)W_{\gamma-2\mu,n}(t)\|_p\}$

$\leq C_2\{(n+2\mu)\|T_{n+2\mu}(t)W_{\gamma-2\mu,n}(t)\|_p + \|T_n(t)W_{\gamma-2\mu,n}(t)\sin^{2\mu-1}t\|_p$

$\quad + n^{-2\mu+1}\|T_n(t)W_{\gamma-2\mu,n}(t)\|_p\}$

$\leq C_3 n\|T_n(t)W_{\gamma,n}\|_p,$

as $\sin^{2\mu}t$, $n^{-1}\sin^{2\mu-1}t$, and $n^{-2\mu}$ are smaller than $C_4(|\sin t| + (1/n))^{2\mu}$. □

Lemma 8.4.2. *For $\gamma_1, \gamma_2 > -1$, $W_{\gamma_1,n}(t)$ and $\overline{W}_{\gamma_2,n}(t)$ given in (8.4.1) and (8.4.2), and $1 \leq p \leq \infty$ we have*

$$\|T_n'(t)W_{\gamma_1,n}(t)\overline{W}_{\gamma_2,n}(t)\|_{L_p(T)} \leq Cn\|T_n(t)W_{\gamma_1,n}(t)\overline{W}_{\gamma_2,n}(t)\|_{L_p(T)}, \quad (8.4.5)$$

where $C \equiv C(\gamma_1,\gamma_2)$ is independent of n and $T_n \in \tau_n$.

PROOF. The definition of V_2 implies $y^{\gamma_2}V_2(y) \geq Ay^{m-1}$ if $m - 1 > \gamma_2$ and $0 < y < 2$. Therefore, $\overline{W}_{\gamma_2,n}(t) \geq A(|\cos t| + (1/n))^{m-1}$, and since on $(-\pi/4, \pi/4)$ $\sqrt{2}/2 \leq \cos t \leq 1$, we have, using Lemma 8.4.1,

$\|T_n'W_{\gamma_1,n}\overline{W}_{\gamma_2,n}\|_{L_p[-\pi/4,\pi/4]}$

$\leq B\left\|T_n'(t)\left(\cos t + \frac{1}{n}\right)^m W_{\gamma_1,n}(t)\right\|_{L_p[-\pi/4,\pi/4]}$

$\leq B\left\|T_n'(t)\left(\cos t + \frac{1}{n}\right)^m W_{\gamma_1,n}(t)\right\|_{L_p[-\pi,\pi]}$

$$\leq B\left\{\left\|\frac{d}{dt}\left(T_n(t)\left(\cos t + \frac{1}{n}\right)^m\right)W_{\gamma_1,n}(t)\right\|_{L_p[-\pi,\pi]}\right.$$

$$\left. + m\left\|T_n(t)\sin t\left(\cos t + \frac{1}{n}\right)^{m-1}W_{\gamma_1,n}(t)\right\|_{L_p[-\pi,\pi]}\right\}$$

$$\leq B_1(n+m)\left\|T_n\left(|\cos t| + \frac{1}{n}\right)^m W_{\gamma_1,n}(t)\right\|_p$$

$$+ B_1\left\|T_n(t)\left(|\cos t| + \frac{1}{n}\right)^{m-1}W_{\gamma_1,n}(t)\right\|_p$$

$$\leq B_2 n\|T_n(t)W_{\gamma_1,n}(t)W_{\gamma_2,n}(t)\|_{L_p[-\pi,\pi]}.$$

In the intervals $[\pi/4, 3\pi/4]$, $[-3\pi/4, -\pi/4]$, and $[3\pi/4, 5\pi/4]$ we use the translations $y \to y - (\pi/2)$, $y \to y + (\pi/2)$, and $y \to y - \pi$ respectively, and hence the problem is reduced to the estimates on $[-\pi/4, \pi/4]$ which we have just discussed. (In $[\pi/4, 3\pi/4]$ and $[-3\pi/4, -\pi/4]$ V_1 replaces V_2 and $\sin t$ replaces $\cos t$.) $\square$

Lemma 8.4.3. *For $\rho > -1$ and arbitrary α there exists small enough $L > 0$ for which*

$$\max_t |T_n(t)|\left(|\sin t| + \frac{1}{n}\right)^\rho$$

$$\leq Cn^{1/p}\left\{\int_{T\setminus E}|T_n(t)|^p|\sin t|^\alpha\left(|\sin t| + \frac{1}{n}\right)^{\rho p - \alpha}dt\right\}^{1/p}, \quad 1 \leq p < \infty \quad (8.4.6)$$

and

$$\left\|T_n(t)\left(|\sin t| + \frac{1}{n}\right)^\rho\right\|_{L_\infty(T)} \leq C\left\|T_n(t)\left(|\sin t| + \frac{1}{n}\right)^\rho\right\|_{L_\infty(T\setminus E)}, \quad (8.4.7)$$

where $T = [-\pi, \pi]$, E is any measurable set satisfying $m(E) < L/n$ and C is independent of n and $T_n \in \tau_n$.

This is a slight variation of the lemma of Potapov [1], more general in some sense and less in another, which seems to be what we need here.

PROOF. Using Lemma 8.4.1 with $W_{\gamma,n}(t) = (|\sin t| + (1/n))^\rho$ and $p = \infty$, we have

$$\left\|T_n'(t)\left(|\sin t| + \frac{1}{n}\right)^\rho\right\|_{L_\infty(T)} \leq C_1 n\left\|T_n(t)\left(|\sin t| + \frac{1}{n}\right)^\rho\right\|_{L_\infty(T)}.$$

Writing $Q(t) = T_n(t)(|\sin t| + (1/n))^\rho$, we observe that for $t \neq k\pi$

8.4. Proof of Some Crucial Inequalities for $w \in J_p^*$

$$|Q'(t)| \leq \left\| T_n'(t)\left(|\sin t| + \frac{1}{n}\right)^p \right\|_{L_\infty(T)} + |\rho| \left\| T_n(t)\left(|\sin t| + \frac{1}{n}\right)^{p-1} \cos t \right\|_{L_\infty(T)}$$

$$\leq (C_1 + |\rho|)n \left\| T_n(t)\left(|\sin t| + \frac{1}{n}\right)^p \right\|_{L_\infty(T)}.$$

We now choose $L \leq 1/8(C_1 + \rho)$. For t_0 satisfying

$$\mu \equiv |T_n(t_0)|\left(|\sin t_0| + \frac{1}{n}\right)^p = \left\| T_n(t)\left(|\sin t| + \frac{1}{n}\right)^p \right\|_{L_\infty(T)}$$

we have either $t_0 \in [0, \pi]$ or $t_0 \in [-\pi, 0]$, and with no loss of generality, we can assume $t_0 \in [0, \pi]$. For $C_2 = C_1 + |\rho|$ and

$$I = \left[t_0 - \frac{1}{2C_2 n}, t_0 + \frac{1}{2C_2 n}\right] \cap [0, \pi]$$

we have for $t \in I$,

$$\left| T_n(t)\left(|\sin t| + \frac{1}{n}\right)^p \right| \geq \mu - C_2 n \mu \frac{1}{2C_2 n} \geq \frac{\mu}{2},$$

and since $m(I) \geq 1/2C_2 n > m(E)$, we have proved (8.4.7) with $L = 1/8C_2$ and $C = 2$. We can also write for $L = 1/8C_2$, $m(E) < L/n$ and $E_1 = (-\delta, \delta) \cup (\pi - \delta, \pi + \delta)$, where $\delta = 1/16C_2 n$,

$$I(n) \equiv \int_{T \setminus E} \left| T_n(t)\left(|\sin t| + \frac{1}{n}\right)^p \right|^p (|\sin t|/(|\sin t| + n^{-1}))^\alpha \, dt$$

$$\geq M \int_{I \setminus E \cup E_1} |T_n(t)|^p \left(|\sin t| + \frac{1}{n}\right)^{pp} dt,$$

as for $t \notin E_1$, $|\sin t|/(|\sin t| + n^{-1}) \geq A > 0$. Since $M(I \setminus E \cup E_1) \geq 1/4C_2 n$, we have

$$I(n) \geq M \left(\frac{\mu}{2}\right)^p \frac{1}{4C_2 n} \geq M_1 n^{-1} \left(\max_t |T_n(t)|\left(|\sin t| + \frac{1}{n}\right)^p\right)^p,$$

which is what we wanted to show. □

Lemma 8.4.4. Suppose $\gamma_i > -1$, $1 \leq p \leq \infty$, and the weights $W_{\gamma_1, n}(t)$ and $\overline{W}_{\gamma_2, n}(t)$ are defined in (8.4.1) and (8.4.2). Then there exist $a > 0$ and $C = C(a)$ big enough, such that

$$\|T_n W_{\gamma_1, n} \overline{W}_{\gamma_2, n}\|_{L_p(T)} \leq C \|T_n W_{\gamma_1, n} \overline{W}_{\gamma_2, n}\|_{L_p(T \setminus E_n(a))}, \tag{8.4.8}$$

where $E_n(a) = \bigcup_{k=-2}^{2}((k\pi/2) - (a/n), (k\pi/2) + (a/n))$ and both a and C are independent of n and $T_n \in \tau_n$.

PROOF. We choose a such that $mE_n(a) = 10a/n < L/n$, where L is that of Lemma 8.4.3. Obviously, we have to estimate only $\|T_n W_{\gamma_1, n} \overline{W}_{\gamma_2, n}\|_{L_p(E_n(a))}$, and because of symmetry and simple transformations used already in Lemma 8.4.2, it is enough to estimate $\|T_n W_{\gamma_1,n}\overline{W}_{\gamma_2,n}\|_{L_p[-a/n,a/n]}$. Writing $T_n^*(t) = T_n(t)(\cos t + 1/n)^m$ with m satisfying $y^{\gamma_2} V_2(y) \geq Ay^m$ ($m > \gamma_2, 0 < y < \sqrt{2}$), and therefore $\overline{W}_{\gamma_2,n}(t) \geq A(|\cos t| + 1/n)^m$, and $0 < \delta < 1$ such that $\delta + \gamma_1 > 0$, we have

$$\|T_n(t) W_{\gamma_1,n}(t) \overline{W}_{\gamma_2,n}(t)\|_{L_p[-a/n,a/n]}$$

$$\leq M \left\| T_n(t) \left(\cos t + \frac{1}{n}\right)^m W_{\gamma_1,n}(t) \right\|_{L_p[-a/n,a/n]}$$

$$\leq M \left\| T_n^*(t) W_{\gamma_1+\delta,n}(t) \left(|\sin t| + \frac{1}{n}\right)^{-\delta} \right\|_{L_p[-a/n,a/n]}$$

$$\leq M W_{\gamma_1+\delta,n}\left(\frac{a}{n}\right) \left\| T_n^*(t) \left(|\sin t| + \frac{1}{n}\right)^{-\delta} \right\|_{L_p[-a/n,a/n]}$$

$$\leq M_1 W_{\gamma_1+\delta,n}\left(\frac{a}{n}\right) n^{-1/p} \left\| T_n^*(t) \left(|\sin t| + \frac{1}{n}\right)^{-\delta} \right\|_{L_\infty[T]}$$

$$\leq M_2 W_{\gamma_1+\delta,n}\left(\frac{a}{n}\right) \left\| T_n^*(t) \left(|\sin t| + \frac{1}{n}\right)^{-\delta} \right\|_{L_p[T \setminus E_n(a)]},$$

where, in the last step we used Lemma 8.4.3. For $t \notin E_n(a)$, $W_{\gamma_1+\delta,n}(a/n) \leq M W_{\gamma_1+\delta,n}(t)$ and for all t, $(|\cos t| + (1/n))^m \leq (1/A) W_{\gamma_2,n}(t)$, and therefore

$$\|T_n(t) W_{\gamma_1,n}(t) \overline{W}_{\gamma_2,n}(t)\|_{L_p[-a/n,a/n]}$$

$$\leq M_3 \left\| T_n(t) W_{\gamma_1+\delta,n}(t) \overline{W}_{\gamma_2,n}(t) \left(|\sin t| + \frac{1}{n}\right)^{-\delta} \right\|_{L_p[T \setminus E_n(a)]}$$

$$\leq M_3 \|T_n(t) W_{\gamma_1,n}(t) \overline{W}_{\gamma_2,n}(t)\|_{L_p[T \setminus E_n(a)]}. \qquad \square$$

Consequently we have the following corollary.

Corollary 8.4.5. *For $\gamma_i > -1$ and $W_{\gamma_1,n}$ and $\overline{W}_{\gamma_2,n}$ as above,*

$$\|T_n W_{\gamma_1,n} \overline{W}_{\gamma_2,n}\|_{L_p(T)} \leq C \|T_n W_{\gamma_1} \overline{W}_{\gamma_2}\|_{L_p(T)}. \qquad (8.4.9)$$

PROOF. We use Lemma 8.4.4 and the fact that for $t \notin E_n(a)$, $W_{\gamma,n}(t) \sim W_{\gamma,n}(t)$ and $\overline{W}_{\gamma,n}(t) \sim \overline{W}_{\gamma,n}(t)$. (The conditions look more general than those in the following theorems but the norm on the right of (8.4.9) is not always finite.) $\square$

Lemma 8.4.6. *Suppose $\gamma_1, \gamma_2 > -1/p$; $W_{\gamma_1}, \overline{W}_{\gamma_2}, W_{\gamma_1,n}$, and $\overline{W}_{\gamma_2,n}$ are given in (8.4.1) and (8.4.2), and if $p = \infty$, we may have $\gamma_1 = 0$ or $\gamma_2 = 0$, in which case $V_1(y)$ or $V_2(y)$ are nondecreasing for small y. Then*

$$\|T_n(t) W_{\gamma_1}(t) \overline{W}_{\gamma_2}(t)\|_{L_p(T)} \leq C \|T_n(t) W_{\gamma_1,n}(t) \overline{W}_{\gamma_2,n}(t)\|_{L_p(T)}. \qquad (8.4.10)$$

PROOF. The additional condition for $p = \infty$ leads trivially to (8.4.10) and otherwise the interesting situation is $\gamma_i \leq 0$. Define $E_n = E_n(1) = \bigcup_k ((k\pi/2) - 1/n,$

$(k\pi/2) + 1/n)$ and the estimate on E_n^c is clear. Using the technique of Lemma 8.4.2, it follows that it is enough to estimate $\|T_n W_{\gamma_1} \overline{W}_{\gamma_2}\|_{L_p[-1/n, 1/n]}$. We choose m such that $y^{\gamma_2} V_2(y) \geq C_1 y^m, 0 < y < \sqrt{2}$, define $T_n^*(t) = T_n(t)(\cos t + (1/n))^m$, choose $\rho < 1/p$ such that $\rho + \gamma_1 > 0$, and following Lemma 8.4.3 with $E = E_n(a)$ for some a (where ρ there is $-\rho$ here), we write

$$\|T_n(t) W_{\gamma_1}(t) \overline{W}_{\gamma_2}(t)\|_{L_p[-1/n, 1/n]}$$

$$\leq A \left\| T_n(t) W_{\gamma_1+\rho}(t)\left(\cos t + \frac{1}{n}\right)^m (|\sin t| + \frac{1}{n})^{-\rho} \left(\frac{|\sin t| + 1/n}{|\sin t|}\right)^\rho \right\|_{L_p[-1/n, 1/n]}$$

$$\leq A W_{\gamma_1+\rho}\left(\frac{1}{n}\right) \left\| T_n^*(t)(|\sin t| + \frac{1}{n})^{-\rho} \right\|_{L_\infty[-1/n, 1/n]} \left\| \left(\frac{|\sin t| + 1/n}{|\sin t|}\right)^\rho \right\|_{L_p[-1/n, 1/n]}$$

$$\leq A_2 W_{\gamma_1+\rho}\left(\frac{1}{n}\right) n^{1/p} \left\| T_n^*(t)(|\sin t| + \frac{1}{n})^{-\rho} \right\|_{L_p(T \setminus E_n(a))} \left(\frac{1}{n}\right)^\rho \left\{ \int_0^{1/n} \left(\frac{1}{t}\right)^{\rho p} dt \right\}^{1/p}$$

$$\leq A_3 \left\| T_n^*(t) W_{\gamma_1+\rho, n}(t)(|\sin t| + \frac{1}{n})^{-\rho} \right\|_{L_p(T)} \leq A_4 \|T_n W_{\gamma_1, n} \overline{W}_{\gamma_2, n}\|_{L_p(T)}. \quad \square$$

We are now ready to prove inequality (8.1.3), and in fact, as $w(x) \in J_p^*$ implies $(1-x^2)^{1/2} w(x) \in J_p^*$, it is sufficient to prove (8.1.3) for $r = 1$ which is the following weighted Bernstein-type inequality.

Theorem 8.4.7. *For $P_n \in \Pi_n$ and $w \in J_p^*$ we have*

$$\|w(x)(1-x^2)^{1/2} P_n'(x)\|_{L_p[-1,1]} \leq Cn \|w(x) P_n(x)\|_{L_p[-1,1]}.$$

PROOF. Let $T_n(t) = P_n(\cos t)$ and therefore $|T_n'(t)| = |P_n'(x)|(1-x^2)^{1/2}|_{x=\cos t}$. We have

$$w(x) = w_-(\sqrt{1+x}) w_+(\sqrt{1-x})$$
$$= (\sqrt{1+x})^{\gamma_2} v_-(\sqrt{1+x})(\sqrt{1-x})^{\gamma_1} v_+(\sqrt{1-x}),$$
$$\sqrt{1-x} = \sqrt{2}\left|\sin \frac{t}{2}\right|, \quad \sqrt{1+x} = \sqrt{2}\left|\cos \frac{t}{2}\right|,$$

where γ_i, v_-, and v_+ satisfy conditions (b)–(d) in Definition 8.1.1. We can now write

$$\|w(x)(1-x^2)^{1/2} P_n'(x)\|_{L_p[-1,1]}$$

$$= \left\| T_n'(t) 2^{\gamma_1+\gamma_2+(1/p)} \left|\sin \frac{t}{2}\right|^{\gamma_1+(1/p)} v_+\left(\sqrt{2}\left|\sin \frac{t}{2}\right|\right) \right.$$
$$\left. \times \left|\cos \frac{t}{2}\right|^{\gamma_2+(1/p)} v_-\left(\sqrt{2}\left|\cos \frac{t}{2}\right|\right) \right\|_{L_p[0,\pi]}$$

$$= C_1 \|T_n'(2t)|\sin t|^{\gamma_1+(1/p)} v_+(\sqrt{2}|\sin t|)|\cos t|^{\gamma_2+(1/p)} v_-(\sqrt{2}|\cos t|)\|_{L_p[0,(\pi/2)]}.$$

Recalling that for $w \in J_p^*$, $\gamma_i > -2/p$, we now set $V_1(y) = v_+(\sqrt{2}y)$, $V_2(y) = v_-(\sqrt{2}y)$, and $\gamma_i + (1/p) > -1/p$ (with appropriate modification for $p = \infty$), and $T_n(2t) = T_{2n}^*(t)$. Making use of (8.4.5), (8.4.9), and (8.4.10), we obtain

$$\|w(x)(1-x^2)^{1/2} P_n'(x)\|_{L_p[-1,1]} \le C_1 \|T_{2n}^{*\prime}(t) W_{\gamma_1+(1/p)}(t) \overline{W}_{\gamma_2+(1/p)}(t)\|_{L_p(T)}$$

$$\le C_2 \|T_{2n}^{*\prime}(t) W_{\gamma_1+(1/p), 2n}(t) \overline{W}_{\gamma_2+(1/p), 2n}(t)\|_{L_p(T)}$$

$$\le C_3 n \|T_{2n}^*(t) W_{\gamma_1+(1/p), 2n}(t) \overline{W}_{\gamma_2+(1/p), 2n}(t)\|_{L_p(T)}$$

$$\le C_4 n \|T_{2n}^*(t) W_{\gamma_1+(1/p)}(t) \overline{W}_{\gamma_2+(1/p)}(t)\|_{L_p(T)}.$$

We now recall that $T_{2n}^*(t)$ is even and so are $|\sin t|$ and $|\cos t|$; moreover, we recall that since $T_{2n}^*(t) = T_n(2t)$, we have $T_{2n}^*(\pi - t) = T_n(-2t) = T_n(2t) = T_{2n}^*(t)$, and that type of identity is valid for $|\sin t|$ and $|\cos t|$ as well. Consequently,

$$\|w(x)(1-x^2)^{1/2} P_n'(x)\|_{L_p[-1,1]} \le 4C_4 n \|T_{2n}^*(t) W_{\gamma_1+(1/p)}(t) \overline{W}_{\gamma_2+(1/p)}(t)\|_{L_p[0, \pi/2]}$$

$$\le C_5 n \left\| T_n(t) \left|\sin \frac{t}{2}\right|^{\gamma_1+(1/p)} v_+\left(\sqrt{2}\left|\sin \frac{t}{2}\right|\right) \right.$$

$$\left. \times \left|\cos \frac{t}{2}\right|^{\gamma_2+(1/p)} v_-\left(\sqrt{2}\left|\cos \frac{t}{2}\right|\right) \right\|_{L_p[0, \pi]}$$

$$\le C_6 n \|P_n(x) w(x)\|_{L_p[-1,1]}. \quad \square$$

We are now ready to prove the second important inequality, that is, (8.1.4).

Theorem 8.4.8. *For $w \in J_p^*$ and $c > 0$ we have for $n^2 > c$*

$$\|P_n(x) w(x)\|_{L_p[-1,1]} \le M(c) \|P_n(x) w(x)\|_{L_p[-1+cn^{-2}, 1-cn^{-2}]},$$

where $M(c)$ does not depend on n and $P_n \in \Pi_n$.

PROOF. With the notations of the proof of the preceding theorem we will need the estimate of Lemma 8.4.4 applied to

$$\|T_{2n}^*(t) W_{\gamma_1+(1/p), 2n}(t) \overline{W}_{\gamma_2+(1/p), 2n}(t)\|_{L_p(T)}$$

which yields

$$\|T_{2n}^*(t) W_{\gamma_1+(1/p), 2n}(t) \overline{W}_{\gamma_2+(1/p), 2n}(t)\|_{L_p(T)} \qquad (8.4.11)$$

$$\le C \|T_{2n}^*(t) W_{\gamma_1+(1/p)}(t) \overline{W}_{\gamma_2+(1/p)}(t)\|_{L_p(T \setminus E_n(a))},$$

where a and C depend on $W_{\gamma_1+(1/p)}$, $\overline{W}_{\gamma_2+(1/p)}$, and p but not on n. We now use (8.4.11) and substitutions as in the proof of Theorem 8.4.7 to obtain for some a

$$\|P_n(x) w(x)\|_{L_p[-1,1]} \le C_1 \|T_{2n}^*(t) W_{\gamma_1+(1/p)}(t) \overline{W}_{\gamma_2+(1/p)}(t)\|_{L_p[0, (\pi/2)]}$$

(and using (8.4.10))

$$\le C_2 \|T_{2n}^*(t) W_{\gamma_1+(1/p), 2n}(t) \overline{W}_{\gamma_2+(1/p), 2n}(t)\|_{L_p(T)}$$

(and using (8.4.8))

$$\leq C_3 \| T_{2n}^*(t) W_{\gamma_1+(1/p), 2n}(t) \overline{W}_{\gamma_2+(1/p), 2n}(t) \|_{L_p(T \setminus E_n(a))}$$

$$\leq C_4 \| T_{2n}^*(t) W_{\gamma_1+(1/p)}(t) \overline{W}_{\gamma_2+(1/p)}(t) \|_{L_p(T \setminus E_n(a))}$$

$$\leq C_5 \| P_n(x) w(x) \|_{L_p[\cos(\pi - (a/n)), \cos(a/n)]}$$

$$\leq C_5 \| P_n(x) w(x) \|_{L_p[-1 + a_1 n^{-2}, 1 - a_1 n^{-2}]},$$

where $a_1 = a^2/2$. If we need just any c for which the theorem is valid, we have it already. To increase c we use the change of variable $x_1 = x/(1 - a_1 n^{-2})$, $P_{n,1}(x_1) = P_n(x)$, $\sqrt{1 \pm x_1} + n^{-1} \sim \sqrt{1 \pm x}$ for $x \in (-1 + a_1 n^{-2}, 1 - a_1 n^{-2})$, and substitute as before to obtain

$$\| P_n(x) w(x) \|_{L_p[-1 + a_1 n^{-2}, 1 - a_1 n^{-2}]}$$

$$\leq C_1 \left\| P_{n,1}(x) \left(\sqrt{1 - x_1} + \frac{1}{n} \right)^{\gamma_1} v_+ \left(\sqrt{1 - x_1} + \frac{1}{n} \right) \left(\sqrt{1 + x_1} + \frac{1}{n} \right)^{\gamma_2} \right.$$

$$\left. \times v_- \left(\sqrt{1 + x_1} + \frac{1}{n} \right) \right\|_{L_p[-1,1]}$$

$$\leq C_2 \| T_{2n,1}^*(t) W_{\gamma_1, 2n}(t) \overline{W}_{\gamma_2, 2n}(t) |\sin t|^{1/p} |\cos t|^{1/p} \|_{L_p(T)}$$

$$\leq C_3 \| T_{2n,1}^*(t) W_{\gamma_1+(1/p), 2n}(t) \overline{W}_{\gamma_2+(1/p), 2n}(t) \|_{L_p(T)}$$

(and using (8.4.8) and the earlier process)

$$\leq C_4 \| P_{n,1}(x_1) w(x_1) \|_{L_p[-1 + a_1 n^{-2}, 1 - a_1 n^{-2}]}$$

$$\leq C_5 \| P_n(x) w(x) \|_{L_p[-1 + a_2 n^{-2}, 1 - a_2 n^{-2}]},$$

where a_2 satisfies $1 - a_2 n^{-2} \geq (1 - a_1 n^{-2})^2$, for example $a_2 = 2a_1(1 - (1/4))$ (the case $n \geq \sqrt{2c}$), as $a_1 < c$. (If $a_1 \geq c$, the earlier part is sufficient, and for $\sqrt{c} < n < \sqrt{2c}$ the whole result is trivial.) This process will lead to the result of the theorem, in finitely many steps. □

8.5. Applications, Calculations, and Specific Examples

In Section 3.4 computations of some particular (nonweighted) moduli were made. We shall use those computations as well as some further calculations of weighted moduli to obtain asymptotic estimates on $E_n(f)_{w,p}$ and $\| w \varphi^l P_n^{(l)} \|_p$.

1. For $f(x) = (1 - x^2)^\rho$ with $\rho \neq 0$ and $-1/p < \rho < 1 - (1/p)$ the first example of Section 3.4 (taking $\varphi(x) = (1 - x^2)^{1/2}$ instead of $x^{1/2}$ and using the interval $[-1, 1]$ instead of R_+), we have for $\rho \neq 0$ $\omega_\varphi^2(f, t)_p \sim t^{2\rho + 2/p}$. Therefore, Corollarys 7.2.5 and 7.3.3 imply for the above ρ

$$E_n(f)_p \sim n^{-2\rho - (2/p)} \quad \text{and} \quad \| (1 - x^2) P_n''(x) \|_{L_p[-1, 1]} = O(n^{-2 - 2\rho - (2/p)}).$$

(For $\rho = 1 - (1/p)$, $\rho \neq 0, 1$ we have $E_n(f)_p \leq Mn^{-2\rho-(2/p)}$ and $\|(1 - x^2)P_n''(x)\|_{L_p[-1, 1]} \leq M_1 \log n$.)

2. Let
$$f(x) = x^\delta [\log x/2]^\gamma, \quad x \in (0, 1), \quad \delta > -1/p.$$
From the second example of Section 3.4 we can deduce via Theorems 7.2.1 and 7.2.4
$$E_n(f)_{L_p(0, 1)} \sim \begin{cases} n^{-2\delta-(2/p)}(\log n)^\gamma & \text{if } \delta \neq 0, 1, \ldots \\ n^{-2\delta-(2/p)}(\log n)^{\gamma-1} & \text{if } \delta = 0, 1, \ldots \text{ but } \gamma \neq 0. \end{cases}$$

3. Similar computation yields the asymptotic estimate of $E_n(f)_p$ for f given by
$$f(x) = \prod_{i=1}^n |x - x_i|^{\delta_i}, \quad \delta_i > -1/p,$$
where $-1 \leq x_1 < x_2 < \cdots < x_n \leq 1$, $n \geq 2$. We define
$$\varepsilon_1 = \begin{cases} \delta_1 + (1/p) & \text{if } x_1 > -1 \\ 2\delta_1 + (2/p) & \text{if } x_1 = -1 \text{ but } \delta_1 \neq 0, 1, \ldots \\ \infty & \text{if } x_1 = -1 \text{ and } \delta_1 = 0, 1, \ldots \end{cases}$$
and ε_n is defined similarly around $x = 1$. For ε given by
$$\varepsilon = \min\{\varepsilon_1, \varepsilon_n, \delta_i + 1/p\}_{1 < i < n}$$
we now obtain
$$E_n(f)_p \sim n^{-\varepsilon}.$$

4. The following constitutes a relatively simple example of weighted polynomial approximation. Let $f(x) = \log(1 - x^2)$, $w(x) = (1 - x^2)^{-1/3}$ (and therefore $p < 3$), and $\varphi(x) = \sqrt{1 - x^2}$. We compute $\Omega_\varphi^r(f, h)_{w, p}$ for $r > (2/p) - (2/3)$ by estimating
$$\left\{ \int_{-1+2h^2}^{1-2h^2} |(1 - x^2)^{-1/3} f^{(r)}(\xi(x))(h\sqrt{1 - x^2})^r|^p \, dx \right\}^{1/p}.$$
Recalling $\xi(x) \in (x - (h/2)\varphi(x), x + (h/2)\varphi(x))$, we derive
$$\Omega_\varphi^r(f, h)_{w, p} \sim h^r \left\{ \int_{-1+2h^2}^{1-2h^2} ((1 - x^2)^{-1/3}(1 - x^2)^{-r}(1 - x^2)^{r/2})^p \, dx \right\}^{1/p}$$
$$\sim h^{2/p-2/3}.$$
Therefore, by Theorem 8.2.1
$$E_n(f)_{w, L_p[-1, 1]} \sim n^{2/3-2/p}$$
for p satisfying $1 \leq p < 3$ and by Corollary 8.3.2
$$\|w\varphi^r P_n^{(r)}\|_{L_p[-1, 1]} = O(n^{r+2/3-2/p})$$

8.5. Applications, Calculations, and Specific Examples

for any $r > (2/p) - (2/3)$ (P_n satisfies $\|w(P_n - f)\|_p = E_n(f)_{w,p}$). For $r \le (2/p) - (2/3)$ the situation is different. For instance for $r = 1$ we have $\Omega^1_\varphi(f,h)_{w,6/5} \sim h|\log h|^{5/6}$ and $\Omega^1_\varphi(f,h)_{w,p} \sim h$ for $p < 6/5$. Therefore for P_n satisfying $\|w(f - P_n)\|_p = E_n(f)_{w,p}$, we have

$$\|(1-x^2)^{1/6} P'_n(x)\|_{L_{6/5}[-1,1]} = O((\log n)^{5/6})$$

and

$$\|(1-x^2)^{1/6} P'_n(x)\|_{L_p[-1,1]} \sim 1 \quad \text{for } p < \tfrac{6}{5}.$$

In the next example $w(x)$ is a J_p^* weight but is not a Jacobi weight.

5. Let $f(x) = (1-x^2)^\alpha$, $w(x) = |\log(1-x^2)|$, and $\varphi(x) = \sqrt{1-x^2}$. Obviously $w \in J_p^*$ for $1 \le p < \infty$. To calculate $\Omega^r_\varphi(f,h)_{w,p}$ for large r it is enough to estimate for $\alpha \ne 0, 1, 2, \ldots$ the expression

$$h^r \left\{ \int_{-1+2h^2}^{1-2h^2} ||\log(1-x^2)| f^{(r)}(\xi(x))(1-x^2)^{r/2}|^p \, dx \right\}^{1/p}$$

$$\sim h^r \left\{ \int_{-1+2h^2}^{1-2h^2} ((1-x^2)^{\alpha-r} |\log(1-x^2)| (1-x^2)^{r/2})^p \, dx \right\}^{1/p}$$

$$\sim h^{2\alpha + 2/p} \log 1/h$$

if $\alpha \ne 0, 1, \ldots$. Therefore, for $\alpha > -1/p$

$$E_n(f)_{w, L_p[-1,1]} \sim n^{-2\alpha - (2/p)} \log n \quad (\alpha \ne 0, 1, \ldots).$$

For other r the results are somewhat different. For $r = 1$, for example, if $\alpha > (1/2) - (1/p)$ and $\alpha \ne 0$, $\Omega^1_\varphi(f,h)_{w,p} \sim h$; if $\alpha = (1/2) - (1/p)$ and $\alpha \ne 0$, $\Omega^1_\varphi(f,h)_{w,p} \sim h|\log h|^{1+1/p}$, and if $-1/p < \alpha < (1/2) - (1/p)$ and $\alpha \ne 0$ ($\alpha > -1/p$ is necessary for $wf \in L_p[-1,1]$), $\Omega^1_\varphi(f,h) \sim h^{2\alpha + 2/p} |\log h|$. Suppose P_n satisfies $\|w(f - P_n)\|_p = E_n(f)_{w,p}$, and define

$$I_n = \|\log(1-x^2)(1-x^2)^{1/2} P'_n(x)\|_p,$$

then the above implies for $\alpha \ne 0$ and $\alpha > (1/2) - (1/p)$, $I_n = O(1)$, for $\alpha \ne 0$ and $\alpha = (1/2) - (1/p)$, $I_n = O((\log n)^{1+(1/p)})$, and for $\alpha \ne 0$ and $-(1/p) < \alpha < (1/2) - (1/p)$, $I_n = O(n^{1-2\alpha - (2/p)} \log n)$ (for $\alpha = 0$, $I_n = 0$).

These examples show that we can actually calculate $\Omega^r_\varphi(f,t)_{w,p}$ and get estimates for $E_n(f)_{w,p}$ in reasonably simple fashion.

CHAPTER 9
EXPONENTIAL-TYPE OR BERNSTEIN-TYPE OPERATORS

Relations between the rate of convergence of several well-known and much-studied approximation operators and the modulus presented in this book will be studied. Earlier partial results on the subject were important for motivating the investigation of $\omega_\varphi^r(f,t)_p$. Results given in detail in this chapter are new.

9.1. Background and Notations, Positive Operators on $C(D)$

For the Bernstein operator on $C[0,1]$ given by

$$B_n(f,x) = \sum_{k=0}^{n} \binom{n}{k} x^k (1-x)^{n-k} f\left(\frac{k}{n}\right) \qquad (9.1.1)$$

it was proved by Lorentz [2, p. 102] ($\alpha = 2$), Berens and Lorentz [1] ($\alpha < 2$), and DeVore [1, p. 267] ($\alpha < 2$) that for $f \in C[0,1]$

$$|B_n(f,x) - f(x)| < K\left(\frac{x(1-x)}{n}\right)^{\alpha/2} \text{ if and only if } \omega^2(f,h) = O(h^\alpha). \qquad (9.1.2)$$

This result had much influence on investigations of the relation between the rate of approximation of specific operators and the smoothness of the function approximated.

The various directions of research that started from (9.1.2) can be classified as follows.

(a) Investigation of the class of functions for which $\|B_n(f,x) - f(x)\| = O(\psi(n))$ (or $(x(1-x))^\eta |B_n(f,x) - f(x)| = O(\psi(n))$.
(b) Investigation of operators similar to $B_n(f,x)$ or classes of these operators and obtaining results of type (9.1.2) or (a).

9.1. Background and Notations, Positive Operators on C(D)

(c) Modification of $B_n(f,x)$ and other similar operators to obtain bounded operators on L_p and investigation of relations between rate of convergence and smoothness in L_p.

(d) Combinations of the operators above that yield faster convergence in relation to a higher degree of smoothness and the investigation of this relation.

We will give a brief survey of past results in these directions that will set up the framework for results of this chapter, as well as explain the necessity for investigating classes of functions determined by smoothness different from $\omega^r(f,h)_p = O(\psi(1/h))$. The survey of results achieved on the subjects in (a) and (b) above will be presented in this section and that on the subjects in (c) and (d) in Section 9.2.

Investigation of $\|B_n f - f\|_{C[0,1]}$ led to

$$\|B_n f - f\|_{C[0,1]} = O(n^{-\alpha/2}) \quad \text{if and only if}$$
$$\|(x(1-x))^{\alpha/2}\Delta_h^2 f(x)\|_{C[h,1-h]} = O(h^\alpha), \tag{9.1.3}$$

for $\alpha < 2$ (see Becker and Nessel [2] and Ditzian [1]). It was also shown (Ditzian [4, p. 322]) that for $\varphi(x) = \sqrt{x(1-x)}$ and $\alpha < 2$

$$\|B_n f - f\|_{C[0,1]} = O(n^{\alpha/2}) \Leftrightarrow \|\Delta_{h\varphi}^2 f\|_{C[0,1]} = O(h^\alpha). \tag{9.1.4}$$

In fact many expressions were explored and for $0 \leq \eta \leq 2$, $0 < \alpha < 2$, $\varphi(x) = \sqrt{x(1-x)}$, and either $\eta \leq \alpha$ or $\alpha + \eta \leq 2$,

$$\varphi(x)^{-\eta}|B_n(f,x) - f(x)| \leq Mn^{-\alpha/2} \Leftrightarrow \varphi(x)^{-\eta}|\Delta_h^2 f(x)| \leq M_1(h/\varphi(x))^\alpha \tag{9.1.5}$$

was proved in Ditzian [4] and [7], which contains both (9.1.3) and (9.1.2). This settled the problem of Bernstein polynomial approximation in a way very similar to that of best polynomial approximation, and it was K.G. Ivanov [4] who observed that for $\alpha < 2$ and

$$E_n(f)_{C[0,1]} = \inf_{P_n \in \Pi_n} \|f - P_n\|_{C[0,1]}$$

we obtain

$$\|B_n f - f\|_{C[0,1]} = O(n^{-\alpha/2}) \Leftrightarrow E_n(f)_{C[0,1]} = O(n^{-\alpha}). \tag{9.1.6}$$

We shall treat other relations with best polynomial approximation later.

Operators similar to Bernstein polynomials and relating their rate of convergence in a way similar to that of (9.1.2) were treated extensively. In particular we mention the following operators:

Szász–Mirakian
$$S_n(f,x) = \sum_{k=0}^{\infty} e^{-nx}\frac{(nx)^k}{k!}f\left(\frac{k}{n}\right) \tag{9.1.7}$$

Baskakov
$$V_n(f,x) = \sum_{k=0}^{\infty}\binom{n+k-1}{k}x^k(1+x)^{-n-k}f\left(\frac{k}{n}\right) \tag{9.1.8}$$

Post–Widder
$$P_n(f,x) = \frac{(n/x)^n}{(n-1)!} \int_0^\infty e^{-nu/x} u^{n-1} f(u)\, du \qquad (9.1.9)$$

Meier–König and Zeller
$$M_n(f,x) = (1-x)^n \sum_{k=0}^\infty f\left(\frac{k}{n+k}\right)\binom{n+k-1}{k} x^k. \qquad (9.1.10)$$

One can observe that the Post–Widder operator, which actually is the Post–Widder real inversion formula for the Laplace transform, appeared and was investigated extensively also in a slightly changed form, as the Gamma operator which is given by

$$G_n(f,x) = \frac{x^{n+1}}{n!}\int_0^\infty e^{-ux} u^n f\left(\frac{n}{u}\right) du. \qquad (9.1.11)$$

It was C.P. May ([1] and [2]) who pointed out that the condition

$$\frac{d}{dx}\{L_n(f(t),x)\} = \frac{n}{\varphi(x)^2} L_n(f(t)(t-x),x) \qquad (9.1.12)$$

is dominant in the above, and yields many estimates on the operators, in particular

$$L_n((t-x)^2,x) = \frac{\varphi(x)^2}{n} \qquad (9.1.13)$$

which controls their rate of convergence. In fact he pointed out that in (9.1.12) and (9.1.13), $\varphi(x)^2 = x(1-x)$ for $L_n = B_n$ the Bernstein polynomial, $\varphi(x)^2 = x$ for $L_n = S_n$ the Szász–Mirakian operator, $\varphi(x)^2 = x(1+x)$ for $L_n = V_n$, the Baskakov operator, $\varphi(x)^2 = x^2$ for $L_n = P_n$ the Post–Widder operator (or for $L_n = G_{n-1}$, where G_n is the Gamma operator), and $\varphi(x) = 1$ for the Weierstrass operator given by

$$W_n(f,x) = \sqrt{\frac{n}{2\pi}} \int_{-\infty}^\infty \exp(-n(u-x)^2/2) f(u)\, du. \qquad (9.1.14)$$

Later, Ismail and May [1] observed that if $\varphi(x)^2$ is a polynomial of degree at most 2 with real zeros, the above operators are, up to a change of variables, all the possible cases, but for $\varphi(x)^2 = 1 + x^2$ they discovered a new operator on $f \in L_1(R)$ (which is, except for a minor transformation, the only one with two imaginary roots) given by

$$T_n(f,x) = \frac{2^{n-2}n}{\pi(n-1)!}(1+x^2)^{-n/2}\int_{-\infty}^\infty e^{nu \arctan x}\left|\Gamma\left(\frac{n}{2}+\frac{inu}{2}\right)\right|^2 f(u)\, du. \qquad (9.1.15)$$

C.P. May in [1] and [2] investigated the saturation and inverse results locally. Actually the function $\varphi(x)$ relates to the step-weight function of this text. Global results analogous to (9.1.2), (9.1.3), and (9.1.4) were achieved later and for these one can refer for details to Becker [1], Becker and Nessel [1], [2],

Ditzian [2], and especially to Totik [1], [2], [7], [12], and [13] (the last two covering virtually all results with regard to (9.1.4)), where most of the above-mentioned analogous problems were settled for the L_∞ norm. Analogues of (9.1.2), (9.1.3), and (9.1.4) are valid for the operators mentioned above and at this point preference for any one of the above directions (i.e., (9.1.2), (9.1.3), and (9.1.4)) over the other is still a matter of taste. We will see in the next section that in fact, when dealing with higher smoothness and the L_p norm ($p \neq \infty$), only analogues of (9.1.4) are possible.

9.2. Operators on $L_p(D)$, Higher Degree of Smoothness

The Bernstein polynomials are not defined for $f \in L_p[0, 1]$. A modification of Bernstein polynomials due to L.V. Kantorovich [1] (see also Lorentz [1, Ch. II, p. 30]), and hence the name Kantorovich polynomials, given by

$$B_n^*(f, x) \equiv \sum_{k=0}^n p_{n,k}(x)(n+1) \int_{k/n+1}^{k+1/n+1} f(u)\, du,$$

$$p_{n,k}(x) \equiv \binom{n}{k} x^k (1-x)^{n-k}$$

(9.2.1)

alleviated this difficulty. In analogy to (9.2.1) the Szász–Kantorovich operator was defined (see Butzer [3]) by

$$S_n^*(f, x) \equiv \sum_{k=0}^\infty s_{n,k}(x) n \int_{k/n}^{k+1/n} f(u)\, du,$$

$$s_{n,k}(x) \equiv e^{-nx} \frac{(nx)^k}{k!}.$$

(9.2.2)

Likewise the Baskakov–Kantorovich operator can be defined by

$$V_n^*(f, x) \equiv \sum_{k=0}^\infty v_{n,k}(x) n \int_{k/n}^{k+1/n} f(u)\, du,$$

$$v_{n,k}(x) \equiv \binom{n+k-1}{k} x^k (1-x)^{-n-k}.$$

(9.2.3)

A similar definition can be used for extending the Meier–König and Zeller operator, but the Post–Widder and Weierstrass operators were defined and bounded on $L_p(D)$ to begin with.

The problems of characterizations of $\|B_n^* f - f\|_p = O(n^{-\alpha/2})$ as well as similar problems for the modification of the Szasz, Baskakov, Meier–König, and Zeller operators and for the Gamma operator were solved by Totik [3–9], [11], and [14]. The results for the Kantorovich polynomials ($\varphi(x)^2 = x(1 - x)$) were of the form

$$\|B_n^* f - f\|_{L_p[0,1]} = O(n^{-\alpha/2}) \Leftrightarrow \omega_\varphi^2(f,h)_p = O(h^\alpha), \qquad \alpha < 2 \quad (9.2.4)$$

(for $\alpha = 2$ the problem was resolved by V. Maier [1], [2] and S.D. Riemenschneider [1]) and a similar result was achieved for other operators. It should be noted that several counterexamples were displayed (Totik [8]) to show that for $p \neq \infty$ neither an analogue of (9.1.2) nor of (9.1.3) is possible.

However, neither Bernstein-type operators nor their Kantorovich modification can be a tool for the investigation of higher degrees of smoothness, and the rate of convergence for these operators cannot be faster than $O(1/n)$. To overcome this problem for Bernstein polynomials P.L. Butzer [1] used combinations of Bernstein polynomials, $B_n(f,r,x)$, defined inductively by

$$(2^r - 1)B_n(f,r,x) = 2^r B_{2n}(f, r-1, x) - B_n(f, r-1, x)),$$
$$B_n(f,0,x) \equiv B_n(f,x). \qquad (9.2.5)$$

This type of combinations was extended by C.P. May ([1] and [2]) who used it also for other exponential-type operators. We will use for the operator $L_n(f,x)$ the combination $L_{n,r}(f,x)$ given by

$$L_{n,r}(f,x) = \sum_{i=0}^{r-1} C_i(n) L_{n_i}(f,x), \qquad (9.2.6)$$

where n_i and $C_i(n)$ satisfy

(a) $n = n_0 < \cdots < n_{r-1} \le Kn$,   (b) $\sum_{i=0}^{r-1} |C_i(n)| < C$,

(c) $\sum_{i=0}^{r-1} C_i(n) = 1$,   (d) $\sum_{i=0}^{r-1} C_i(n) n_i^{-\rho} = 0$, $\qquad (9.2.7)$

$$\text{for} \quad \rho = 1, 2, \ldots, r-1.$$

Clearly (9.2.6) with $C_i(n)$ restricted by (9.2.7) contains (9.2.5) as a special case. Combinations of the type given by (9.2.6) and (9.2.7) for Bernstein polynomials were already used by Z. Ditzian [1] to achieve for $\alpha < 2r$ and $\varphi(x)^2 = x(1-x)$

$$\|B_{n,r} f - f\|_{C[0,1]} = O(n^{-\alpha/2}) \Leftrightarrow \|\varphi^\alpha \Delta_h^{2r} f\|_{C[rh, 1-rh]} = O(h^\alpha) \quad (9.2.8)$$

(and, using Ditzian [4], the above was also shown to be equivalent to $\omega_\varphi^{2r}(f,h)_\infty = O(h^\alpha)$).

The conditions

(c)' $\qquad \sum_{i=0}^{r-1} C_i(n) = 1 + o(n^{-r})$

and

(d)' $\qquad \sum_{i=0}^{r-1} C_i(n) n^{-\rho} = o(n^{-r}), \quad \text{for } \rho = 1, 2, \ldots, r-1,$

can replace (c) and (d) in many cases. (See Remark 9.5.6 in Section 9.5 where the only results that depend on a particular choice of $C_i(n)$ are proved.)

We will denote by $B_{n,r}$, $B^*_{n,r}$, $S_{n,r}$, $S^*_{n,r}$, $V_{n,r}$, $V^*_{n,r}$, $G_{n,r}$, and $P_{n,r}$ the combinations of the above type where in place of L_n we have B_n, B^*_{n-1}, S_n, S^*_n, V_n, V^*_n, G_n, and P_n, respectively. We will obtain direct and converse theorems for these operators, which, apart from those on $B_{n,r}$, were known only locally, if at all.

9.3. Direct and Converse Results

We will state and prove, pending the proof of three inequalities, the direct and converse results for exponential-type operators and their Kantorovich analogue.

To make the statement of various results in the present and following sections easier, we introduce a table.

Definition 9.3.1. Relations between Bernstein-type operators $L_n(f, x)$, the step-weight function $\varphi(x)$, the space B on which they are defined, and the set E_n are given in:

$L_n(f,x)$	$B_n(f,x)$	$B^*_n(f,x)$	$S_n(f,x)$	$S^*_n(f,x)$	$V_n(f,x)$	$V^*_n(f,x)$	$P_n(f,x)$
defined	(9.1.1)	(9.2.1)	(9.1.7)	(9.2.2)	(9.1.8)	(9.2.3)	(9.1.9)
$\varphi(x)^2$	$x(1-x)$	$x(1-x)$	x	x	$x(1+x)$	$x(1+x)$	x^2
B	$C[0,1]$	$L_p[0,1]$	$C[0,\infty)$	$L_p(R_+)$	$C[0,\infty)$	$L_p(R_+)$	$L_p(R_+)$
E_n	$\left[\frac{A}{n}, 1-\frac{A}{n}\right]$	$\left[\frac{A}{n}, 1-\frac{A}{n}\right]$	$\left[\frac{A}{n}, \infty\right)$	$\left[\frac{A}{n}, \infty\right)$	$\left[\frac{A}{n}, \infty\right]$	$\left[\frac{A}{n}, \infty\right]$	R_+

where $A > 0$ is a fixed number, $1 \le p < \infty$ (for $P_n f$ $1 \le p \le \infty$), $C[0,\infty) \subset L_\infty[0,\infty)$ (i.e., only bounded continuous functions are discussed) and $B(E_n)$ is $C(E_n)$ or $L_p(E_n)$ when B is $C(D)$ or $L_p(D)$.

The set E_n is related to the domain of the main part moduli of smoothness with the given $\varphi(x)$ and $t = n^{-1/2}$. It is crucial in the proof of the direct and converse results and is included in the above table with other pertinent data.

The following theorems yield the direct and converse result for all these operators.

Theorem 9.3.2. Suppose $L_n(f, x)$, B, and φ are those given in Definition 9.3.1, $L_{n,r}(f, x)$ is given by (9.2.6) and (9.2.7) and $f \in B$. Then

$$\|L_{n,r}f - f\|_B \le M[\omega^{2r}_\varphi(f, n^{-1/2})_B + n^{-r}\|f\|_B], \tag{9.3.1}$$

$$K_{2r,\varphi}(f, n^{-r})_B \le \|L_{k,r}f - f\|_B + M\left(\frac{k}{n}\right)^r K_{2r,\varphi}(f, k^{-r})_B, \tag{9.3.2}$$

and

$$\|L_{n,r}f - f\|_B = O(n^{-\alpha/2}) \Leftrightarrow \omega^{2r}_\varphi(f, h)_B = O(h^\alpha), \quad \alpha < 2r. \tag{9.3.3}$$

Remark 9.3.3 (a) For B_n^*, S_n^*, and V_n^* (9.3.2) and (9.3.3) are valid when $B = L_\infty(D)$ too, but (9.3.1) is not. Remarks on the proof of these facts will be made at the pertinent points.

(b) The inequality (9.3.2) constitutes a converse of (9.3.1).

(c) We were not able to settle the saturation case, i.e., $\alpha = 2r$ (for $r > 1$) in (9.3.3).

PROOF. For the proof of the theorem we need the following inequalities:

$$\|L_{n,r}f\|_B \leq M\|f\|_B, \tag{9.3.4}$$

$$\|\varphi^{2r} L_n^{(2r)} f\|_B \leq Mn^r \|f\|_B, \tag{9.3.5}$$

$$\|L_{n,r}f - f\|_{B(E_n)} \leq Mn^{-r}(\|\varphi^{2r}f^{(2r)}\|_B + \|f\|_B) \quad \text{for } f^{(2r-1)} \in \text{A.C.}_{\text{loc}}, \tag{9.3.6}$$

and

$$\|\varphi^{2r} L_n^{(2r)} f\|_B \leq M\|\varphi^{2r}f^{(2r)}\|_B \quad \text{for } f^{(2r-1)} \in \text{A.C.}_{\text{loc}}. \tag{9.3.7}$$

(For $L_\infty(D)$ and the operators B_n^*, S_n^*, and V_n^* (9.3.6) cannot be proved. However, the implication "⇐" in (9.3.3), is a corollary of Ditzian [8, Th. 5.1].)

Inequality (9.3.4) follows from the definition of $L_{n,r}(f,x)$ and in particular (9.2.7) (b) and the boundedness of $L_n(f,x)$ given in Definition 9.3.1. In fact, in all cases except $V_n^*(f,x)$, we have $\|L_n f\|_B \leq \|f\|_B$ and for V_n^* we have

$$\|V_n^* f\|_{L_p(R_+)} \leq \left(\frac{n}{n-1}\right)^{1/p} \|f\|_{L_p(R_+)}.$$

The estimate $\|L_n f\|_B \leq M_1 \|f\|_B$ is well known and will be achieved in Section 9.4 as a side product, following Lemma 9.4.2. Inequalities (9.3.5), (9.3.6), and (9.3.7) will be proved in Sections 9.4, 9.5, 9.6, and 9.7, respectively.

To prove (9.3.1) we choose $g_n \in B(D)$ such that $\varphi^{2r} g_n^{(2r)} \in B$ and $g_n^{(2r-1)} \in$ A.C.$_{\text{loc}}$, which satisfies

$$\|f - g_n\|_B \leq 2K_{2r,\varphi}(f, n^{-r})_B$$

and

$$\|\varphi^{2r} g_n^{(2r)}\|_B \leq 2n^r K_{2r,\varphi}(f, n^{-r})_B.$$

Using (9.3.4) and (9.3.6) and writing $f = f - g_n + g_n$, we estimate

$$\|L_{n,r}(f - g_n) - (f - g_n)\|_B \leq C\|f - g_n\|_B$$

and

$$\|L_{n,r} g_n - g_n\|_{B(E_n)} \leq Cn^{-r} \|\varphi^{2r} g_n^{(2r)}\|_B.$$

We have (see Theorem 2.1.1)

$$K_{2r,\varphi}(f, n^{-r})_B \sim \omega_\varphi^{2r}(f, n^{-1/2})_B$$

and therefore for $L_n(f,x) = P_n(f,x)$ this already completes the proof of (9.3.1) as $E_n = R_+$. For $L_n(f,x) = B_n(f,x)$ or $L_n(f,x) = B_n^*(f,x)$ we can, because of Theorems 7.2.1 and 7.3.1, choose g_n to be $P_{[\sqrt{n}]}(x)$ the best $[\sqrt{n}]$-th degree

9.3. Direct and Converse Results

polynomial approximation in $C[0,1]$ or $L_p[0,1]$ respectively. We have the estimates (see (9.3.6))

$$\|B_{n,r}^* P_{[\sqrt{n}]} - P_{[\sqrt{n}]}\|_{L_p[A/n,\,1-(A/n)]} \le M K_{2r,\varphi}(f, n^{-r})_p$$

and

$$\|B_{n,r} P_{[\sqrt{n}]} - P_{[\sqrt{n}]}\|_{C[A/n,\,1-(A/n)]} \le M K_{2r,\varphi}(f, n^{-r})_\infty.$$

Since for $k < n$, $B_n(P_k, x)$ and $B_n^*(P_k, x)$ are polynomials of degree at most k, we can use Theorem 8.4.8 translated from $[-1,1]$ to $[0,1]$ and n to $[\sqrt{n}]$, to obtain the estimate

$$\|B_{n,r}^* P_{[\sqrt{n}]} - P_{[\sqrt{n}]}\|_{L_p[0,1]} \le M_1 \|B_{n,r}^* P_{[\sqrt{n}]} - P_{[\sqrt{n}]}\|_{L_p[A/n,\,1-(A/n)]}$$
$$\le M_2 K_{2r,\varphi}(f, n^{-r})_p$$

and similarly for $B_{n,r}$. This implies (9.3.1) for the Bernstein and Kantorovich operators. To complete the proof of (9.3.1) for the Szász and Baskakov operators and their Kantorovich modifications, we will also use best polynomial approximation. We choose the best $[\sqrt{n}]$-th degree polynomial approximation of f in $L_p[0,2]$ or $C[0,2]$, and recall, using Theorems 7.2.1 and 7.3.1, that we have

$$\|f - P_{[\sqrt{n}]}\|_{L_p[0,1]} \le M K_{2r,\varphi}(f, n^{-r})_p$$

and

$$\|\varphi^{2r} P_{[\sqrt{n}]}^{(2r)}\|_{L_p[0,1]} \le M n^r K_{2r,\varphi}(f, n^{-r})_p, \qquad (9.3.8)$$

as in $[0,1]$ $\varphi(x) \sim (x(2-x))^{1/2}$. We now define $g_n^*(x)$ by

$$g_n^*(x) = P_{[\sqrt{n}]}(x)\psi(x) + g_n(x)(1 - \psi(x)),$$

where $\psi(x)$ is decreasing, $\psi(x) \in C^\infty$, $\psi(x) = 1$ for $x \le 1/4$, and $\psi(x) = 0$ for $x \ge 3/4$. The (now) standard technique implies

$$\|f - g_n^*\|_B \le M_* K_{2r,\varphi}(f, n^{-r})_B \quad \text{and} \quad \|\varphi^{2r} g_n^{*(2r)}\|_B \le M_* n^r K_{2r,\varphi}(f, n^{-r})_B.$$

With no loss of generality we may assume $A/n < 1/8$. Using (9.3.4), we now write

$$\|L_{n,r} f - f\|_B \le \|L_{n,r}(f - g_n^*) - (f - g_n^*)\|_B + \|L_{n,r} g_n^* - g_n^*\|_B$$
$$\le (M+1)\|f - g_n^*\|_B + \|L_{n,r} g_n^* - g_n^*\|_{L_p[1/8,\infty)}$$
$$+ \|L_{n,r} P_{[\sqrt{n}]} - P_{[\sqrt{n}]}\|_{L_p[0,1/8]} + \|L_{n,r}(P_{[\sqrt{n}]} - g_n^*)\|_{L_p[0,1/8]}$$
$$+ \|P_{[\sqrt{n}]} - g_n^*\|_{L_p[0,1/8]}.$$

Having chosen g_n^* to satisfy $\|f - g_n^*\|_B \le M_* K_{2r,\varphi}(f, n^{-r})_B$ and using also (9.3.6),

$$\|L_{n,r} g_n^* - g_n^*\|_{L_p[1/8,\infty)} \le M_1 [K_{2r,\varphi}(f, n^{-r})_B + n^{-r}\|g_n^*\|_B],$$

where $\|g_n^*\|_B \le 4\|f\|_B$.

The fact that for $k < n$, $S_n(P_k, x)$, $S_n^*(P_k, x)$, $V_n(P_k, x)$, and $V_n^*(P_k, x)$ are polynomials of degree at most k and Theorem 8.4.8, imply for L_n standing for these operators

$$\|L_{n,r}P_{[\sqrt{n}]} - P_{[\sqrt{n}]}\|_{L_p[0, 1/8]} \leq C\|L_{n,r}P_{[\sqrt{n}]} - P_{[\sqrt{n}]}\|_{L_p[(A/n), 1/8]},$$

where for S_n and V_n $p = \infty$. We will complete the proof of (9.3.1) when we show that

$$\|L_{n,r}P_{[\sqrt{n}]} - L_{n,r}g_n^*\|_{L_p[0, 1/8]} \leq M\|f\|_p n^{-l}$$

for any integer l and

$$\|L_{n,r}P_{[\sqrt{n}]} - P_{[\sqrt{n}]}\|_{L_p[A/n, 1/8]} \leq CK_{2r,\varphi}(f, n^{-r})_p.$$

We will begin by demonstrating the first of these estimates. Using the definition of $L_{n,r}$, it is enough to show

$$\|L_s P_{[\sqrt{n}]} - L_s g_n^*\|_{L_p[0, 1/8]} = M\|f\|_p n^{-l}, \quad s = n_i.$$

As L_n is a positive operator, it is enough to show for $\chi(x)$ given by $\chi(x) = 0$ for $x \leq 1/4$ and $\chi(x) = 1$ for $x > 1/4$, that for $s = n_i$

$$\|L_s(\chi|g_n^*|, x)\|_{L_p[0, 1/8]} \leq M_2\|f\|_p n^{-l}$$

and

$$\|L_s(\chi|P_{[\sqrt{n}]}|, x)\|_{L_p[0, 1/8]} = M_3\|f\|_p n^{-l}.$$

The proof of the first inequality, which is quite standard, will be shown first. For $1 \leq p < \infty$ and $q^{-1} + p^{-1} = 1$ we write

$$L_s(\chi|g_n^*|, x) \leq 8^{2m} L_s((\cdot - x)^{2m}|g_n^*(\cdot)|, x)$$
$$\leq 8^{2m} L_s(|\cdot - x|^{2mq}, x)^{1/q}(L_s(|g_n^*(\cdot)|^p, x))^{1/p}$$

and for $p = \infty$

$$L_s(\chi|g_n^*|, x) \leq 8^{2m} L_s(|\cdot - x|^{2m}, x)\|g_n^*(x)\|_{L_\infty}.$$

We now use (9.3.4) to obtain

$$\|L_s(|g_n^*(\cdot)|^p, x)^{1/p}\|_{L_p[0, 1/8]} \leq \|L_s(|g_n^*(\cdot)|^p, x)\|_{L_1[0, \infty)}^{1/p} \leq C\|g_n^*\|_p \leq 4C\|f\|_p$$

which, together with the inequality

$$L_s(|\cdot - x|^{2mq}, x)^{1/q} \leq C_1 n^{-m} \quad \text{for } x \in [0, 1], s = n_i \text{ and } 1 \leq q < \infty$$

(see May [2, 1227 Prop. 3.2(2)] or Lemma 9.4.4 below for example), completes the estimate of $\|L_s(\chi|g_n^*|, x)\|$ in $L_p[0, 1/8]$. For the second inequality, that is $\|L_s(\chi|P_{[\sqrt{n}]}|, x)\|_{L_p[0, 1/8]} \leq M\|f\|_p n^{-l}$, we recall that, as $P_{[\sqrt{n}]}$ approximates f in $[0, 2]$, $\|P_{[\sqrt{n}]}\|_{L_p[0, 2]} \leq 2\|f\|_{L_p[0, 2]}$, and therefore (see Timan [1, Sec. 4.9.6])

$$|P_{[\sqrt{n}]}(x)| \leq M\|f\|_{L_p[0, 2]} n^{1/p} \quad \text{for } 0 \leq x \leq 2.$$

9.3. Direct and Converse Results

Using a theorem due to Bernstein (see Lorentz [2, p. 42, Th. 7]), we have

$$|P_{[\sqrt{n}]}(u)| \leq M_1 \|f\|_p n^{1/p} |4u|^{[\sqrt{n}]} \quad \text{for } u \geq 1/4$$

and therefore

$$\left|P_{[\sqrt{n}]}\left(\frac{k}{n}\right)\right| \leq M_1 \|f\|_p n^{1/p} \left(\frac{4k}{n}\right)^{[\sqrt{n}]}, \quad k \geq n/4$$

and

$$\left|n\int_{k/n}^{k+1/n} P_{[\sqrt{n}]}(u)\,du\right| \leq M_2 \|f\|_p n^{1/p} \left(\frac{4k}{n}\right)^{[\sqrt{n}]}, \quad k \geq n/4.$$

Consequently, we have to estimate for $S_n(f,x)$ and $S_n^*(f,x)$ the sum

$$I_n = \sum_{k \geq n/4} e^{-nx} \frac{(nx)^k}{k!} n^{1/p} \left(\frac{4k}{n}\right)^{\sqrt{n}}$$

and for $V_n(f,x)$ and $V_n^*(f,x)$ the sum

$$J_n = \sum_{k \geq n/4} \binom{n+k-1}{k} x^k (1+x)^{-n-k} \left(\frac{4k}{n}\right)^{[\sqrt{n}]} n^{1/p}.$$

Using Stirling's formula and the fact that for $x < 1/8$ the terms of these sums decrease geometrically we have

$$I_n \leq 3e^{-n/8} \frac{(n/8)^{[n/4]}}{[n/4]!} n^{1/p} \left(\frac{4[n/4]}{n}\right)^{\sqrt{n}} = O(n^{-l})$$

and

$$J_n \leq 3\binom{n+[n/4]-1}{[n/4]} (1/8)^{[n/4]} \left(1+\frac{1}{8}\right)^{-[n/2]-n} \left(\frac{4[n/4]}{n}\right)^{\sqrt{n}} n^{1/p} = O(n^{-l}).$$

To complete the proof of (9.3.1) for all operators of Definition 9.3.1 we have to show

$$\|L_{n,r} P_{[\sqrt{n}]} - P_{[\sqrt{n}]}\|_{L_p[A/n,\,1/8]} \leq CK_{2r,\varphi}(f, n^{-r})_p,$$

where L_n is S_n^*, V_n^*, S_n, or V_n (the latter two for $p = \infty$). To achieve this estimate we construct a function $\Phi_n(x)$ such that

$$\Phi_n(x) = P_{\sqrt{n}}(x), \quad \text{for } x \leq \tfrac{1}{4}, \qquad \Phi_n(x) = 0 \quad \text{for } x > 1,$$

$$\|\Phi_n(x)\|_p \leq \|P_{[\sqrt{n}]}(x)\|_{L_p(0,1)},$$

and

$$\|\varphi^{2r} \Phi_n^{(2r)}\|_p \leq M(\|\varphi^{2r} P_{[\sqrt{n}]}^{(2r)}\|_{L_p[0,1]} + \|P_{[\sqrt{n}]}\|_{L_p[0,1]}).$$

Using (9.3.6) and (9.3.8) we have

$$\|L_{n,r}\Phi_n - \Phi_n\|_{L_p[A/n,\,1/8]} \le C_1 n^{-r}(\|\varphi^{2r}\Phi_n^{(2r)}\|_p + \|\Phi_n\|_p)$$
$$\le C_2(K_{2r,\varphi}(f,n^{-r})_p + n^{-r}\|f\|_p).$$

To complete the proof it is enough to estimate for $s = n_i$, as was already done earlier,

$$\|L_s(\chi|\Phi_n|,x)\|_{L_p[0,\,1/8]} \le M\|f\|_p n^{-l}$$

and

$$\|L_s(\chi|P_{[\sqrt{n}]}|,x)\|_{L_p[0,\,1/8]} \le M\|f\|_p n^{-l},$$

where l is a big integer and $\chi(x)$ is the characteristic function of $[0, 1/4]$.

To prove (9.3.2) we write $f = f - L_{k,r}f + L_{k,r}f$ and therefore,

$$K_{2r,\varphi}(f,n^{-r})_B \le \|f - L_{k,r}f\|_B + n^{-r}\|\varphi^{2r}L_{k,r}^{(2r)}f\|_B,$$

which, using (9.3.5) and (9.3.7) on the second term with $f - g$ and g respectively, implies (9.3.2). The equivalence result (9.3.3) follows now from (9.3.1), (9.3.2), and a well-known lemma by Berens and Lorentz [1, p. 969] restated below. □

Lemma 9.3.4 (Berens–Lorentz). *Suppose for a fixed r and α satisfying $r > \alpha$, the positive sequence $\Psi(n)$ satisfies*

$$\Psi(n) \le An^{-\alpha} + M(k/n)^r \Psi(k) \quad \text{and} \quad \Psi(n_0) \le A_1 \quad \text{for } n \ge n_0 \quad \text{and all } k.$$

Then we have

$$\Psi(n) \le M_1 n^{-\alpha} \quad \text{for all } n \ge n_0.$$

We also obtain the following corollary by combining the results of Chapter 7 with Theorem 9.3.2.

Corollary 9.3.5. *For $f \in L_p[0,1]$, $1 \le p \le \infty$, $\alpha < 2r$ and*

$$E_n(f)_p \equiv \inf_{P_n \in \Pi_n} \|f - P_n\|_{L_p[0,1]}$$

we have

$$\|B_{n,r}^* f - f\|_{L_p[0,1]} = O(n^{-\alpha/2}) \Leftrightarrow E_n(f)_p = O(n^{-\alpha}). \tag{9.3.9}$$

For $f \in C[0,1]$ and $\alpha < 2r$ we have

$$\|B_{n,r}f - f\|_{C[0,1]} = O(n^{-\alpha/2}) \quad \text{if and only if} \quad E_n(f)_\infty = O(n^{-\alpha}). \tag{9.3.10}$$

We can also derive the following weak-type estimate for $\omega_\varphi^{2s}(f,t)_p$.

Theorem 9.3.6. *Under the assumptions of Theorem 9.3.2 we have for any $s \ge 1$*

$$\omega_\varphi^{2s}(f, 1/\sqrt{n})_B \le M_s n^{-s}\left(\sum_{k=1}^n k^{s-1}\|L_{k,r}f - f\|_B + \|f\|_B\right). \tag{9.3.11}$$

9.3. Direct and Converse Results

PROOF. We denote $E_k = \|L_{k,r} f - f\|_B$ and recall that (9.3.2) follows from (9.3.5) and (9.3.7) and not from properties of $L_{n,r} f$, and therefore,

$$\omega_\varphi^{2m}(f, 1/\sqrt{n}) \le M_m\left(E_k + \left(\frac{k}{n}\right)^m \omega_\varphi^{2m}(f, 1/\sqrt{k})\right) \tag{9.3.12}$$

for every m. Iterating (9.3.12) and using the standard technique (see Totik [7, p. 469]), we obtain for $\rho < m$

$$\omega_\varphi^{2m}(f, 1/\sqrt{k})_B \le C\left(k^{-\rho} \sum_{l=1}^{k} l^{\rho-1} E_k + k^{-\rho} \|f\|_B\right). \tag{9.3.13}$$

We now recall the Marchaud-type inequality (Theorem 4.3.1)

$$\omega_\varphi^m(f, t)_B \le Ct^m \left(\int_t^c \frac{\omega_\varphi^{m+1}(f, u)_B}{u^{m+1}} du + \|f\|_B\right)$$

which easily implies

$$\omega_\varphi^{2s}(f, t)_B \le Ct^{2s} \left(\int_t^c \frac{\omega_\varphi^{2s+2}(f, u)_B}{u^{2s+1}} du + \|f\|_B\right).$$

The above written in series form is given by

$$\omega_\varphi^{2s}(f, 1/\sqrt{n})_B \le Cn^{-s} \left(\sum_{k=1}^{[\sqrt{n}]} k^{2s-1} \omega_\varphi^{2s+2}(f, 1/k)_B + \|f\|_B\right)$$

$$\le Cn^{-s} \left(\sum_{k=1}^{n} k^{s-1} \omega_\varphi^{2s+2}(f, 1/\sqrt{k})_B + \|f\|_B\right).$$

Using the estimate of $\omega_\varphi^{2m}(f, 1/\sqrt{k})_B$ from (9.3.13) with $m = s + 1$ and $\rho = s + 1/2$ in the above, we obtain

$$\omega_\varphi^{2s}(f, 1/\sqrt{n})_B \le Cn^{-s}\left(\sum_{k=1}^{n} k^{s-1} k^{-\rho} \sum_{l=1}^{k} l^{\rho-1} E_l + \|f\|_B\right)$$

$$= C\left(n^{-s} \sum_{l=1}^{n} E_l l^{\rho-1} \sum_{k=l}^{n} k^{s-1-\rho} + \|f\|_B\right)$$

$$\le Cn^{-s}\left(\sum_{l=1}^{n} l^{s-1} E_l + \|f\|_B\right). \qquad \square$$

Remark 9.3.7. (a) For $r = 1$ and $L_k = B_k$ or $L_k = B_k^*$ the term $\|f\|_B$ in (9.3.11) can be dropped. To show this we observe that for $L_1 f = B_1 f$ or $L_1 f = B_1^* f$,

$$\omega_\varphi^{2m}(f, t) = \omega_\varphi^{2m}(f - L_1, t).$$

For $L_n = B_n$ we now substitute $f - B_1 f$ in place of f in (9.3.11). For $L_n = B_n^*$ we observe that $(d/dx)^2 B_1^*(f, x) = 0$, and hence the term $\|f\|_B$ can be dropped in (9.3.13). The Marchaud-type inequality is then applied to $f - B_1^* f$.

(b) For $s = 1$ Theorem 9.3.6 was proved by E. Wickern [1], [2].

For the given approximation operator Theorem 9.3.6 forms an analogue of the Bernstein–Stechkin estimates for best trigonometric polynomial approximation. Being a common way of expressing estimates of moduli of smoothness, this, i.e., (9.3.11) appears to be preferable to (9.3.2). However, in some cases we actually lost some information while deriving (9.3.11) from (9.3.2) and one such case is given in the following corollary.

Corollary 9.3.8. *Suppose there is a constant $C < 2^{2r}$ for which*

$$\omega_\varphi^{2r}(f, 2t)_B \leq C\omega_\varphi^{2r}(f, t)_B \quad \text{for } t \leq t_0. \tag{9.3.14}$$

Then

$$\|L_{n,r}f - f\|_B \sim \omega_\varphi^{2r}(f, 1/\sqrt{n})_B \quad \text{for } n \geq n_0. \tag{9.3.15}$$

PROOF. We have to show that (9.3.14) implies

$$\|L_{k,r}f - f\|_B \geq L\omega_\varphi^{2r}(f, 1/\sqrt{k})_B \qquad (L > 0, k \geq k_0).$$

In fact (9.3.14) implies

$$C^{-m}\omega_\varphi^{2r}(f, k^{-1/2})_B \leq \omega_\varphi^{2r}(f, 2^{-m}k^{-1/2})_B = \omega_\varphi^{2r}(f, (2^{2m}k)^{-1/2})_B.$$

Using (9.3.2) for $n = 2^m k$ and $\omega_\varphi^{2r}(f,t)_B \sim K_{2r\varphi}(f, t^r)_B$ (the main equivalence theorem), we have

$$C^{-m}\omega_\varphi^{2r}(f, k^{-1/2})_B \leq \omega_\varphi^{2r}(f, (2^{2m}k)^{-1/2})_B$$

$$\leq M_1\|L_{k,r}f - f\|_B + M_1 2^{-2mr}\omega_\varphi^{2r}(f, k^{-1/2})_B.$$

We now choose m so that $C^{-m} > 2M_1 2^{-2mr}$, which is possible since $C^{-1} > 2^{2r}$, and obtain

$$2^{-2mr}\omega_\varphi^{2r}(f, k^{-1/2})_B \leq M_1\|L_{k,r}f - f\|_B,$$

which completes the proof. □

Corollary 9.3.8 provides us with an effective method of calculating the asymptotic behavior of the operator approximation if $\omega_\varphi^{2r}(f, t)_B$ is not too close to t^{2r}. For example, for $f(x) = x^\delta (\log x)^\gamma$, where $-1/p < \delta < 1 - (1/p)$ and $1 \leq p \leq \infty$, we have, using the calculation of $\omega_\varphi^{2s}(f, t)_p$ in the second example of Section 3.4,

$$\|B_n^* f - f\|_p \sim \begin{cases} n^{-\delta - 1/p}(\log n)^\gamma & \text{if } \delta \neq 0, \\ n^{-1/p}(\log n)^{\gamma - 1} & \text{if } \delta = 0, \end{cases}$$

and when $\delta > 1 - (1/p)$, we have $\|B_n^* f - f\|_p \sim n^{-1}$.

9.4. The Bernstein-Type Inequality $\|\varphi^{2r} L_n^{(2r)} f\|_p \leq Mn^r \|f\|_p$

As promised, we will prove the inequality (9.3.5) in this section. The inequality is a Bernstein-type inequality and is of importance in itself. We shall state the result in the following theorem.

9.4. The Bernstein-Type Inequality $\|\varphi^{2r}L_n^{(2r)}f\|_p \leq Mn^r\|f\|_p$

Theorem 9.4.1. *For $f \in B$ we have*

$$\|\varphi^{2r}L_n^{(2r)}f\|_B \leq Mn^r\|f\|_B, \quad (9.4.1)$$

where L_n, φ, and B are those of Definition 9.3.1 and M is independent of n and f.

The proof consists of two parts estimating (9.4.1) on the domain E_n^c (complement of E_n) and E_n respectively, where E_n is given in Definition 9.3.1. Of course the proof for P_n involves only the estimate on E_n (as for $P_n(f,x)$ $E_n = R_+$). For the estimate on E_n^c we will need the following elementary lemma.

Lemma 9.4.2. *Suppose $P_{n,k}(x) \geq 0$, $\sum_k P_{n,k}(x) = 1$, and for all k $\int_D P_{n,k}(x)\,dx = \Phi(n)$. Then*

$$\left\|\sum b_k P_{n,k}\right\|_{L_p(D)} \leq \Phi(n)^{1/p}\|\{b_k\}\|_{l_p} \quad \text{for } 1 \leq p \leq \infty. \quad (9.4.2)$$

Actually for $p = \infty$ the condition $\int_D P_{n,k}(x)\,dx = \Phi(n)$ is redundant.

PROOF. Using Jensen's inequality,

$$\int_D \left|\sum b_k P_{n,k}(x)\right|^p dx \leq \int_D \sum |b_k|^p P_{n,k}(x)\,dx = \Phi(n)\sum|b_k|^p.$$

For $p = \infty$ we have

$$\sup\left|\sum b_k P_{n,k}(x)\right| \leq \sup|b_k|\sum P_{n,k}(x) = \sup|b_k| = \|\{b_k\}\|_{l_\infty}. \quad \square$$

PROOF OF (9.4.1) ON E_n^c. We will need the following expressions of derivatives of L_n given by Martini [1] or simple calculations.

$$B_n^{(m)}(f,x) = \frac{n!}{(n-m)!}\sum_{k=0}^{n-m}\vec{\Delta}_{1/n}^m f\left(\frac{k}{n}\right)P_{n-m,k}(x)$$

$$S_n^{(m)}(f,x) = n^m\sum_{k=0}^{\infty}\vec{\Delta}_{1/n}^m f\left(\frac{k}{n}\right)S_{n,k}(x) \quad (9.4.3)$$

$$V_n^{(m)}(f,x) = \frac{(n+m-1)!}{(n-1)!}\sum_{k=0}^{\infty}\vec{\Delta}_{1/n}^m f\left(\frac{k}{n}\right)v_{n+m,k}(x)$$

and

$$B_n^{*(m)}(f,x) = \frac{n!}{(n-m)!}\sum_{k=0}^{n-m}\Delta^m a_k(n+1)p_{n-m,k}(x)$$

$$S_n^{*(m)}(f,x) = n^m\sum_{k=0}^{\infty}\Delta^m a_k(n)s_{n,k}(x) \quad (9.4.4)$$

$$V_n^{*(m)}(f,x) = \frac{(n+m-1)!}{(n-1)!}\sum_{k=0}^{\infty}\Delta^m a_k(n)v_{n+m,k}(x),$$

where $a_k(n) = n\int_{k/n}^{k+1/n} f(u)\,du$, $\Delta a_k = a_{k+1} - a_k$, and $\Delta^m a_k \equiv \Delta(\Delta^{m-1}a_k)$. Using

$$\frac{n!}{(n-2r)!} \sim n^{2r} \sim \frac{(n+2r-1)!}{(n-1)!} \quad \text{and} \quad \|\varphi^{2r}\|_{L_\infty(E_n^c)} \sim n^{-r},$$

we have

$$\|\varphi^{2r} L_n^{(2r)} f\|_{L_p(E_n^c)} \leq C n^r \|\sum P_{n,k}(x) b_k\|_{L_p(D)},$$

where $P_{n,k}(x)$, b_k, D, and p are given by:

(a) $P_{n,k}(x)$ is $p_{n-m,k}(x)$, $D = [0,1]$ and $b_k = \Delta^{2r} a_k(n+1)$ or $b_k = \overline{\Delta}_{1/n}^{2r} f(k/n)$ for $L_n = B_n^*$ $(1 \leq p \leq \infty)$ or $L_n = B_n$ $(p = \infty)$.
(b) $P_{n,k}(x)$ is $s_{n,k}(x)$, $D = R_+$ and $b_k = \Delta^{2r} a_k(n)$ or $b_k = \overline{\Delta}_{1/n}^{2r} f(k/n)$ for $L_n = S_n^*$ $(1 \leq p \leq \infty)$ or $L_n = S_n$ $(p = \infty)$.
(c) $P_{n,k}(x)$ is $v_{n+m,k}(x)$, D and b_k are as in (b) for $L_n = V_n^*$ or $L_n = V_n$.

Using

$$\int_0^1 p_{n,k}(x)\, dx = \frac{1}{n+1}, \quad \int_0^\infty s_{n,k}(x)\, dx = \frac{1}{n}, \quad \text{and} \quad \int_0^\infty v_{n,k}(x)\, dx = \frac{1}{n-1},$$

the estimate

$$|a_k(n)|^p \leq \left| n \int_{k/n}^{k+1/n} f(u)\, du \right|^p \leq n \int_{k/n}^{k+1/n} |f(u)|^p\, du, \quad 1 \leq p < \infty,$$

and Lemma 9.4.2 we deduce

$$\|\sum P_{n,k} b_k\|_{L_p(D)} \leq 2^{2r}(1 + o(1)) \|f\|_{L_p(D)}, \quad n \to \infty$$

which completes the proof of (9.4.1) on E_n^c. Actually the expression $(1 + o(1))$ as $n \to \infty$ is 1 for $p_{n,k}(x)$ and $s_{n,k}(x)$, and $(n/(n-1))^{1/p}$ for $v_{n,k}(x)$. We have therefore shown that $B_n(f,x)$, $B_n^*(f,x)$, $S_n(f,x)$, $S_n^*(f,x)$, and $V_n(f,x)$ are contractions on their respective space B (of Definition 9.3.1). The operator $V_n^*(f,x)$ is not a contraction for $1 \leq p < \infty$ but we have the estimate

$$\|V_n^* f\|_{L_p(R_+)} \leq \left(\frac{n}{n-1}\right)^{1/p} \|f\|_{L_p(R_+)}. \qquad \square$$

If we want a variant of V_n^* that is a contraction, we may define

$$V_n^{**}(f,x) = \sum_{k=0}^{\infty} a_k(n-1) v_{n,k}(x)$$

for which Theorem 9.3.1 will also be valid.

To prove our estimate on E_n, we will need to discuss further derivatives of $L_n(f,x)$. This discussion will be needed also in the proof of (9.3.7). Following C.P. May, we have (9.1.12) which applies to $B_n(f,x)$, $S_n(f,x)$, $V_n(f,x)$, and $P_n(f,x)$, and for their Kantorovich modification we have the variants

$$\frac{d}{dx} B_n^*(f,x) = \frac{n}{x(1-x)} \sum_{k=0}^{n} p_{n,k}(x) \left(\frac{k}{n} - x\right) a_k(n+1), \tag{9.4.5}$$

9.4. The Bernstein-Type Inequality $\|\varphi^{2r} L_n^{(2r)} f\|_p \leq M n^r \|f\|_p$

$$\frac{d}{dx} S_n^*(f, x) = \frac{n}{x} \sum_{k=0}^{n} s_{n,k}(x) \left(\frac{k}{n} - x\right) a_k(n), \qquad (9.4.6)$$

and

$$\frac{d}{dx} V_n^*(f, x) = \frac{n}{x(1 + x)} \sum_{k=0}^{n} v_{n,k}(x) \left(\frac{k}{n} - x\right) a_k(n). \qquad (9.4.7)$$

Differentiating the above and using derivatives of $P_{n,k}$, $S_{n,k}$, and $V_{n,k}$, we have the following expressions for the various operators.

$$B_n^{*(2r)}(f, x) = (x(1 - x))^{-2r} \sum_{i=0}^{2r} Q_i^B(x, n) n^i \sum_{k=0}^{n} p_{n,k}(x) \left(\frac{k}{n} - x\right)^i a_k(n + 1) \quad (9.4.8)$$

with $Q_i^B(x, n)$ a polynomial in $nx(1 - x)$ of degree $[(2r - i)/2]$ with non-constant bounded coefficients. Therefore,

$$|(x(1 - x))^{-2r} Q_i^B(x, n) n^i| \leq C \left(\frac{n}{x(1 - x)}\right)^{r+(i/2)} \quad \text{for } x \in E_n.$$

$$S_n^{*(2r)}(f, x) = x^{-2r} \sum_{i=0}^{2r} Q_i^S(nx) n^i \sum_{k=0}^{\infty} s_{n,k}(x) \left(\frac{k}{n} - x\right)^i a_k(n), \qquad (9.4.9)$$

where $Q_i^S(nx)$ is a polynomial in nx of degree $[(2r - i)/2]$ with constant coefficients; and therefore

$$|x^{-2r} Q_i^S(xn) n^i| \leq C \left(\frac{n}{x}\right)^{r+(i/2)} \quad \text{for } x \in E_n.$$

$$V_n^{*(2r)}(f, x) = (x(1 + x))^{-2r} \sum_{i=0}^{2r} Q_i^V(x, n) n^i \sum_{k=0}^{\infty} v_{n,k}(x) \left(\frac{k}{n} - x\right)^i a_k(n), \quad (9.4.10)$$

where $Q_i^V(x, n)$ is a polynomial in $n(x(1 + x))$ of degree $[(2r - i)/2]$ with non-constant polynomial coefficients, and therefore

$$|((1 + x)x)^{-2r} Q_i^V(x, n) n^i| \leq C \left(\frac{n}{x(1 + x)}\right)^{r+(i/2)} \quad \text{for } x \in E_n \cap [0, 2].$$

To investigate $(x(1 + x))^{-2r} Q_i^V(x, n) n^i$ for $x \in [2, \infty)$ we recall that it is generated by taking the $2r$-th derivative of V_n^*. We denote by j the number of times a derivative of the term $v_{n,k}(x)$ in V_n^* is taken creating a form of the type $(n/x(1 + x))^j (x - (k/n))^j v_{n,k}$. To reduce the power of $(x - (k/n))$ to i, we take $j - i$ times the derivative of the power of $(x - (k/n))$ terms. We also take $2r - j - (j - i)$ times the derivatives of terms like $(1/x(1 + x))^m$. (One should note that we do not restrict ourselves in the order of taking the derivatives or part of them.) For $x > 1$ we have $(x(1 + x))^{-1} = \sum_{l=0}^{\infty} x^{-l-2}(-1)^l$, and taking derivatives and multiplying this type of function, we can observe that they remain analytic in $1/x$ with easily computable lowest power. Therefore $(x(1 + x))^{-2r} Q_i^V(x, n) n^i$ is a combination of terms of the type $n^j x^{-2j-2r+2j-i} A_{i,j}(x)$, where $A_{i,j}(x)$ is analytic in x^{-1} and therefore bounded in

[2, ∞) and $2i \le 2j \le 2r + i$. Since $x^{-2} < 2/x(1 + x)$ in $[1, \infty)$, we have

$$|(x(1 + x))^{-2r} Q_i^V(x, n) n^i| \le C_i (n/(x(1 + x)))^{r+(i/2)} \quad \text{for } x \in [2, \infty).$$

For $B_n^{(2r)}(f, x)$, $f(k/n)$ replaces $a_k(n + 1)$ in (9.4.8), and for $S_n^{(2r)}(f, x)$ and $V_n^{(2r)}(f, x)$, $f(k/n)$ replaces $a_k(n)$ in (9.4.9) and (9.4.10) respectively. For the Post–Widder operator we have

$$P_n^{(2r)}(f, x) = \sum_{i=0}^{2r} Q_i(n, x) P_n((\cdot - x)^i f(\cdot), x), \tag{9.4.11}$$

where $Q_i(n, x) = \sum_{2j+l-i=2r} C(i, l) n^j / x^{2j+l}$, and therefore

$$x^{2r} |Q_i(n, x)| \le C n^{r+(i/2)} / x^i \quad \text{for } x \in R_+.$$

From (9.1.12) we can deduce the following lemmas (see May [2]).

Lemma 9.4.3. *If $L_n(f, x)$ is one of the operators $B_n(f, x)$, $S_n(f, x)$, $V_n(f, x)$, or $P_n(f, x)$, and $A_{m,n}(x)$ is given by*

$$A_{m,n}(x) \equiv n^m L_n((\cdot - x)^m, x), \tag{9.4.12}$$

then the recursion relation

$$A_{m+1,n}(x) = mn\varphi(x)^2 A_{m-1,n}(x) + \varphi(x)^2 \frac{d}{dx} A_{m,n}(x),$$
$$A_{0,n}(x) = 1 \quad \text{and} \quad A_{1,n}(x) = 0 \tag{9.4.13}$$

holds with $\varphi(x)$ given in Definition 9.3.1.

From this we can deduce the following lemma.

Lemma 9.4.4. *If $A_{m,n}(x)$ is as in Lemma 9.4.3, then*

$$A_{2m,n}(x) \le C n^m \varphi(x)^{2m} \quad \text{for } x \in E_n, \tag{9.4.14}$$

where E_n, φ, and L_n are related by Definition 9.3.1.

PROOF. We can use induction on m and the recursion relation (9.4.13). For Bernstein polynomials we use (9.4.13) to observe that $A_{m,n}(x)$ is a polynomial in $n(x(1 - x))$ of degree $[m/2]$ with coefficients that are bounded polynomials in x (bounded uniformly for all n). Since on E_n $nx(1 - x) \ge C$, we have (9.4.14). For the Szász operator we show, using (9.4.13), that $A_{m,n}(x)$ is a polynomial in nx of degree $[m/2]$ with constant coefficients (in x and n). Since on E_n $nx \ge A$, we have (9.4.14). For the Baskakov operator we deduce, using (9.4.13), that $A_{m,n}(x)$ is a polynomial in $n(x(1 + x))$ with coefficients that are polynomials in x whose coefficients depend on n but are bounded for all n. This implies for $x \in E_n \cap [0, 1]$, $A_{2m,n}(x) \le C n^m \varphi^{2m}(x)$. For $x \ge 1$ we observe that $A_{m,n}(x)$ is a polynomial in x of degree m. Also $A_{2m,n}(x)$ is a polynomial in n (with coefficients in x) with leading term n^m. Therefore,

$$A_{2m,n}(x) \le C n^m x^{2m} \le C_1 n^m (x(1 + x))^m \quad \text{for } x \in [1, \infty).$$

9.4. The Bernstein-Type Inequality $\|\varphi^{2r}L_n^{(2r)}f\|_p \le Mn^r\|f\|_p$

For the Post–Widder operator the formula (9.4.13) implies, using induction on m, $A_{m,n}(x) = C(m)n^{[m/2]}x^m$, where $C(m,n) = \sum_{j=0}^{[m/2]-1} C_j(m)n^{-j}$ from which (9.4.14) follows. $\square$

To prove our theorem for L_1 we will need the following two technical lemmas which we will prove after completing the proof of Theorem 9.4.1 (since they are needed only for L_1).

Lemma 9.4.5. *Suppose $P_{n,k}(x)$ is $p_{n,k}(x)$, $s_{n,k}(x)$, or $v_{n,k}(x)$ of B_n, S_n, or V_n, and both $\varphi(x)$ and E_n are related to them via Definition 9.3.1; then for all integers m we have*

$$\int_{E_n} \varphi(x)^{-2m} P_{n,k}(x)\left(\frac{k}{n} - x\right)^{2m} dx \le Cn^{-m-1}. \tag{9.4.15}$$

Lemma 9.4.6. *For any integer i we have*

$$\frac{n^n}{(n-1)!}\int_0^\infty e^{-nu/x}\left(\frac{u}{x}\right)^{n-1}\left|\frac{u}{x} - 1\right|^i \frac{dx}{x} \le Mn^{-i/2}. \tag{9.4.16}$$

PROOF OF (9.4.1) ON E_n. In case L_n is B_n^*, S_n^*, V_n^*, B_n, S_n, and V_n, we recall (9.4.8), (9.4.9), and (9.4.10) and estimates on Q_i there, to see that it is enough to prove

$$\left\|\left(\frac{n^{1/2}}{\varphi(x)}\right)^i \sum P_{n,k}(x)\left|\frac{k}{n}-x\right|^i |a_k|\right\|_{L_p(E_n)} \le \|f\|_{L_p(D)},$$

where a_k is $a_k(n+1)$, $a_k(n)$, or $f(k/n)$ in the appropriate places. For $1 < p < \infty$ and $q^{-1} + p^{-1} = 1$ we have

$$\left\|\left(\frac{n^{1/2}}{\varphi(x)}\right)^i \sum P_{n,k}(x)\left|\frac{k}{n}-x\right|^i |a_k|\right\|_{L_p(E_n)}$$

$$\le \left\|\left(\frac{n^{1/2}}{\varphi(x)}\right)^i \left(\sum P_{n,k}(x)\left|\frac{k}{n}-x\right|^{iq}\right)^{1/q} (\sum P_{n,k}(x)|a_k|^p)^{1/p}\right\|_{L_p(E_n)}$$

$$\le \left\|\left(\frac{n^{1/2}}{\varphi(x)}\right)^i \left(\sum P_{n,k}(x)\left|\frac{k}{n}-x\right|^{iq}\right)^{1/q}\right\|_{L_\infty(E_n)} \left(\sum\left\{\int_D P_{n,k}(x)\,dx\right\}|a_k|^p\right)^{1/p}.$$

Using

$$\sum\left\{\int_D P_{n,k}(x)\,dx\right\}|a_k|^p \le (1 + o(1))\|f\|_{L_p(D)}^p,$$

and Lemma 9.4.4 we have for $2m \ge iq$

$$\left\|\left(\frac{n^{1/2}}{\varphi(x)}\right)^i \left(\sum P_{n,k}(x)\left|\frac{k}{n}-x\right|^{iq}\right)^{1/q}\right\|_{L_\infty(E_n)}$$

$$\le \|(\varphi(x)n^{1/2})^{-i}(A_{2m,n}(x))^{i/2m}\|_{L_\infty(E_n)} \le C,$$

which completes the proof for $1 < p < \infty$. For $p = \infty$,

$$\left\| \left(\frac{n^{1/2}}{\varphi(x)} \right)^i \sum P_{n,k}(x) \left| \frac{k}{n} - x \right|^i |a_k| \right\|_{L_\infty(E_n)}$$

$$\leq \|\{a_k\}\|_{l_\infty} \left\| (n^{1/2}/\varphi(x))^i \sum P_{n,k}(x) \left| \frac{k}{n} - x \right|^i \right\|_{L_\infty(E_n)}$$

$$\leq \|f\|_{L_\infty(D)} \left\| (n^{1/2}/\varphi(x))^i \left\{ \sum P_{n,k}(x) \left| \frac{k}{n} - x \right|^{2i} \right\}^{1/2} \right\|_{L_\infty(D)} \leq \|f\|_{L_\infty(D)}.$$

For L_1 we use Lemma 9.4.5 to obtain

$$\left\| \left(\frac{n^{1/2}}{\varphi(x)} \right)^i \sum P_{n,k}(x) \left| \frac{k}{x} - x \right|^i |a_k| \right\|_{L_1(E_n)}$$

$$\leq \sum |a_k| n^{i/2} \int_{E_n} \varphi(x)^{-i} P_{n,k}(x) \left| \frac{k}{n} - x \right|^i dx$$

$$\leq \sum |a_k| n^{i/2} \left\{ \int_{E_n} \varphi(x)^{-2i} P_{n,k}(x) \left| \frac{k}{n} - x \right|^{2i} dx \right\}^{1/2} \left\{ \int_{E_n} P_{n,k}(x) dx \right\}^{1/2}$$

$$\leq C \sum |a_k| n^{-1} \leq C \|f\|_{L_1}.$$

For the Post-Widder operator the proof is very similar. We have to estimate

$$\|n^{i/2} x^{-i} P_n(|\cdot - x|^i f(\cdot), x)\|_{L_p(R_+)}$$

and this can be done recalling

$$|P_n(|\cdot - x|^i f(\cdot), x)| \leq P_n(|\cdot - x|^{iq}, x)^{1/q} P_n(|f|^p, x)^{1/p} \quad \text{for } 1 < p < \infty$$

and

$$|P_n(|\cdot - x|^i f(\cdot), x)| \leq \|f\|_\infty P_n(|\cdot - x|^i, x) \quad \text{for } p = \infty.$$

For $p = 1$ we use Lemma 9.4.6 and obtain

$$\int_0^\infty n^{i/2} x^{-i} \frac{1}{(n-1)!} \left(\frac{n}{x} \right)^n \int_0^\infty e^{-nu/x} |u - x|^i u^{n-1} |f(u)| \, du \, dx$$

$$= \int_0^\infty |f(u)| \left\{ n^{i/2} \frac{n^n}{(n-1)!} \int_0^\infty e^{-nu/x} \left| \frac{u}{x} - 1 \right|^i \left(\frac{u}{x} \right)^{n-1} \frac{dx}{x} \right\} du \leq C \|f\|_{L_1(R_+)}. \quad \square$$

PROOF OF LEMMA 9.4.5. We first recall that $\int_D P_{n,k}(x) \, dx \sim n^{-1}$ which is our lemma for $m = 0$. We assume, using induction, that

$$\int_{E_n} P_{n,k}(x) \varphi(x)^{-2l} \left(\frac{k}{n} - x \right)^{2l} dx = O(n^{-l-1}).$$

We now recall

$$\frac{d}{dx} P_{n,k}(x) = \frac{n}{\varphi(x)^2} P_{n,k}(x) \left(\frac{k}{n} - x \right).$$

9.4. The Bernstein-Type Inequality $\|\varphi^{2r}L_n^{(2r)}f\|_p \le Mn^r\|f\|_p$

We may also observe that:
(a) For $l < k < n - l$ $p_{n,k}(x)/(x(1-x))^l \equiv e_1(n,k,l)p_{n-2l,k-l}(x)$.
(b) For $l < k$ $s_{n,k}(x)/x^l \equiv e_2(n,k,l)s_{n,k-l}(x)$.
(c) For $l < k$ $v_{n,k}(x)/(x(1+x))^l \equiv e_3(n,k,l)v_{n+2l,k-l}(x)$.

We now have for the Szász operator and $k > l + 2$

$$\int_{E_n} \frac{s_{n,k}(x)}{x^l}\left(\frac{k}{n}-x\right)^{2l} dx = \frac{n}{2l+1}\int_{E_n} \frac{s_{n,k}(x)}{x^{l+1}}\left(\frac{k}{n}-x\right)^{2l+1}\left(\frac{k-l}{n}-x\right)dx$$

$$+ \frac{1}{2l+1}\frac{s_{n,k}\left(\frac{A}{n}\right)}{\left(\frac{A}{n}\right)^l}\left(\frac{k}{n}-\frac{A}{n}\right)^{2l+1}$$

or

$$\left|\int_{E_n} \frac{s_{n,k}(x)}{x^{l+1}}\left(\frac{k}{n}-x\right)^{2l+1}\left(\frac{k-l}{n}-x\right)dx\right|$$

$$\le O(n^{-l-2}) + M\frac{1}{n}s_{n,k}\left(\frac{A}{n}\right)\left(\frac{A}{n}\right)^{-l}\left(\frac{k}{n}-\frac{A}{n}\right)^{2l+1}.$$

We first observe that $s_{n,k}(A/n)((k/n)-(A/n))^{2l+1}$ is decreasing in k for $k \ge k_0(l)$ for which

$$\frac{1}{k+1}A\left(\frac{k+1}{n}-\frac{A}{n}\right)^{2l+1} \le \left(\frac{k}{n}-\frac{A}{n}\right)^{2l+1} \qquad (k > \max(A+1, A2^{2l+1}))$$

will do). For small k

$$s_{n,k}\left(\frac{A}{n}\right)\left(\frac{k}{n}-\frac{A}{n}\right)^{2l+1}\left(\frac{A}{n}\right)^{-l} \le Cn^{-l-1},$$

and therefore, this latter inequality is valid for all k. We have now proved that

$$\left|\int_{E_n} \frac{s_{n,k}(x)}{x^{l+1}}\left(\frac{k}{n}-x\right)^{2l+1}\left(\frac{k-l}{n}-x\right)dx\right| = O(n^{-l-2}).$$

We define

$$G_{n,k} = \left\{x : \frac{k-l-1}{n} < x < \frac{k+1}{n}\right\},$$

on $G_{n,k}^c$

$$\left(\frac{k}{n}-x\right)\left(\frac{k-l}{n}-x\right) > 0 \quad \text{and} \quad \left|\frac{k}{n}-x\right| \le C\left|\frac{k-l}{n}-x\right|.$$

We have

$$\int_{E_n} \frac{s_{n,k}(x)}{x^{l+1}}\left|\frac{k}{n}-x\right|^{2l+2} dx = \int_{E_n \setminus G_{n,k}} \frac{s_{n,k}(x)}{x^{l+1}}\left|\frac{k}{n}-x\right|^{2l+2} dx$$

$$+ \int_{G_{n,k}} \frac{s_{n,k}(x)}{x^{l+1}}\left|\frac{k}{n}-x\right|^{2l+2} dx \equiv I_1 + I_2$$

and

$$I_1 \leq C \int_{E_n \setminus G_{n,k}} \frac{s_{n,k}(x)}{x^{l+1}} \left(\frac{k}{n} - x\right)^{2l+1} \left(\frac{k-l}{n} - x\right) dx$$

$$\leq C \left| \int_{E_n} \frac{s_{n,k}(x)}{x^{l+1}} \left(\frac{k}{n} - x\right)^{2l+1} \left(\frac{k-l}{n} - x\right) dx \right|$$

$$+ C \int_{G_{n,k}} \frac{s_{n,k}(x)}{x^{l+1}} \left|\frac{k}{n} - x\right|^{2l+1} \left|\frac{k-l}{n} - x\right| dx = O(n^{-l-2}) + CI_3.$$

The integrals I_2 and I_3 are defined on $G_{n,k}$, and for $x \in G_{n,k}$ one has

$$\left|\frac{k}{n} - x\right|^{2l+2} = O(n^{-2l-2}) \quad \text{and} \quad \left|\frac{k}{n} - x\right|^{2l+1} \left|\frac{k-l}{n} - x\right| = O(n^{-2l-2}),$$

for any $x \in E_n$, $x^{-l-1} = O(n^{l+1})$, and $m(G_{n,k}) = O(1/n)$. Therefore, $I_2 = O(n^{-l-2})$ and $I_3 = O(n^{-l-2})$. To estimate our result for small k ($k \leq \max(A+1, A2^{2l+1})$) we use $|(k/n) - x| < C|x|$ ($C = C(A)$) for $x \in E_n$, and therefore

$$\int_{E_n} \frac{s_{n,k}(x)}{x^{l+1}} \left(\frac{k}{n} - x\right)^{2l+2} dx \leq C^{2l+2} \int_{E_n} s_{n,k}(x) x^{l+1} dx \leq C^{2l+2} \frac{(k+l+1)!}{k!} n^{-l-2}.$$

For $p_{n,k}(x)$ and $v_{n,k}(x)$ we proceed using

$$\int_{E_n} \frac{p_{n,k}(x)}{(x(1-x))^l} \left(\frac{k}{n} - x\right)^{2l} dx$$

$$= \frac{n-2l}{2l+1} \int_{E_n} \frac{p_{n,k}(x)}{(x(1-x))^{l+1}} \left(\frac{k}{n} - x\right)^{2l+1} \left(\frac{k-l}{n-2l} - x\right) dx$$

$$+ \frac{1}{(2l+1)} \frac{p_{n,k}\left(\frac{A}{n}\right)}{\left(\frac{A}{n}\right)^l \left(1 - \frac{A}{n}\right)^l} \left(\frac{k}{n} - \frac{A}{n}\right)^{2l+1}$$

$$- \frac{1}{(2l+1)} \frac{p_{n,k}\left(1 - \frac{A}{n}\right)}{\left(\frac{A}{n}\right)^l \left(1 - \frac{A}{n}\right)^l} \left(\frac{k}{n} - \frac{A}{n}\right)^{2l+1};$$

and

$$\int_{E_n} \frac{v_{n,k}(x)}{(x(1+x))^l} \left(\frac{k}{n} - x\right)^{2l} dx$$

$$= \frac{n+l}{2l+1} \int_{E_n} \frac{v_{n,k}(x)}{(x(1+x))^{l+1}} \left(\frac{k}{n} - x\right)^{2l+1} \left(\frac{k-l}{n+2l} - x\right) dx$$

$$+ \frac{1}{(2l+1)} \frac{v_{n,k}\left(\frac{A}{n}\right)}{\left(\frac{A}{n}\right)^l \left(1 + \frac{A}{n}\right)^l} \left(\frac{k}{n} - \frac{A}{n}\right)^{2l+1}.$$

9.4. The Bernstein-Type Inequality $\|\varphi^{2r}L_n^{(2r)}f\|_p \leq Mn^r\|f\|_p$

The proof follows similarly, changing $G_{n,k}$ to

$$\left\{\frac{k-1}{n} \leq x \leq \frac{k-l+1}{n-2l}\right\} \cup \left\{\frac{k-l-1}{n-2l} \leq x \leq \frac{k+1}{n}\right\} \quad \text{and}$$

$$\left\{\frac{k-l-1}{n+2l} \leq x \leq \frac{k+1}{n}\right\}$$

for $p_{n,k}(x)$ and $v_{n,k}(x)$ respectively, and some other computational modification. For the estimate of the analogues of I_2 and I_3 in case $p_{n,k} = v_{n,k}$ we use on the revised $G_{n,k}$

$$\left|\frac{k}{n} - x\right| = O\left(\frac{1}{n}\left(1 + \frac{k}{n}\right)\right),$$

$$\left|\frac{k-l}{n+2l} - x\right| = O\left(\frac{1}{n}\left(1 + \frac{k}{n}\right)\right), \quad \text{and} \quad \varphi(x)^2 \geq B\left(\frac{k}{n}\left(1 + \frac{k}{n}\right)\right)$$

while

$$\int_{G_{n,k}} v_{n,k}(x)\,dx \leq \int_0^\infty v_{n,k}(x)\,dx \leq \frac{c}{n}. \qquad \square$$

PROOF OF LEMMA 9.4.6. Set $z = nu/x$ and write

$$\frac{n^n}{(n-1)!}\int_0^\infty e^{-nu/x}\left(\frac{u}{x}\right)^{n-1}\left|\frac{u}{x}-1\right|^i \frac{dx}{x}$$

$$= \frac{n^{-i+1}}{(n-1)!}\int_0^\infty e^{-z}z^{n-2}|z-n|^i\,dz$$

$$\leq \frac{n^{-i+1}}{(n-1)}\left\{\frac{1}{(n-2)!}\int_0^\infty e^{-z}z^{n-2}(z-n)^{2i}\,dz\right\}^{1/2}\left\{\frac{1}{(n-2)!}\int_0^\infty e^{-z}z^{n-2}\,dz\right\}^{1/2}$$

$$= n^{-i}\left(\frac{n}{n-1}\right)\left\{\frac{1}{(n-2)!}\int_0^\infty e^{-z}z^{n-2}(z-n)^{2i}\,dz\right\}^{1/2}.$$

Since $|z-n|^{2i} \leq 2^{2i}(|z-(n-2)|^{2i} + 2^{2i})$ we will complete the proof when we show

$$\frac{1}{n!}\int_0^\infty e^{-z}z^n(z-n)^{2i}\,dz \leq M_i n^i \qquad \text{(for all } n\text{)}.$$

We prove the above by induction on i. Integration by parts yields

$$\frac{1}{n!}\int_0^\infty e^{-z}z^n(z-n)^{2i}\,dz = \frac{1}{2i+1}\frac{1}{n!}\int_0^\infty e^{-z}z^{n-1}(z-n)^{2i+2}\,dz$$

or for all n

$$\frac{1}{(n-1)!}\int_0^\infty e^{-z}z^{n-1}(z-n)^{2i+2}\,dz \leq (2i+1)M(i)n^{i+1},$$

which completes the proof as the last inequality implies

$$\frac{1}{(n-1)!} \int_0^\infty e^{-z} z^{n-1} (z-(n-1))^{2i+2} \, dz \leq M_{i+1}(n-1)^{i+1}. \qquad \square$$

9.5. Rate of Convergence for Smooth Functions

To estimate $L_{n,r}(f,x) - f(x)$ we operate $L_{n,r}(f,x)$ on the Taylor expansion with integral remainder

$$f(u) = \sum_{i=0}^{m-1} \frac{(u-x)^i}{i!} f^{(i)}(x) + R_m(f,u,x),$$

where

$$R_m(f,u,x) = \frac{1}{(m-1)!} \int_x^u (u-v)^{m-1} f^{(m)}(v) \, dv$$

to write

$$L_{n,r}(f,x) - f(x) = f(x)\{L_{n,r}(1,x) - 1\} + \sum_{i=1}^{2r-1} \frac{1}{i!} f^{(i)}(x) L_{n,r}((\cdot - x)^i, x) \qquad (9.5.1)$$
$$+ L_{n,r}(R_{2r}(f, \cdot, x), x).$$

For calculating the rate of convergence we will need the following two lemmas.

Lemma 9.5.1.

(a) *For L_n given by Definition 9.3.1 we have*

$$L_{n,r}((\cdot - x)^i, x) = 0 \quad \text{for } 0 < i < r \quad \text{and all } x. \qquad (9.5.2)$$

(b) *For L_n given by B_n, B_{n-1}^*, S_n, and S_n^* we have*

$$L_{n,r}((\cdot - x)^{2r-i}, x) = \varphi(x)^{2r-2i} O(n^{-r}) \quad \text{for } i \leq r \quad \text{and } x \in E_n. \qquad (9.5.3)$$

(c) *For L_n given by V_n, V_n^*, and P_n we have*

$$L_{n,r}((\cdot - x)^{2r-i}, x) = \varphi(x)^{2r} x^{-i} O(n^{-r}) \quad \text{for } i \leq r \quad \text{and } x \in E_n \qquad (9.5.4)$$

(*for S_n and S_n^* (9.5.4) is identical with (9.5.3)*).

(d) *For L_n given by B_n, S_n, V_n, or P_n and $0 < i \leq r$,*

$$L_{n,r}((\cdot - x)^i, x) = 0. \qquad (9.5.5)$$

Lemma 9.5.2. *For L_n given by Definition 9.3.1 we have*

$$\|L_n(R_{2r}(f, \cdot, x), x)\|_{L_p(E_n)} \leq Mn^{-r}(\|\varphi^{2r} f^{(2r)}\|_{L_p(D)} + \|f\|_{L_p(D)}). \qquad (9.5.6)$$

9.5. Rate of Convergence for Smooth Functions

We will prove the estimate of $L_{n,r}(f, x) - f(x)$ in E_n pending these lemmas and later prove Lemma 9.5.1 in this section. Lemma 9.5.2, which is easy for $1 < p \leq \infty$ and difficult for $p = 1$, will be the subject of Section 9.6.

Theorem 9.5.3. *For L_n of Definition 9.3.1 we have*

$$\|L_n(f, x) - f(x)\|_{L_p(E_n)} \leq Mn^{-r}(\|\varphi^{2r} f^{(2r)}\|_p + \|f\|_p). \tag{9.5.7}$$

PROOF. Using (9.5.1), it follows from Lemma 9.5.2 and the definition of $L_{n,r}$, especially (9.2.7) (a) and (b), that

$$\|L_{n,r}(R_{2r}(f, \cdot, x), x)\|_{L_p(E_n)} \leq Mn^{-r}(\|\varphi^{2r} f^{(2r)}\|_{L_p(D)} + \|f\|_{L_p(D)}).$$

We recall again this definition of $L_{n,r}$ and in particular (9.2.7) (c) to obtain $L_{n,r}(1, x) - 1 = 0$. Using Lemma 9.5.1 and the expansion (9.5.1), we observe that it is enough to prove the following assertions.

(a) For $\varphi(x)^2 = x(1 - x)$, $\varphi(x)^2 = x(1 + x)$ or $\varphi(x)^2 = x$ we have

$$\|f^{(i)}\|_{L_p[0, 1]} \leq C(\|\varphi^{2r} f^{(2r)}\|_{L_p(D)} + \|f\|_{L_p(D)}),$$

where $i \leq r$ in case $1 \leq p < \infty$ and $i < r$ in case $p = \infty$.

(b) For $\varphi(x)^2 = x(1 - x)$, $\varphi(x)^2 = x(1 + x)$, or $\varphi(x)^2 = x$ and $i < r$ we have

$$\|\varphi^{2r-2i} f^{(2r-i)}\|_{L_p[0, 1]} \leq C(\|\varphi^{2r} f^{(2r)}\|_{L_p[0, 1]} + \|f\|_{L_p[0, 1]})$$

where $1 \leq p \leq \infty$.

(c) For $\varphi(x)^2 = x(1 + x)$ and $i \leq 2r$ or $\varphi(x)^2 = x$ and $i \leq r$ we have

$$\|\varphi(x)^{2r} x^{-i} f^{(2r-i)}(x)\|_{L_p[1, \infty]} \leq C(\|\varphi^{2r} f^{(2r)}\|_{L_p(R_+)} + \|f\|_{L_p(R_+)})$$

where $1 \leq p \leq \infty$.

(d) For $\varphi(x) = x$ and $i \leq 2r$ we have

$$\|\varphi^{2r-i} f^{(2r-i)}\|_{L_p(R_+)} \leq C(\|\varphi^{2r} f^{(2r)}\|_{L_p(R_+)} + \|f\|_{L_p(R_+)}) \quad \text{where } 1 \leq p \leq \infty.$$

(e) For $\varphi(x) = \sqrt{x}$ and $i \leq r$ we have

$$\|f^{(i)}\|_{L_p[1, \infty)} \leq C(\|\varphi^{2r} f^{(2r)}\|_{L_p(R_+)} + \|f\|_{L_p(R_+)}) \quad \text{where } 1 \leq p \leq \infty.$$

We separate the proof of the assertions for $1 \leq p < \infty$ and $p = \infty$ and prove the results for $1 \leq p < \infty$ first. We recall the Hardy inequality for $1 \leq p < \infty$, $g \geq 0$, and $\beta > 0$ given by

$$\left(\int_0^\infty \left(\int_x^\infty g(y)\,dy\right)^p x^{\beta-1}\,dx\right)^{1/p} \leq \frac{p}{\beta}\left(\int_0^\infty (yg(y))^p y^{\beta-1}\,dy\right)^{1/p}. \tag{9.5.8}$$

We will use this inequality for S_n^* ($\varphi(x)^2 = x$) and P_n ($\varphi(x) = x$) first, with $g(y) = |f^{(2r)}(y)|$ and with $\beta = (r-1)p + 1$ and $\beta = (2r-1)p + 1$ for S_n^* and P_n respectively, to obtain

$$\|\varphi(x)^{2r} x^{-1} f^{(2r-1)}(x)\|_p \leq MQ \quad \text{where } Q = \|\varphi^{2r} f^{(2r)}\|_p + \|f\|_p$$

(since $|f^{(2r-1)}(x)| \leq \int_x^\infty g(y)\,dy$). We continue by induction and after we have shown $\|\varphi(x)^{2r}x^{-i}f^{(2r-i)}(x)\| \leq M_i Q$ for $i < r$, we define $g(x) = |f^{(2r-i)}(x)|$ with $\beta - 1 = (r - i - 1)p$ (for S_n^*) or $\beta - 1 = (2r - i - 1)p$ (for P_n) to obtain $\|\varphi(x)^{2r}x^{-i-1}f^{(2r-i-1)}(x)\| \leq M_{i+1}Q$, which concludes the appropriate parts of (c) and (d) for $i \leq r$ and the corresponding part of (a) for $i = r$. For V_n^* and B_n^* the situation is similar but somewhat more involved. We define $f(x) = \psi(x)f(x) + (1 - \psi(x))f(x) \equiv f_1(x) + f_2(x)$, where $\psi(x) = 1$ for $x \leq 1/4$, $\psi(x) = 0$ for $x \geq 3/4$, and $\psi(x) \in C^\infty$. For both V_n^* and B_n^* we use

$$\|\varphi^{2r}f_j^{(2r)}\|_{L_p(D)} \leq \|\varphi^{2r}f^{(2r)}\|_{L_p(D)} + C(\|f^{(2r)}\|_{L_p[1/4,3/4]} + \|f\|_{L_p[1/4,3/4]})$$

$$\leq C_1(\|\varphi^{2r}f^{(2r)}\|_{L_p(D)} + \|f\|_{L_p(D)}).$$

We now proceed with the above method (used for P_n and S_n^*) for f_1 in relation to V_n^* and B_n^* with β chosen as we did for S_n^*, and for f_2 in relation to V_n^* with β chosen as we did for P_n. We obtain

$$\|x^{r-i}f_1^{(2r-i)}(x)\|_{L_p(R_+)} \leq M(\|\varphi^{2r}f^{(2r)}\|_{L_p(D)} + \|f\|_{L_p(D)})$$

which implies

$$\|\varphi(x)^{2r}x^{-i}f^{(2r-i)}(x)\|_{L_p[0,1/4]} \leq M(\|\varphi^{2r}f^{(2r)}\|_p + \|f\|_p)$$

($\varphi(x)^{2r-2i} \sim \varphi(x)^{2r}x^{-i}$ for $x \in [0, 1/4]$). Similarly we estimate for V_n^*

$$\|\varphi(x)^{2r}x^{-i}f^{(2r-i)}(x)\|_{L_p[3/4,\infty)} \leq \|\varphi(x)^{2r}x^{-i}f_2^{(2r-i)}(x)\|_{L_p(R_+)}$$

$$\leq M(\|\varphi(x)^{2r}f^{(2r)}\|_p + \|f\|_p).$$

The estimate in $[1/4, 3/4]$ is easy as $\varphi(x) \sim 1$ there. For B_n^* the estimate on $[3/4, 1]$ is symmetric to that on $[0, 1/4]$. From the result for $i = r$ and $1 \leq p < \infty$ we derive (a) for $1 \leq p < \infty$ and $i < r$, using Landau-type estimates or the estimate (2.2.14) with $c = 0$ and $d = 1$.

The above technique does not work for $p = \infty$; in fact for $i = r$ the result is not valid, as can be seen for $r = 1$, $\varphi(x)^2 = x(1 - x)$ and $f(x) = x \log x$ for $x \in (0, 1)$, where $f' \notin L_\infty$ while $\|f\|_\infty$ and $\|\varphi^2 f''\|_\infty$ are finite. We observe that in (a) $i = r$ is not required for $p = \infty$. For B_n, S_n, V_n, and P_n $L_{n,r}((\cdot - x)^r, x) = 0$ and therefore for the cases in question $i = r$ causes no difficulty. For proving the rest of the result for $p = \infty$ we observe first that

$$|f^{(j)}(\tfrac{1}{2})| \leq M(\|f^{(2r)}\|_{L_\infty[1/4,3/4]} + \|f\|_{L_\infty[1/4,3/4]}) \leq M_1(\|\varphi^{2r}f^{(2r)}\|_\infty + \|f\|_\infty).$$

Near 0, $(x \leq 1/2)$ $x^\eta f^{(j)}(x) \in L_\infty$ and $\eta > 1$ implies

$$\left|f^{(j-1)}(x) - f^{(j-1)}\left(\frac{1}{2}\right)\right| \leq \int_x^{1/2} |f^{(j)}(x)|\,dx \leq \|x^\eta f^{(j)}(x)\|_{L_\infty[0,1/2]} \int_x^{1/2} \frac{du}{u^\eta}$$

which implies

$$\|x^{\eta-1}f^{(j-1)}(x)\|_{L_\infty[0,1/2]} \leq M(\|\varphi^{2r}f^{(2r)}\|_\infty + \|f\|_\infty) + \|x^\eta f^{(j)}(x)\|_{L_\infty[0,1/2]}$$

from which (b) (near 0) follows by induction. For $x \geq C$ we use $x^\eta f^{(j)}(x) \in L_\infty$ for $\eta > 1$ to show $\lim_{x \to \infty} f^{(j-1)}(x) = f^{(j-1)}(\infty) = 0$ as $\|f\|_\infty < \infty$ and

9.5. Rate of Convergence for Smooth Functions

$$|f^{(j-1)}(x) - f^{(j-1)}(\infty)| \leq \|x^\eta f^{(j)}(x)\|_\infty \int_x^\infty u^{-\eta}\,du$$

and so $x^{\eta-1}f^{(j-1)}(x) \in L_\infty$. We again proceed by induction. Near 1 for B_n the situation is similar to that near zero and actually this case has already been settled (see Ditzian [1, p. 280]).

Assertion (a) with regard to $p = \infty$, $i < r$, and $x \leq 1/2$ follows from the estimate $|f^{(r)}(x)| \leq C(\log x)(\|\varphi^{2r}f^{(2r)}\| + \|f\|)$ which implies $\|f^{(r-1)}\| \leq C(\|\varphi^{2r}f^{(2r)}\| + \|f\|)$ and hence

$$\|f^{(i)}\| \leq C(\|f\| + \|f^{(r-1)}\|) \leq C_1(\|f\| + \|\varphi^{2r}f^{(2r)}\|) \quad \text{for } i < r-1$$

using (2.2.14) again. Clearly we can use similar treatment for B_n or B_n^* and $1/2 \leq x \leq 1$.

For $x \geq 1$ we utilize

$$\|f^{(i)}\|_{L_p[2^k, 2^{k+1}]} \leq M(h^{2r-i}\|f^{(2r)}\|_{L_p[2^k, 2^{k+1}]} + h^{-i}\|f\|_{L_p[2^k, 2^{k+1}]})$$

for $h < 2^k$ (see (6.3.4)), and therefore we set $h = C\varphi(2^k) \sim \varphi(x)$ for $x \in [2^k, 2^{k+1}]$ and obtain

$$\|\varphi^i f^{(i)}\|_{L_p[2^k, 2^{k+1}]} \leq M_1(\|\varphi^{2r}f^{(2r)}\|_{L_p[2^k, 2^{k+1}]} + \|f\|_{L_p[2^k, 2^{k+1}]})$$

from which the result (c) follows for $L_\infty(1, \infty)$ and actually for $L_p(1, \infty)$ too. (Observe that if $\varphi(x)^2 \sim x$, $\|\varphi(x)^{2r-j}f^{(2r-j)}\|_{L_p[1,\infty)} < \infty$ implies $\|\varphi(x)^{2r}x^{-j}f^{(2r-j)}\|_{L_p[1,\infty)} < \infty$ which essentially proves (e)). This method can be applied to the proof of (d) with $k \in Z$. $\square$

Remark 9.5.4. For L_n given by B_n^*, S_n^*, or V_n^* an earlier result (Ditzian [8, Th. 5.1]) implies for $\alpha < 2r$

$$\omega_\varphi^{2r}(f, h)_\infty = O(h^\alpha) \Rightarrow \|L_{n,r}(f, \cdot) - f(\cdot)\|_\infty = O(n^{-\alpha/2}). \tag{9.5.9}$$

The result in Lemma 9.5.1 can be used to guarantee the numerous conditions of the theorem mentioned above.

We will now prove Lemma 9.5.1 for the various operators.

PROOF OF LEMMA 9.5.1 FOR POST-WIDDER OPERATOR. We recall the description $P_n((\cdot - x)^m, x)$, i.e.,

$$P_n((\cdot - x)^m, x) = C(m, n)n^{[m/2]-m}x^m \quad \text{where } C(m, n) = \sum_{j=0}^{[m/2]-1} c_j(m)n^{-j}.$$

Using (9.2.7) (d) for $\rho = m - [m/2] + j < r$, we get $P_{n,r}((\cdot - x)^m, x) = 0$ for $m \leq r$. Using (9.2.7) (b) and the above, we have

$$P_{n,r}((\cdot - x)^m, x) = M(m, n, r)n^{-r}x^m \quad \text{where } M(m, n, r) \leq M(m). \quad \square$$

For the proof of Lemma 9.5.1 for the other operators we need the following result about moments which one proves using induction, the recursion relation (9.4.13), and the definition of $A_{m,n}(x)$ in (9.4.12).

Lemma 9.5.5. *For* $L_n = B_n$, $L_n = S_n$, *or* $L_n = V_n$ *and the related* φ *we have*

$$L_n((\cdot - x)^{2j}, x) = \sum_{m=0}^{j-1} \frac{\varphi(x)^{2j-2m}}{n^{j+m}} q_m(x) \qquad (9.5.10)$$

and

$$L_n((\cdot - x)^{2j+1}, x) = \sum_{m=0}^{j-1} \frac{\varphi(x)^{2j-2m}}{n^{j+m+1}} p_m(x), \qquad (9.5.11)$$

where q_m *and* p_m *are constants for* $L_n = S_n$, *fixed bounded polynomials for* $L_n = B_n$, *and fixed polynomials of degree $2m$ and $2m+1$ respectively, for* $L_n = V_n$.

PROOF OF LEMMA 9.5.1 (For the other operators). For L_n equal to B_n, S_n, or V_n and $i \leq r$ we use Lemma 9.5.5 and observe that in the sum (9.5.10) for $i = 2j$ or (9.5.11) for $i = 2j + 1$ terms always have as coefficients $n^{-\rho}$ with $1 \leq \rho \leq r - 1$. We now use (9.2.7) (d) to obtain

$$B_{n,r}((\cdot - x)^i, x) = S_{n,r}((\cdot - x)^i, x) = V_{n,r}((\cdot - x)^i, x) = 0 \quad \text{for } i \leq r.$$

We recall that (9.2.7) implies with $C_s = C_s(n)$, $\sum_{s=0}^{r-1} C_s n_s^{-\rho} \equiv d_\rho(n) = O(n^{-\rho})$ for $\rho \geq r$ and $\sum_{s=0}^{r-1} C_s n_s^{-\rho} = 0$ for $1 \leq s < r$. Using Lemma 9.5.5, we have for L_n equal to B_n, S_n, or V_n

$$L_{n,r}((\cdot - x)^{2r-2j}, x) = \sum_{m=0}^{r-j-1} \varphi(x)^{2r-2j-2m} q_m(x) \sum_{s=0}^{r-1} C_s n_s^{-r+j-m}$$

$$= \sum_{m=j}^{r-j-1} \varphi(x)^{2r-2j-2m} q_m(x) d_{r-j+m}(n) \qquad (9.5.12)$$

$$= n^{-r} \varphi(x)^{2r-4j} \sum_{l=0}^{r-2j-1} (n(\varphi(x))^2)^{-l} q_{l+j}(x) d_{r+l}(n) n^{r+l},$$

and

$$L_{n,r}((\cdot - x)^{2r-2j+1}, x)$$

$$= \sum_{m=0}^{r-j-1} \varphi(x)^{2r-2j-2m} p_m(x) \sum_{s=0}^{r-1} C_s n_s^{-r+j-m-1}$$

$$= \sum_{m=j-1}^{r-j-1} \varphi(x)^{2r-2j-2m} p_m(x) d_{r-j+m+1}(n) \qquad (9.5.13)$$

$$= n^{-r} \varphi(x)^{2r-4j+2} \sum_{l=0}^{r-2j+1} (n(\varphi(x))^2)^{-l} p_{l+j-1}(x) d_{r+l}(n) n^{r+l}.$$

Using $n(\varphi(x))^2 \geq A_1$ for $x \in E_n$ and the behavior of $p_m(x)$ and $q_m(x)$ for the Baskakov operator, the above implies our lemma for B_n, S_n, and V_n.

We have to estimate the moments of B_n^*, S_n^*, and V_n^*. We will use a result similar to the formula $B_n^*(f, x) = (d/dx) B_{n+1}(F, x)$ and $S_n^*(f, x) = (d/dx) S_n(F, x)$ for $F(x) = \int_0^x f(u) du$ and $V_n^*(f, x) = (d/dx) V_{n-1}(F, x)$ for $F(x) = \int_0^{x(1-1/n)} f(u) du$.

9.5. Rate of Convergence for Smooth Functions

Here we use a variant of the above, namely,

$$B_n^*((\cdot - x)^i, x) = \frac{d}{dx}\left[B_{n+1}\left(\frac{(\cdot - \zeta)^{i+1}}{i+1}, x\right)\right]\bigg|_{\zeta=x} = \frac{d}{dx}\left[B_{n+1}\left(\frac{(\cdot - x)^{i+1}}{i+1}, x\right)\right]$$
$$+ B_{n+1}((\cdot - x)^i, x) \qquad (9.5.14)$$

$$S_n^*((\cdot - x)^i, x) = \frac{d}{dx}\left[S_n\left(\frac{(\cdot - x)^{i+1}}{i+1}, x\right)\right] + S_n((\cdot - x)^i, x),$$

but for V_n^* the expression is slightly more complicated and we will use a different method that will work on S_n^* as well. (S_n^* is the easiest case.) Following (9.5.14) and the definition of $B_{n,r}^*(f, x)$ (i.e., $B_{n,r}^*(f, x) = \sum_{s=0}^{r-1} C_s B_{n_s-1}^*(f, x)$) we may study

$$\sum_{s=0}^{r-1} C_s B_{n_s}((\cdot - x)^i, x) = B_{n,r}((\cdot - x)^i, x)$$

and

$$\sum_{s=0}^{r-1} C_s \frac{d}{dx} B_{n_s}((\cdot - x)^{i+1}, x) = \frac{d}{dx} B_{n,r}((\cdot - x)^{i+1}, x)$$

separately to find out the behavior of $B_{n,r}^*((\cdot - x)^i, x)$. As we have already proved, $B_{n,r}((\cdot - x)^i, x)$ satisfies (a) and (b) of our lemma. We use (9.5.5) for $B_{n,r}(f, x)$ to observe that for $i < r$ $B_{n,r}((\cdot - x)^{i+1}, x) = 0$ and therefore $(d/dx)B_{n,r}((\cdot - x)^{i+1}, x) = 0$ which completes the proof of part (a) of our lemma for $B_{n,r}^*$. To complete the proof of the lemma of $B_{n,r}^*$ (that is part (b)) it remains to show that

$$\frac{d}{dx} B_{n,r}((\cdot - x)^{2r-i+1}, x) = \varphi(x)^{2r-2i} O(n^{-r}) \quad \text{for } i \leq r \quad \text{and } x \in E_n.$$

We now use (9.5.12) and (9.5.13) with $L_n = B_n$ for i odd or even, respectively. Recalling that polynomials are bounded in $[0,1]$, that $(d/dx)\{\varphi(x)^l\} = l(1-2x)\varphi(x)^{l-1}$ for $l > 0$, and that $d_\rho(n) n^\rho = O(1)$, the proof is reduced to examining (9.5.12) and (9.5.13). This can be followed for S_n^* and the estimates of $\sum_{s=0}^{r-1} C_s S_{n_s}((\cdot - x)^i, x)$ and $\sum_{s=0}^{r-1} C_s(d/dx)S_{n_s}((\cdot - x)^{i+1}, x)$. Here we derive $S_n^*((\cdot - x)^i, x) = 0$ for $i < r$, but for $i = r$ we have $S_{n,r}^*((\cdot - x)^r, x) = O(n^{-r})$. In the last step we have to recall that we have constants rather than polynomials for $p_m(x)$ and $q_m(x)$.

For V_n^* and S_n^* we can use

$$n \int_{k/n}^{k+1/n} (u - x)^i \, du = \frac{1}{i+1} \sum_{l=0}^{i} \binom{i+1}{l} \left(\frac{k}{n} - x\right)^l n^{-i+l}.$$

This implies that we may estimate instead of $V_{n,r}^*$ and $S_{n,r}^*$ of $(u-x)^i$ the sums

$$I_n(V, l, i) \equiv \sum_{s=0}^{r-1} C_s V_{n_s}((\cdot - x)^l, x) n_s^{-i+l}$$

and

$$I_n(S, l, i) \equiv \sum_{s=0}^{r-1} C_s S_{n_s}((\cdot - x)^l, x) n_s^{-i+l}$$

for $0 \leq l \leq i$. We will estimate $I_n(V, l, i)$ and $I_n(S, l, i)$ separately for each l. Obviously $I_n(V, 0, i) = I_n(S, 0, i) = \sum_{s=0}^{r-1} C_s n_s^{-i} = d_i(n)$, where $d_i(n) = 0$ for $i < r$ and $d_i(n) = O(n^{-i})$ for $i \geq r$ which clearly satisfy the required estimates ((a), (b), and (c) of our lemmas).

For $i \leq r$ we again use (9.5.10) and (9.5.11) in conjunction with (9.2.7) (d) to obtain

$$I_n(V, l, i) = I_n(S, l, i) = 0 \quad \text{for } 0 < i \leq r \text{ and } 0 < l \leq i.$$

For $i > r$ we follow estimates (9.5.12) and (9.5.13) for l even or odd, respectively. For even $l \leq i \leq 2r$ we set $0 < l = 2r - 2j$ and L for either S or V and write

$I_n(L, 2r - 2j, i)$

$$= \sum_{m=0}^{r-j-1} \varphi(x)^{2r-2j-2m} q_m(x) \sum_{s=0}^{r-1} C_s n_s^{-i+r-j-m}$$

$$= \sum_{2r-i-j \leq m \leq r-j-1} \varphi(x)^{2r-2j-2m} q_m(x) d_{i-r+j+m}(n)$$

$$= n^{-r} \varphi(x)^{2r-2(2r-i)} \sum_{2r-i-j \leq m \leq r-j-1} (n\varphi(x)^2)^{2r-i-j-m} q_m(x) d_{i-r+j+m}(n) n^{i+j+m-r}.$$

Using $n\varphi(x)^2 \geq A$ in E_n, $2r - i - j - m \leq 0$, $d_\rho(n) n^\rho = O(1)$, and $m \leq r - j - 1$, we get the result (c) of our lemma for such l. For odd l the proof follows similarly. □

Remark 9.5.6. Examining closely the above, we can require for B_n^*, S_n^*, V_n^*, and P_n and $1 \leq p < \infty$ the conditions (c′) and (d′) instead of (9.2.7) (c) and (d) (where (c)′ $\sum_{i=0}^{r-1} C_i(n) = 1 + o(n^{-r})$ and (d)′ $\sum_{i=0}^{r-1} C_i(n) n_i^{-\rho} = o(n^{-r})$ $0 < \rho < r$). Conditions (c′) and (d′) would not be adequate for B_n, S_n, and V_n and the C norm.

9.6. Estimate of $\|L_n(R_{2r}(f, \cdot, x), x)\|_{L_p(E_n)}$

In this section we just prove Lemma 9.5.2 and estimate $L_n(R_{2r}(f, \cdot, x), x)$ for L_n given by Definition 9.3.1. In fact the proof for $1 < p \leq \infty$ is relatively easy and will be shown first but the proof for $p = 1$ is more difficult and will occupy most of this section. We first prove a lemma about $R_{2r}(f, u, x)$.

Lemma 9.6.1. *For*

$$R_{2r}(f, u, x) = \frac{1}{(2r-1)!} \int_x^u (u-v)^{2r-1} f^{(2r)}(v) \, dv$$

we have

9.6. Estimate of $\|L_n(R_{2r}(f,\cdot,x),x)\|_{L_p(E_n)}$

$$|R_{2r}(f,u,x)| \leq \frac{|u-x|^{2r-1}}{\varphi(x)^{2r}}\left|\int_u^x \varphi(v)^{2r}|f^{(2r)}(v)|dv\right| \tag{9.6.1}$$

for $\varphi(v)^2 = v(1-v)$ and $x, u \in [0,1]$ or $\varphi(x)^2 = x$ and $x, u \in R_+$; and

$$|R_{2r}(f,u,x)| \leq \frac{|u-x|^{2r-1}}{\varphi(x)^{2r-2}x}\left[\frac{1}{\varphi_1(x)} + \frac{1}{\varphi_1(u)}\right]\left|\int_u^x \varphi(v)^{2r}|f^{(2r)}(v)|dv\right| \tag{9.6.2}$$

for $\varphi(v)^2 = v(1+v)$, $\varphi_1(u) = 1 + u$, and $x, u \in R_+$ or $\varphi(x) = x$, $\varphi_1(x) = x$, and $x, u \in R_+$.

PROOF OF LEMMA 9.6.1. All we have to do is estimate $|u-v|^{2r-1}/\varphi(v)^{2r}$ in the following ways. For $\varphi(v)^2 = v(1-v)$ and v between x and u we have $|v-u|^{2r-1}/(v(1-v))^r \leq |u-x|^{2r-1}/(x(1-x))^r$ as $|v-u|/v(1-v) \leq |u-x|/x(1-x)$, which in turn follows from $(v-u)/v \leq (x-u)/x$, $1/(1-v) \leq 1/(1-x)$ for $u \leq v \leq x$, and $(u-v)/(1-v) \leq (u-x)/(1-x)$, $1/v \leq 1/x$ for $x \leq v \leq u$. For $\varphi(v)^2 = v$, $|v-u|^{2r-1}v^{-r} \leq |x-u|^{2r-1}x^{-r}$ follows from $|v-u|/v \leq |x-u|/x$. For $\varphi(v)^2 = v^2$, $|v-u|^{2r-1}v^{-2r} \leq |x-u|^{2r-1}x^{-2r+1}[(1/x) + (1/u)]$ as $|v-u|/v \leq |x-u|/x$ and $1/v \leq (1/x) + (1/u)$ for v between u and x. For $\varphi(v)^2 = v(1+v)$

$$|v-u|^{2r-1}v^{-r}(1+v)^{-r} \leq |x-u|^{2r-1}x^{-r}(1+x)^{-r+1}\left[\frac{1}{1+x} - \frac{1}{1+u}\right]$$

as $|v-u|/(1+v) \leq |x-u|/(1+x)$ (trivial for $x > u$, otherwise $(u-v)/(1+v) \leq (u-x)/(1+x)$ as $u - v \leq u - x$ and $x \leq v$) and $1/(1+v) \leq 1/(1+x) + 1/(1+u)$ for v between u and x. $\square$

We will need the following lemma on V_n or V_n^*.

Lemma 9.6.2. *For any integer m and $L_n = V_n$ or $L_n = V_n^*$ we have for $\varphi_1(u) = 1 + u$*

$$L_n(\varphi_1(\cdot)^{-m}, x) \leq C(m)\frac{1}{(1+x)^m}. \tag{9.6.3}$$

PROOF. We first observe that

$$n\int_{k/n}^{k+1/n} \frac{1}{(1+x)^m}dx \leq \left(1 + \frac{k}{n}\right)^{-m}$$

and therefore it is sufficient to prove (9.6.3) for V_n. For $V_n(\varphi_1^{-m}, x)$ we have

$$\sum_{k=0}^{\infty} v_{n,k}(x)\left(1 + \frac{k}{n}\right)^{-m}$$

$$= \frac{1}{(1+x)^m}\sum_{k=0}^{\infty} v_{n-m,k}(x)\frac{(n+k-1)!(n-m-1)!}{(n-m+k-1)!(n-1)!}\left(1 + \frac{k}{n}\right)^{-m}$$

$$\leq \frac{C(m)}{(1+x)^m}\sum_{k=0}^{\infty} v_{n-m,k}(x) = \frac{C(m)}{(1+x)^m}. \quad \square$$

142   9. Exponential-Type or Bernstein-Type Operators

We will also need the following lemma about $P_n(f,x)$.

Lemma 9.6.3. *For an integer m and $\varphi(x) = x$ we have*

$$P_n(\varphi^{-m}, x) \leq C\varphi(x)^{-m}. \tag{9.6.4}$$

PROOF. We write

$$P_n(\varphi^{-m}, x) = \frac{(n/x)^n}{(n-1)!} \int_0^\infty e^{-nu/x} u^{n-m-1}\, du$$

$$= x^{-m} \frac{n^m}{(n-1)!} \int_0^\infty e^{-z} z^{n-m-1}\, dz$$

$$= x^{-m} n^m \frac{(n-m-1)!}{n!} \leq c(m)/x^m. \qquad \square$$

PROOF OF LEMMA 9.5.2 (For $1 < p \leq \infty$). We define the maximal function

$$M(G,x) = \sup_u \left| \frac{1}{x-u} \int_u^x G(r)\, dv \right| \quad \text{where } G(r) = \varphi(v)^{2r} f^{(2r)}(v).$$

We use Lemma 9.6.1 to estimate $L_n(R_{2r}(f,\cdot,x),x)$ by the terms

$$\frac{1}{(2r-1)!} L_n((\cdot - x)^{2r} \varphi(x)^{-2r}, x) M(G,x)$$

and

$$\frac{1}{(2r-1)!} L_n((\cdot - x)^{2r-2} \varphi(x)^{-2r+2} x^{-1} \varphi_1(\cdot)^{-1}, x) M(G,x).$$

(The second expression has to be estimated only for the Baskakov and Post–Widder operators.) Taking the $L_p(E_n)$ norm, we have

$$\|L_n((\cdot - x)^{2r} \varphi(x)^{-2r}, x) M(G,x)\|_{L_p(E_n)}$$

$$\leq \left\| L_n\left(\frac{(\cdot - x)^{2r}}{\varphi(x)^{2r}}, x\right) \right\|_{L_\infty(E_n)} \|M(G,x)\|_{L_p(D)}$$

and

$$\left\| L_n\left(\frac{(\cdot - x)^{2r}}{\varphi(x)^{2r-2} x \varphi_1(\cdot)}, x\right) M(G,x) \right\|_{L_p(E_n)}$$

$$\leq \left\| L_n\left(\frac{(\cdot - x)^{2r}}{\varphi(x)^{2r-2} x \varphi_1(\cdot)}, x\right) \right\|_{L_\infty(E_n)} \|M(G,x)\|_{L_p(D)}.$$

Recalling now Lemma 9.4.4, we have

$$\|L_n((\cdot - x)^{2r} \varphi(x)^{-2r}, x)\|_{L_\infty(E_n)} \leq M n^{-r},$$

9.6. Estimate of $\|L_n(R_{2r}(f,\cdot,x),x)\|_{L_p(E_n)}$ 143

and recalling (9.6.2), (9.6.3), and (9.6.4), we have

$$\left\|L_n\left(\frac{(\cdot-x)^{2r}}{\varphi(x)^{2r-2}x\varphi_1(\cdot)},x\right)\right\|_{L_\infty(E_n)}$$

$$\leq \left\|\frac{1}{\varphi(x)^{2r-2}x}\{L_n((\cdot-x)^{4r},x)\}^{1/2}\{L_n(\varphi_1(\cdot)^{-2},x)\}^{1/2}\right\|_{L_\infty(E_n)} \leq Mn^{-r}.$$

We complete the proof observing that for $p > 1$ we may use the inequality $\|M(G)\|_p \leq M_1\|G\|_p$ and obtain

$$\|M(G,x)\|_{L_p(D)} \leq M_1\|\varphi(x)^{2r}f^{(2r)}\|_{L_p(D)}. \qquad \square$$

We are now in the position to prove Lemma 9.5.2 for $p = 1$ for B_n^* and S_n^*; however, for V_n^* and P_n we will need two additional lemmas, which we will prove first.

Lemma 9.6.4. *For* $v_{n,k}(x) = \binom{n+k-1}{k}x^k(1+x)^{-n-k}$ *and any integer* l *we have*

$$\int_{n^2}^\infty \left\{\sum_{k=0}^{[nx/3]} + \sum_{[2nx]+1}^\infty\right\} v_{n,k}(x) n \int_{k/n}^{k+1/n} |f(u)|\,du\,dx \leq Mn^{-l}\|f\|_{L_1(R_+)} \quad (9.6.5)$$

and

$$\int_1^\infty \left\{\sum_{k=0}^{[nx/3]} + \sum_{[2nx]+1}^\infty\right\} v_{n,k}(x)\left|\frac{k}{n}-x\right|^i |f^{(i)}(x)|\,dx \leq Mn^{-l}\|x^i f^{(i)}(x)\|_{L_1[1,\infty)}. \quad (9.6.6)$$

PROOF. To prove (9.6.6) we use Lemma 9.4.4 and observe that for $x \geq 1$

$$\left\{\sum_{k=0}^{[nx/3]} + \sum_{[2nx]+1}^\infty\right\} v_{n,k}(x)\left|\frac{k}{n}-x\right|^i \leq (2x/3)^m \sum_{k=0}^\infty v_{n,k}(x)\left|\frac{k}{n}-x\right|^{m+i}$$

$$\leq Cx^{-m}\left(\frac{x(1+x)}{n}\right)^{(m+i)/2}$$

$$\leq C_1 n^{-(m+i)/2} x^i.$$

To prove (9.6.5) we write

$$\int_{n^2}^\infty \left\{\sum_{k=0}^{[nx/3]} + \sum_{[2nx]+1}^\infty\right\} v_{n,k}(x) n \int_{k/n}^{k+1/n} |f(u)|\,du\,dx$$

$$\leq \sum_{k=0}^\infty \left\{\int_{\max(n^2,3k/n)}^\infty v_{n,k}(x)\,dx\right\} n \int_{k/n}^{k+1/n} |f(u)|\,du$$

$$+ \sum_{k=2n^3}^\infty \left\{\int_0^{k/2n} v_{n,k}(x)\,dx\right\} n \int_{k/n}^{k+1/n} |f(u)|\,du.$$

We now estimate all coefficients of $n\int_{k/n}^{k+1/n}|f(u)|\,du$ to be $O(n^{-l-1})$. Using the Stirling formula $n! \sim n^{n+1/2}e^{-n}$, and recalling that $\xi^k/(1+\xi)^{n+k}$ is increasing for $\xi < k/n$, we write for $k > 2n^3$

$$\int_0^{k/2n} v_{n,k}(x)\,dx = \binom{n+k-1}{k}\int_0^{k/2n} \frac{x^k}{(1+x)^{n+k}}\,dx$$

$$= \binom{n+k-1}{k}\frac{k}{2n}\frac{\xi^k}{(1+\xi)^{n+k}}$$

$$\leq C\frac{(n+k-1)^{n+k-1/2}}{k^{k+1/2}(n-1)^{n-1/2}}\frac{k}{2n}\frac{\zeta^k}{(1+\zeta)^{n+\zeta}}$$

$$\leq C_1 n^{-1/2}\binom{n+k-1}{k}^k\binom{n+k-1}{n-1}^{n-1}\frac{(k/2n)^k}{(1+(k/2n))^{n+k}}\frac{k}{2n}$$

$$\leq C_2 n^{-1/2} e^n \left(\frac{k}{n}\right)^{n-1} \frac{(k/2n)}{\left(1+\frac{2n}{k}\right)^k \left(\frac{k}{2n}\right)^n} \leq C_2 n^{-1/2} e^{-n} 2^n.$$

For the other terms we have

$$J_{k,n} \equiv \binom{n+k-1}{k}\int_{\max(n^2,3k/n)}^\infty \frac{x^k}{(1+x)^{n+k}}\,dx \leq \binom{n+k-1}{k}\frac{1}{n}\frac{1}{(1+y)^{n-1}},$$

where $y = \max(n^2, 3k/n)$. For $k \geq n^3/3$ we have

$$J_{k,n} \leq Cn^{-3/2}e^{n-1}\left(\frac{k}{n}\right)^{n-1}\left(\frac{n}{3k}\right)^{n-1} \leq Cn^{-3/2}\left(\frac{e}{3}\right)^{n-1}.$$

For $n < k < n^3/3$ we have for $n \geq n_0$

$$J_{k,n} \leq Cn^{-3/2}\left(1+\frac{1}{k/(n-1)}\right)^k \binom{n+k-1}{n-1}^{n-1}\left(\frac{1}{1+n^2}\right)^{n-1}$$

$$\leq Cn^{-3/2}e^{n-1}((1+3/n^2)/3(1+n^{-2}))^{n-1} \leq Cn^{-3/2}\left(\frac{e}{3}\right)^{n-1}.$$

For $k \leq n$ and $n > 3$

$$J_{k,n} \leq (2n)^k n^{-1}(1+n^2)^{-n+1} \leq 2(2n/(1+n^2))^{n-1} \leq 2(3/5)^{n-1}. \quad \square$$

Lemma 9.6.5. *For* $j = -1, 0, 1, 2, \ldots$

$$\frac{1}{(n-1)!}\left(\frac{n}{x}\right)^n \left\{\int_0^{x/2} + \int_{2x}^\infty\right\} e^{-nu/x} u^{n-1} u^j\,du \leq Cx^j (0.9)^n. \qquad (9.6.7)$$

9.6. Estimate of $\|L_n(R_{2r}(f, \cdot, x), x)\|_{L_p(E_n)}$

PROOF. Actually we separate (9.6.7) into the following two estimates:

I. $\dfrac{1}{(n-1)!}\left(\dfrac{n}{x}\right)^n \displaystyle\int_0^{x/2} e^{-nu/x} u^{n-1} u^j \, du = \dfrac{n^n}{(n-1)!} x^j \displaystyle\int_0^{1/2} e^{-nz} z^{n+j-1} \, dz$

$\leq \dfrac{n^n}{(n-1)!} x^j e^{-n/2} (1/2)^{n+j}$

$\leq C n^n e^{n-1} e^{-n/2} (1/2)^n x^j / ((n-1)^{n-1} n^{1/2})$

$\leq C_1 n^{1/2} (e^{1/2}/2)^n x^j$

$\leq C_2 n^{1/2} (0.85)^n x^j$

$\leq C_3 (0.9)^n x^j.$

II. $\dfrac{1}{(n-1)!}\left(\dfrac{n}{x}\right)^n \displaystyle\int_{2x}^\infty e^{-nu/x} u^{n-1} u^j \, du = \dfrac{x^j n^n}{(n-1)!} \displaystyle\int_2^\infty e^{-nz} z^{n+j-1} \, dz$

$\leq \dfrac{x^j n^n}{(n-1)!} 2^{n+j-1} e^{-(n-1)2} \displaystyle\int_2^\infty e^{-z} \, dz$

$\leq C \dfrac{x^j n^n}{(n-1)!} 2^{n+j-1} e^{-2n}$

$\leq C_1 n \left(\dfrac{2}{e}\right)^n x^j$

$\leq C_1 (0.9)^n x^j \qquad \text{for } n > j.$ □

PROOF OF LEMMA 9.5.2 FOR $p = 1$. This estimate for $p = 1$ is one of the most involved in this chapter. We will first prove the result for B_n^* and then discuss the alterations needed and overcome the additional difficulties for S_n^*, V_n^*, and P_n. We will try to use notations and steps which are mutual to all cases. In fact the major additional difficulties for V_n^* and P_n are the reasons for Lemmas 9.6.4 and 9.6.5 respectively.

For
$$g(v) = \varphi(v)^{2r} |f^{(2r)}(v)| \equiv v^r (1-v)^r |f^{(2r)}(v)|$$

we have, using Lemma 9.6.1,

$\displaystyle\int_{E_n} |B_n^*(R_{2r}(f, \cdot, x), x)| \, dx$

$\leq \dfrac{1}{(2r-1)!} \displaystyle\int_{E_n} \sum_{k=0}^n p_{n,k}(x)(n+1) \displaystyle\int_{k/n+1}^{k+1/n+1} \dfrac{|u-x|^{2r-1}}{x^r(1-x)^r} \left|\displaystyle\int_x^u g(v) \, dv\right| du \, dx.$

Using

$(n+1)\displaystyle\int_{k/n+1}^{k+1/n+1} |(u-x)|^{2r-1} \, dx \leq C_1 \left(\left|\dfrac{k}{n} - x\right|^{2r-1} + n^{-2r+1}\right),$

we write

$$\int_{E_n} |B_n^*(R_{2r}(f,\cdot,x),x)|\,dx$$

$$\leq C \int_{E_n} \sum_{k=0}^{n} p_{n,k}(x)\varphi(x)^{-2r}\left(\left|\frac{k}{n}-x\right|^{2r-1}+n^{-2r+1}\right)\left|\int_x^{k^*/n+1} g(v)\,dv\right|dx$$

$$\equiv C \cdot I_n, \tag{9.6.8}$$

where k^* is either k or $k+1$ such that

$$\left|\int_x^{k^*/n+1} g(v)\,dv\right| = \max_{j=k,k+1}\left|\int_x^{j/n+1} g(v)\,dv\right|.$$

We now define $D(l,n,x)$ (the definition holds for S_n^* and V_n^* as well) by

$$D(l,n,x) = \left\{k;\, l\varphi(x)n^{-1/2} \leq \left|\frac{k}{n}-x\right| < (l+1)\varphi(x)n^{-1/2}\right\}. \tag{9.6.9}$$

This enables us to rewrite I_n (of (9.6.8)) which is the term we have to evaluate, as follows:

$$I_n = \int_{E_n} \varphi(x)^{-2r} \sum_{l=0}^{\infty} \sum_{k \in D(l,n,x)} p_{n,k}(x)\left(\left|\frac{k}{n}-x\right|^{2r-1}+n^{-2r+1}\right)$$

$$\times \left|\int_x^{k^*/n+1} \varphi(v)^{2r}|f^{(2r)}(v)|\,dv\right|dx. \tag{9.6.10}$$

We observe that for $P_{n,k}$ equal to $p_{n,k}$, $s_{n,k}$, or $v_{n,k}$ with $\varphi(x)^2$ equal to $x(1-x)$, x or $x(1+x)$ respectively, we have for $x \in E_n$

$$\sum_{k \in D(l,n,x)} P_{n,k}(x)\left(\left|\frac{k}{n}-x\right|^{2r-1}+n^{-2r+1}\right) \leq \frac{C}{(l+1)^4}(\varphi(x)n^{-1/2})^{2r-1}. \tag{9.6.11}$$

This follows for $l \geq 1$ and $x \in E_n$, from (cf. Lemma 9.4.4)

$$\sum_{k \in D(l,n,x)} P_{n,k}(x)\left(\left|\frac{k}{n}-x\right|^{2r-1}+n^{-2r+1}\right)$$

$$\leq \frac{n^2}{l^4\varphi(x)^4} \sum P_{n,k}(x)\left(\left|\frac{k}{n}-x\right|^{4+2r-1}+n^{-2r+1}\left|\frac{k}{n}-x\right|^4\right)$$

$$\leq C_1 \frac{1}{l^4} \frac{n^2}{\varphi(x)^4}(\varphi(x)n^{-1/2})^4((\varphi(x)n^{-1/2})^{2r-1}+n^{-2r+1})$$

$$\leq C_2 \frac{1}{l^4}(\varphi(x)n^{-1/2})^{2r-1} \leq C_1 \frac{1}{(l+1)^4}(\varphi(x)n^{-1/2})^{2r-1}.$$

As $x \in E_n$, we have

$$\sum_{k \in D(0,n,x)} P_{n,k}(x)\left(\left|\frac{k}{n}-x\right|^{2r-1}+n^{-2r+1}\right) \leq C_4[(\varphi(x)n^{-1/2})^{2r-1}+n^{-2r+1}]$$

$$\leq C_5(\varphi(x)n^{-1/2})^{2r-1}.$$

9.6. Estimate of $\|L_n(R_{2r}(f,\cdot,x),x)\|_{L_p(E_n)}$

We now define

$$F(l,x) \equiv \left\{v; v \in D, |v-x| \leq (l+1)\varphi(x)n^{-1/2} + \frac{1}{n}\right\} \quad \text{and} \quad (9.6.12)$$

$$G(l,v) \equiv \{x; x \in E_n, v \in F(l,x)\}$$

and obtain

$$I_n \leq Cn^{-r} \sum_{l=0}^{\infty} \frac{1}{(l+1)^4} \int_{E_n} \frac{n^{1/2}}{\varphi(x)} \int_{F(l,x)} \varphi(v)^{2r} |f^{(2r)}(v)| \, dv \, dx$$

$$\leq C_1 n^{-r} \sum_{l=0}^{\infty} \frac{1}{(l+1)^4} \int_0^1 \varphi(v)^{2r} |f^{(2r)}(v)| n^{1/2} \left\{\int_{G(l,v)} \varphi(x)^{-1} \, dx\right\} dv. \quad (9.6.13)$$

We will now recall for $l \geq n^{1/4}$ that for B_n^* $G(l,v) \subset E_n \subset [0,1]$ and $\int_0^1 \varphi(x)^{-1} \, dx = \int_0^1 (x(1-x))^{-1/2} \, dx \leq B$, and therefore,

$$n^{1/2} \sum_{l \geq n^{1/4}} \frac{1}{(l+1)^4} \int_0^1 \varphi(v)^{2r} |f^{(2r)}(v)| \int_{G(l,v)} \varphi(x)^{-1} \, dx \, dv \leq B_1 \|\varphi(v)^{2r} f^{(2r)}(v)\|.$$

We will show that for $h \leq h_0$

$$\int_{x; |v-x| \leq h\varphi(x)+h^2} \varphi(x)^{-1} \, dx \leq Mh, \quad (9.6.14)$$

where M is independent of v and $h \leq h_0$. We will prove the result for $0 < v \leq 1/2$ and the proof is similar for $1/2 \leq v < 1$. If $v \leq Ch^2$ then $x \leq h\varphi(x) + (C+1)h^2$ and therefore $x \leq 2(C+1)h^2$. Choosing $C = 7$, we have $\int_0^{16h^2} \varphi(x)^{-1} \, dx \leq Mh$. For $v \geq 7h^2$ we have $x + h\varphi(x) \geq 6h^2$ and therefore $x \geq 4h^2$ which implies $2h\varphi(x) < x$. We now have for $x \geq 4h^2$ $x/4 \leq (x/2) - h^2 \leq x - h\varphi(x) - h^2 \leq v \leq x + h\varphi(x) + h^2 \leq 2x$ and therefore $x/4 \leq v \leq \min(2x, (1/2))$ which implies $C_1^{-1}\varphi(x) \leq \varphi(v) \leq C_1\varphi(x)$ and

$$m\{x; |v-x| \leq h\varphi(x) + h^2\} \leq m\{x; |v-x| \leq C_1 h\varphi(v) + h^2\}$$

$$\leq m\{x; |v-x| \leq C_2 h\varphi(v)\} \leq 2C_2 h\varphi(v).$$

Therefore,

$$\int_{x: |v-x| \leq h\varphi(x)+h^2} \varphi(x)^{-1} \, dx \leq \frac{C_1}{\varphi(v)} 2C_2 h\varphi(x) \leq Mh.$$

Choosing $h = (l+1)/n^{1/2}$ and recalling $h^2 \geq 1/n$, we have

$$\int_{G(l,v)} \varphi(x)^{-1} \, dx \leq Mn^{-1/2}(l+1).$$

This estimate substituted in (9.6.13) completes the proof of our result for $L_n = B_n^*$.

148   9. Exponential-Type or Bernstein-Type Operators

For the Szász–Kantorovich operator S_n^* we write

$$I_n \equiv \int_{E_n} |S_n^*(R_{2r}(f,\cdot,x),x)|\,dx$$

$$\leq C \int_{E_n} \sum_{k=0}^{\infty} s_{n,k}(x)\varphi(x)^{-2r} \left[\left|\frac{k}{n} - x\right|^{2r-1} + n^{-2r+1} \right] \left| \int_x^{k^*/n} g(v)\,dv \right| dx, \tag{9.6.15}$$

where $g(v) = \varphi(v)^{2r}|f^{(2r)}(v)|$, $\varphi(v)^2 = v$, k^* is either k or $k+1$ and

$$\left|\int_x^{k^*/n} g(v)\,dv\right| = \max_{j=k,k+1} \left|\int_x^{j/n} g(v)\,dv\right|.$$

We now follow (9.6.9) through (9.6.12) and have

$$I_n \leq Cn^{-r} \sum_{l=0}^{\infty} \frac{1}{(l+1)^4} \int_0^{\infty} g(v) n^{1/2} \int_{G(l,v)} \varphi(x)^{-1}\,dx\,dv \equiv J_1(n) + J_2(n).$$

We now prove (9.6.14) for $\varphi(x) = \sqrt{x}$, which follows in exactly the same fashion as that for $\varphi(x) = \sqrt{x(1-x)}$ and completes the proof in the same way. (Here the whole domain replaces $0 < v \leq 1/2$ in the proof for B_n^*.)

For V_n^*, the Baskakov–Kantorovich operator, we will have to split the estimate of $V_n^*(R_{2r}(f,\cdot,x),x)$ into estimates on two domains, that is $((A/n), Bn^2)$ (with fixed $B \geq 1$) and (n^2, ∞). We define the auxiliary operator

$$\tilde{V}_n^*(f,x) = \sum_{k=[nx/3]}^{[2nx]} v_{n,k}(x) n \int_{k/n}^{k+1/n} f(u)\,du. \tag{9.6.16}$$

For B such that $Bn^2 \geq n_{r-1}^2 \geq n_i^2$ (see Definition (9.2.7)(a) of $L_{n,r}$) we have, using Lemma 9.6.4,

$$\|\tilde{V}_n^*(f,x) - V_n^*(f,x)\|_{L_1[Bn^2,\infty)} \leq Mn^{-r}\|f\|_1,$$

and for $j \leq 2r$

$$\|\tilde{V}_n^*((\cdot - x)^j f^{(j)}(x), x) - V_n^*((\cdot - x)^j f^{(j)}(x), x)\|_{L_1[Bn^2,\infty)}$$
$$\leq Mn^{-r}\|x^j f^{(j)}(x)\|_{L_1[1,\infty)} \leq M_1 n^{-r}(\|f\|_1 + \|\varphi^{2r} f^{(2r)}\|_1).$$

Therefore, using Taylor's formula, we have

$$\|\tilde{V}_n^*(R_{2r}(f,\cdot,x),x) - V_n^*(R_{2r}(f,\cdot,x),x)\|_{L_1[Bn^2,\infty)}$$
$$\leq \|\tilde{V}_n^*(f,x) - V_n^*(f,x)\|_{L_1[Bn^2,\infty)} + \sum_{j=0}^{2r-1} \|f^{(j)}(x)\{\tilde{V}_n^*((\cdot-x)^j,x)$$
$$- V_n^*((\cdot-x)^j,x)\}\|_{L_1[Bn^2,\infty)} \leq M_2 n^{-r}(\|f\|_1 + \|\varphi^{2r} f^{(2r)}\|_1).$$

Following the above, we will have to estimate $\|V_n^*(R_{2r}(f,\cdot,x),x)\|_{L_1[A/n,Bn^2]}$ and $\|\tilde{V}_n^*(R_{2r}(f,\cdot,x),x)\|_{L_1[n^2,\infty)}$, where the overlapping of the domain is intentional and necessary, and at the same time does not cause difficulties. Using Lemma 9.6.1 (9.6.2), the above expressions will be estimated by

9.6. Estimate of $\|L_n(R_{2r}(f,\cdot,x),x)\|_{L_p(E_n)}$

$$M\int \sum v_{n,k}(x)x^{-r}(1+x)^{-r+1}\left[\frac{1}{1+x}+\frac{1}{1+(k/n)}\right]\left[\left|\frac{k}{n}-x\right|^{2r-1}+n^{-2r+1}\right]$$

$$\times \left|\int_x^{k^*/n} g(v)\,dv\right|dx,$$

where the integral is on $[A/n, Bn^2]$ and $[n^2,\infty)$ and the sums are $\sum_{n=0}^{\infty}$ and $\sum_{k=[nx/3]}^{[2nx]}$ for estimating V_n^* and $\tilde{V}_n^*$ respectively, and where $g(v) = \varphi(v)^{2r}|f^{(2r)}(v)|$ and $\varphi(v)^2 = v(1+v)$. Moreover, we have

$$v_{n,k}(x)/1+\frac{k}{n} = v_{n-1,k}(x)\frac{1}{1+x}\frac{n+k-1}{n-1}\frac{n}{n+k} \le 2v_{n-1,k}(x)\frac{1}{1+x}$$

and therefore having $[(1/1+x)+(1/1+(k/n))]$ in the above expression, we use the estimate (9.6.11) and another similar expression with $n-1$ replacing n, that is

$$\sum_{k\in D(l,n,x)} v_{n-1,k}(x)\left(\left|\frac{k}{n}-x\right|^{2r-1}+n^{-2r+1}\right) \le \frac{C}{(l+1)^4}(\varphi(x)n^{-1/2})^{2r-1}. \quad (9.6.11)'$$

This follows very similarly the proof of (9.6.11). For $l \ge 1$ for instance, we assume n is big enough ($n > 2r+3$) and $x \in E'_{n-1} = (A/2(n-1),\infty)$, and therefore, $x \in E_n$. We use the fact that Lemma 9.4.4 is valid for E_n defined for any positive A and show

$$\sum_{k\in D(l,n,x)} v_{n-1,k}(x)\left(\left|\frac{k}{n}-x\right|^{2r-1}+n^{-2r+1}\right)$$

$$\le \frac{n^2}{l^4\varphi(x)^4}\sum_{k=0}^{\infty}v_{n-1,k}(x)\left(\left|\frac{k}{n}-x\right|^{2r+3}+n^{-2r+1}\left|\frac{k}{n}-x\right|^4\right)$$

$$\le C\frac{n^2}{l^4\varphi(x)^4}\sum_{k=0}^{\infty}v_{n-1,k}(x)\left(\left|\frac{k}{n-1}-x\right|^{2r+3}+n^{-2r+1}\left|\frac{k}{n-1}-x\right|^4\right)$$

$$+ \max\left(1,\left(\frac{k}{n}\right)^{2r+3}\right)n^{-2r-3}\right)$$

$$\le C_1\frac{n^2}{(l+1)^4\varphi(x)^4}\left[\left(\frac{\varphi(x)}{(n-1)^{1/2}}\right)^{2r+3}+n^{-2r+1}\left(\frac{\varphi(x)}{(n-1)^{1/2}}\right)^4+n^{-2r-3}\right]$$

$$+ C_1\frac{n^{-2r-1}}{(l+1)^4\varphi(x)^4}\sum_{k=n}^{\infty}v_{n-1,k}(x)\left(\frac{k}{n}\right)^{2r+3}$$

$$\le \frac{C_2}{(l+1)^4}\left(\frac{\varphi(x)}{n^{1/2}}\right)^{2r-1}+C_3\frac{n^{-2r-1}x^{2r+3}}{(l+1)^4\varphi(x)^4}\sum_{k=n}^{\infty}v_{n+2r+2,k-2r-3}(x)$$

$$\le \frac{C_4}{(l+1)^4}\left(\frac{\varphi(x)}{n^{1/2}}\right)^{2r-1},$$

where in the last step we used $(k/n)^j v_{n,k}(x) \le C_5 x^j v_{n+j,k-j}(x)$, $x < \varphi(x)^{1/2}$, and $n\varphi(x) > A$ in E_n. Therefore, following (9.6.9) through (9.6.13) for V_n^* we have

$$\|V_n^*(R_{2r}(f,\cdot,x))\|_{L_1[An^{-1}, Bn^2]}$$

$$\le Cn^{-r} \sum_{l=0}^{\infty} \frac{1}{(l+1)^4} \int_0^{\infty} \varphi(v)^{2r} |f^{(2r)}(v)| n^{1/2} \int_{G(l,v) \cap [A/n, Bn^2]} \varphi(x)^{-1} \, dx \, dv.$$

To estimate the last expression we notice that

$$\int_{G(l,v) \cap [A/n, Bn^2]} \varphi(x)^{-1} \, dx \le \int_0^{Bn^2} \varphi(x)^{-1} \, dx \le M \log n$$

which we will use for $l \ge n^{1/4}$, and the estimate will follow as

$$\sum_{l \ge n^{1/4}} \frac{1}{(l+1)^4} n^{1/2} \log n \le M.$$

For $l \le n^{1/4}$ we actually have (9.6.14) as in this case $(l+1)/n^{1/2} \le h_0$. For the estimate of $\tilde{V}_n$ on $[n^2, \infty)$ we follow (9.6.14) to complete the estimate for $l \le n^{1/4}$. For $l > n^{1/4}$ we split the domain of the index set of k $D(l, n, x) \cap \{k; k \in [[nx/3], 2nx]\}$ into two, $\tilde{D}_1(l, n, x)$ and $\tilde{D}_2(l, n, x)$, which will be subsets of $\{k; k \in [[nx/3], x]\}$ and $\{k; k \in (x, 2nx]\}$ respectively. Following the same argument as before, we have to estimate

$$n^{-r} \sum_{l=0}^{b_j n^{1/2}} \frac{1}{(l+1)^4} \int_0^{\infty} \varphi(v)^{2r} |f^{(2r)}(v)| n^{1/2} \int_{\tilde{G}_j(l,v)} \varphi(x)^{-1} \, dx \, dv,$$

where $b_1 = 2/3$, $b_2 = 1$, $\tilde{F}_1(l, x) = \{v; x - (l+1)n^{-1/2}\varphi(x) - n^{-1} \le v \le x\}$, $\tilde{F}_2(l, x) = \{v; x \le v \le x + (l+1)n^{-1/2}\varphi(x) + n^{-1}\}$, and $\tilde{G}_j(l, v) = \{x; x \in [n^2, \infty), v \in \tilde{F}_j(l, x)\}$. For $x \in \tilde{G}_1(l, v)$ we have $x \ge v$ and $\sup \tilde{G}_1(l, v) \le 4v$ as $v > x - (l+1)n^{-1/2}\varphi(x) - n^{-1} \ge (1/(3+\delta))x$ and hence $\int_{\tilde{G}_1(l,v)} \varphi(x)^{-1} \le (1/v)\int_v^{4v} dx \le 3$. Similarly, for $\tilde{G}_2(l, v)$ $\inf \tilde{G}_2(l, v) \ge v/(2+\delta)$ and $x \in \tilde{G}_2(l, v)$ implies $x \le v$, and therefore $\int_{\tilde{G}_2(l,v)} \varphi(x)^{-1} \, dx \le 2$. We use these estimates for $l \ge n^{1/4}$. This concludes the proof of our lemma for the Baskakov–Kantorovich operator.

To prove our lemma for the Post–Widder operator P_n we have to estimate $P_n(R_{2r}(f, \cdot, x), x)$. We define

$$\tilde{P}_n(f, x) = \frac{1}{(n-1)!} \left(\frac{n}{x}\right)^n \int_{x/2}^{2x} e^{-nu/x} u^{n-1} f(u) \, du.$$

We will show first that

$$\|P_n(R_{2r}(f,\cdot,x), x) - \tilde{P}_n(R_{2r}(f,\cdot,x), x)\|_1 \le C_1 n^{-r}(\|\varphi^{2r} f^{(2r)}\|_1 + \|f\|_1),$$

and therefore, it is enough to estimate $\|\tilde{P}_n(R_{2r}(f,\cdot,x), x)\|_1$. To do this we use Fubini's Theorem and the computation of Lemma 9.6.5 and obtain

9.6. Estimate of $\|L_n(R_{2r}(f,\cdot,x),x)\|_{L_p(E_n)}$ 151

$$\|P_n(f,x) - \tilde{P}_n(f,x)\|_1$$

$$\leq \frac{1}{(n-1)!} \int_0^\infty \left(\frac{n}{x}\right)^n \left\{\int_0^{x/2} + \int_{2x}^\infty\right\} e^{-nu/x} u^{n-1} |f(u)|\, du\, dx$$

$$= \int_0^\infty |f(u)| \frac{1}{(n-1)!} \left\{\int_0^{u/2} + \int_{2u}^\infty\right\} \left(\frac{n}{x}\right)^n e^{-nu/x} u^{n-1}\, dx\, du$$

$$= \int_0^\infty |f(u)| \frac{n^n}{(n-1)!} \left\{\int_2^\infty + \int_0^{1/2}\right\} e^{-nz} z^{n-2}\, dz\, du \leq C_2 (0.9)^n \|f\|_1.$$

On the other hand, using Taylor's formula and Lemma 9.6.5, we write

$$\|P_n(f,x) - \tilde{P}_n(f,x) - \{P_n(R_{2r}(f,\cdot,x),x) - \tilde{P}_n(R_{2r}(f,\cdot,x),x)\}\|_1$$

$$\leq \sum_{i=0}^{2r-1} \int_0^\infty |f^{(i)}(x)| \frac{(n/x)^n}{(n-1)!} \left\{\int_0^{x/2} + \int_{2x}^\infty\right\} e^{-nu/x} u^{n-1} |u-x|^i\, du\, dx$$

$$\leq C_3 \sum_{i=0}^{2r-1} (0.9)^n \int_0^\infty |f^{(i)}(x)| x^i\, dx \leq C_4 (0.9)^n (\|\varphi^{2r} f^{(2r)}\|_1 + \|f\|_1).$$

Therefore, to obtain $\|P_n(R_{2r}(f,\cdot,x),x)\|_1 \leq Mn^{-r}(\|\varphi^{2r} f^{(2r)}\|_1 + \|f\|_1)$ we have only to prove $\|\tilde{P}_n(R_{2r}(f,\cdot,x)\|_1 \leq C_3 n^{-r}(\|\varphi^{2r} f^{(2r)}\|_1 + \|f\|_1)$. Using Lemma 9.6.1, we have to show

$$\int_0^\infty \frac{1}{(n-1)!} \left(\frac{n}{x}\right)^n \int_{x/2}^{2x} e^{-nu/x} u^{n-1} |u-x|^{2r-1} x^{-2r-1} \left[\frac{1}{u} + \frac{1}{x}\right]$$

$$\times \left|\int_u^x v^{2r} |f^{(2r)}(v)|\, dv\right|\, du\, dx \leq Mn^{-r} \|\varphi^{2r} f^{(2r)}\|_1. \qquad (9.6.17)$$

The technique of proving (9.6.17) is similar to earlier estimates, but in the following lemma we will prove a somewhat more general result which will be needed in Section 9.7. □

Lemma 9.6.6. *For* $g \in L_1(R)$, $g \geq 0$ *and any integer* j *we have*

$$I_n \equiv n^{j/2} \int_0^\infty \frac{x^{-j+1}}{(n-1)!} \left(\frac{n}{x}\right)^n \int_{x/2}^{2x} e^{-nu/x} u^{n-1} |u-x|^{j-1} \left[\frac{1}{u} + \frac{1}{x}\right] \left|\int_u^x g(v)\, dv\right| du\, dx$$

$$\leq M \|g\|_1. \qquad (9.6.18)$$

PROOF. The proof follows the proof of Lemma 9.5.2 ($p = 1$) for other operators. We choose $D(l,n,x) = \{u: lxn^{-1/2} \leq |u-x| \leq (l+1)xn^{-1/2}\}$, and $D_j(l,n,x)$ is the intersection of $D(l,n,x)$ with $\{u: x/2 \leq u \leq x\}$ and $\{u: x \leq u \leq 2x\}$ for $j = 1$ and $j = 2$ respectively. We also define $F_1(l,n,x) = \{v: x\max(1-(l+1)n^{-1/2}, 1/2) \leq v \leq x\}$, $F_2(l,n,x) = \{v: x \leq v \leq x\min(2, 1+(l+1)n^{-1/2})\}$, and $G_j(l,n,v) = \{x; v \in F_j(l,n,x)\}$. We recall that for $x/2 \leq u \leq 2x$ we have $(1/u) + (1/x) \leq 3/x$ and obtain for $l \geq 1$

$$\frac{n^{j/2}}{(n-1)!}x^{-j+1}\left(\frac{n}{x}\right)^n \int_{\substack{D(l,n,x)\\x/2<u<2x}} e^{-nu/x}u^{n-1}|u-x|^{j-1}\left(\frac{1}{u}+\frac{1}{x}\right)du$$

$$\leq 3n^{j/2}x^{-j+1}\frac{1}{l^4x^4n^{-2}}\frac{1}{(n-1)!}\left(\frac{n}{x}\right)^n \int_0^\infty e^{-nu/x}u^{n-1}|u-x|^{j+3}x^{-1}\,du$$

$$\leq \frac{3n^{(j+4)/2}x^{-j-4}}{l^4}P_n(|\cdot - x|^{j+3},x) \leq \frac{Cn^{1/2}}{(l+1)^4x},$$

as

$$P_n(|\cdot - x|^s, x) \leq \{P_n((\cdot - x)^{2s}, x)\}^{1/2} \leq C_1 x^s/n^{s/2}.$$

To estimate the integral on $u \in D(0, n, x)$ we replace $D(0, n, x)$ by $[0, \infty)$ and use the above method. It is easy to see, following the technique used for other operators, that

$$I_n \leq M\left\{\sum_{l=0}^{(1/2)n^{1/2}}\frac{1}{(l+1)^4}\int_0^\infty g(v)\left\{n^{1/2}\int_{G_1(l,n,v)}\frac{1}{x}dx\right\}dv\right.$$

$$\left.+\sum_{l=0}^{n^{1/2}}\frac{1}{(l+1)^4}\int_0^\infty g(v)\left\{n^{1/2}\int_{G_2(l,n,v)}\frac{dx}{x}\right\}dv\right\}.$$

For $l \geq n^{1/4}$ we use the estimates

$$\int_{G_1(l,n,v)}\frac{1}{x}dx \leq \frac{1}{v}\int_v^{2v}dx \leq 2 \quad \text{and} \quad \int_{G_2(l,n,v)}\frac{1}{x}dx \leq \frac{2}{v}\int_{v/2}^v dx = 1$$

(which are valid for all l). For $l \leq n^{1/4}$ we write for $G_1(l, v)$,

$$x/2 \leq v \leq x, \quad x_1 = \inf G_1(l, v) = v,$$

$$x_2 = \sup G_1(l, v) \leq v/(1 - (l+1)n^{-1/2}),$$

and

$$\int_{G_1(l,v)}\frac{dx}{x} \sim (l+1)n^{-1/2},$$

which, together with

$$\int_{G_2(l,v)}\frac{dx}{x} \sim (l+1)n^{-1/2},$$

completes the proof of our lemma. $\square$

9.7. The Estimate $\|\varphi(x)^{2r}L_n^{(2r)}(f)\|_{L_p} \leq M\|\varphi^{2r}f^{(2r)}\|_p$

In this section we obtain the estimate (9.3.7). Many of the preliminary results that we need have already been proved. We will prove the results for P_n after we treat the other cases.

9.7. The Estimate $\|\varphi(x)^{2r}L_n^{(2r)}(f)\|_{L_p} \le M\|\varphi^{2r}f^{(2r)}\|_p$

We use (9.4.3) and (9.4.4) as well as definitions of B_n^*, S_n^*, and V_n^*. We obtain for B_n^*

$$x^r(1-x)^r B_n^{*(2r)}(f,x)$$
$$= \frac{n!}{(n-2r)!}\sum_{k=0}^{n-2r} x^r(1-x)^r p_{n-2r,k}(x)\Delta^{2r}a_k(n+1)$$
$$= \sum_{k=0}^{n-2r} p_{n,k+r}(x)(k+r)\cdots(k+1)(n-r-k)\cdots(n-2r-k+1)\Delta^{2r}a_k(n+1).$$

Therefore,

$$|x^r(1-x)^r B_n^{*(2r)}(f,x)| \le C(n^r p_{n,r}(x)|\Delta^{2r}a_0(n+1)|$$
$$+ n^r p_{n,n-r}(x)|\Delta^{2r}a_{n-2r}(n+1)| \quad (9.7.1)$$
$$+ n^{2r}\sum_{k=1}^{n-2r-1} p_{n,k+r}(x)\left(\frac{k}{n}\left(1-\frac{k}{n}\right)\right)^r|\Delta^{2r}a_k(n+1)|),$$

and when B_n replaces B_n^*, $\vec{\Delta}_{1/n}^{2r}f(k/n)$ replaces $\Delta^{2r}a_k(n+1)$. Similarly,

$$x^r S_n^{*(2r)}(f,x) = n^{2r}\sum_{k=0}^{\infty} x^r s_{n,k}(x)\Delta^{2r}a_k(n)$$
$$= n^{2r}\sum_{k=0}^{\infty} s_{n,k+r}(x)\left(\frac{k}{n}+\frac{1}{n}\right)\cdots\left(\frac{k}{n}+\frac{r}{n}\right)\Delta^{2r}a_k(n),$$

and therefore

$$|x^r S_n^{*(2r)}(f,x)| \le C\left(n^r s_{n,r}(x)|\Delta^{2r}a_0(n)|\right.$$
$$\quad (9.7.2)$$
$$\left. + n^{2r}\sum_{k=1}^{\infty} s_{n,k+r}(x)\left(\frac{k}{n}\right)^r|\Delta^{2r}a_k(n)|\right).$$

For $V_n^*(f,x)$ we have

$$x^r(1+x)^r V_n^{*(2r)}(f,x) = \frac{(n+2r-1)!}{(n-1)!}\sum_{k=0}^{\infty}\Delta^{2r}a_k(n)x^r(1+x)^r v_{n+2r,k}(x)$$
$$= n^{2r}\sum_{n=0}^{\infty}\Delta^{2r}a_k(n)v_{n,k+r}(x)\left(\frac{k}{n}+\frac{1}{n}\right)\cdots\left(\frac{k}{n}+\frac{r}{n}\right)$$
$$\times\left(1+\frac{k}{n}+\frac{r}{n}\right)\cdots\left(1+\frac{k}{n}+\frac{2r-1}{n}\right),$$

and therefore,

$$|x^r(1+x)^r V_n^{*(2r)}(f,x)| = C\left(n^r v_{n,r}(x)|\Delta^{2r}a_0(n)|\right.$$
$$\quad (9.7.3)$$
$$\left. + n^{2r}\sum_{k=1}^{\infty} v_{n,k+r}(x)\left(\frac{k}{n}\right)^r\left(1+\frac{k}{n}\right)^r|\Delta^{2r}a_k(n)|\right).$$

In case we have S_n or V_n rather than S_n^* or V_n^*, $\vec{\Delta}_{1/n}^{2r} f(k/n)$ replaces $\Delta^{2r} a_k(n)$ in (9.7.2) and (9.7.3) respectively.

Using Taylor's formula, we write

$$f\left(\frac{k+i}{n} + \theta\right) = f\left(\frac{k}{n} + \theta\right) + \frac{i}{n} f'\left(\frac{k}{n} + \theta\right) + \cdots$$

$$+ \frac{1}{(2r-1)!} \left(\frac{i}{n}\right)^{2r-1} f^{(2r-1)}\left(\frac{k}{n} + \theta\right)$$

$$+ \frac{1}{(2r-1)!} \int_0^{i/n} \left(\frac{i}{n} - y\right)^{2r-1} f^{(2r)}\left(\frac{k}{n} + \theta + y\right) dy.$$

Therefore,

$$|\Delta^{2r} a_k(n)| \leq Cn \int_0^{1/n} \sup_{1 \leq i \leq 2r} \int_0^{i/n} \left|y - \frac{i}{n}\right|^{2r-1} \left|f^{(2r)}\left(\frac{k}{n} + \theta + y\right)\right| dy\, d\theta$$

$$\leq Cn^{-2r+1} \int_{k/n}^{(k+2r+1)/n} |f^{(2r)}(u)|\, du.$$

Moreover, with $k = 2r$ and $i = -1, -2, \ldots, -2r$, we have

$$|\Delta^{2r} a_0(n)| \leq C/n \int_0^{1/n} \sup_{1 \leq i \leq 2r} \int_{-i/n}^0 \left|y + \frac{i}{n}\right|^{2r-1} \left|f^{(2r)}\left(\frac{2r}{n} + y + \theta\right)\right| dy\, d\theta$$

$$\leq Cn \int_0^{1/n} \sup_{1 \leq i \leq 2r} \int_0^{i/n} u^{2r-1} \left|f^{(2r)}\left(\frac{2r-i}{n} + u + \theta\right)\right| du\, d\theta$$

$$\leq C \int_0^{(2r+1)/n} u^{2r-1} |f^{(2r)}(u)|\, du \leq Cn^{-r+1} \int_0^{(2r+1)/n} u^r |f^{(2r)}(u)|\, du.$$

We now estimate $x^r(1+x)^r V_n^{*(2r)}$ using $(k/n)(1+(k/n)) \leq ((k/n) + u) \times (1 + (k/n) + u)$, formula (9.7.3), and the above to obtain

$$\left| x^r(1+x)^r \left(\frac{d}{dx}\right)^{2r} V_n^*(f, x) \right|$$

$$\leq C\left(n^{2r} \sum_{k=1}^{\infty} v_{n,k+r}(x) \left(\frac{k}{n}\right)^r \left(1 + \frac{k}{n}\right)^r |\Delta^{2r} a_k(n)| + n^r v_{n,r}(x) |\Delta^{2r} a_0(n)|\right) \quad (9.7.4)$$

$$\leq C_1 n \sum_{k=0}^{\infty} v_{n,k+r}(x) \int_{k/n}^{(k+2r+1)/n} \varphi(u)^{2r} |f^{(2r)}(u)|\, du.$$

Therefore,

$$\left\| x^r(1+x)^r \left(\frac{d}{dx}\right)^{2r} V_n^*(f, x) \right\|_{L_1}$$

$$\leq C \frac{n}{n-1} \sum_{k=0}^{\infty} \int_{k/n}^{(k+2r+1)/n} \varphi(u)^{2r} |f^{(2r)}(u)|\, du$$

$$\leq C_1 \|\varphi^{2r} f^{(2r)}\|_{L_1}$$

9.7. The Estimate $\|\varphi(x)^{2r}L_n^{(2r)}(f)\|_{L_p} \le M\|\varphi^{2r}f^{(2r)}\|_p$

and

$$\left\|x^r(1+x)^r\left(\frac{d}{dx}\right)^{2r} V_n^*(f,x)\right\|_{L_\infty}$$

$$\le C \sum_{k=0}^{\infty} v_{n,k+r}(x)n \int_{k/n}^{(k+2r+1)/n} \varphi(u)^{2r}|f^{(2r)}(u)|\,du$$

$$\le C_1\|\varphi^{2r}f^{(2r)}\|_{L_\infty}.$$

Using the Riesz–Thorin interpolation theorem, we get

$$\left\|x^r(1+x)^r\left(\frac{d}{dx}\right)^{2r} V_n^*(f,x)\right\|_{L_p} \le C_1\|\varphi^{2r}f^{(2r)}\|_{L_p}, \qquad 1 \le p \le \infty.$$

To complete the proof for the other cases we have to observe

(a) $|\Delta^{2r}a_{n-2r}(n+1)| \le Cn^{-r+1}\int_{1-((2r+1)/(n+1))}^{1}(1-u)^r|f^{(2r)}(u)|\,du$ and $(k/n)\times(1-(k/n)) \le C_1((k/(n+1))+y)(1-(k/(n+1))-y)$ for $0 < k < n-2r$ and $0 < y < (2r+1)/(n+1)$ ($C_1 = 2(2r-1)$ will do) for B_n^*.
(b) $k/n < (k/n) + y$ (and other estimates which we have already achieved) for S_n^*.
(c) For S_n, V_n, and B_n we use

$$|\vec{\Delta}_{1/n}^{2r}f(k/n)| \le Cn^{-2r+1}\int_0^{2r/n}\left|f^{(2r)}\left(\frac{k}{n}+u\right)\right|du,$$

$$|\vec{\Delta}_{1/n}^{2r}f(0)| \le Cn^{-r+1}\int_0^{2r/n} u^r|f^{(2r)}(u)|\,du,$$

and

$$\left|\overleftarrow{\Delta}_{1/n}^{2r}f\left(\frac{n-2r}{n}\right)\right| \le Cn^{-2r+1}\int_{1-2r/n}^{1}(1-u)^r|f^{(2r)}(u)|\,du.$$

Also one has $(k/n)(1-(k/n)) \le C((k/n)+y)(1-(k/n)-y)$ for $0 < k < n-2r$ and $0 < y < 2r/n$.

From these observations we obtain

$$\left|x^r(1+x)^r\left(\frac{d}{dx}\right)^{2r} B_n^*(f,x)\right| \le Cn \sum_{k=0}^{n-2r} p_{n,k+r}(x) \int_{k/n+1}^{(k+2r+1)/n+1} \varphi(u)^{2r}|f^{(2r)}(u)|\,du \qquad (9.7.5)$$

and

$$\left|x^r\left(\frac{d}{dx}\right)^{2r} S_n^*(f,x)\right| \le Cn \sum_{k=0}^{\infty} s_{n,k+r}(x) \int_{k/n}^{(k+2r+1)/n} \varphi(u)^{2r}|f^{(2r)}(u)|\,du. \qquad (9.7.6)$$

The rest follows in the same way as the above consideration for V_n^*.

The operators V_n, B_n, and S_n can replace V_n^*, B_n^*, and S_n^* on the left hand side of (9.7.4), (9.7.5), and (9.7.6) respectively, in which case the internal integral on the right hand side of these equations is

$$\int_{k/n}^{(k+2r)/n} \varphi(u)^{2r}|f^{(2r)}(u)|\,du.$$

To conclude, we have obtained the following result:

Theorem 9.7.1. *For L_n, being one of $B_n^*, S_n^*, V_n^*, B_n, S_n,$ and V_n, for $f^{(2r-1)} \in A.C._{\text{loc}}$ and $f^{(2r)} \in L_p(D)$ (for $B_n, S_n,$ and V_n we have $p = \infty$ only and assume $f \in C$), we have*

$$\|\varphi(x)^{2r} L_n^{(2r)}(f,x)\|_{L_p(D)} \le C\|\varphi^{2r} f^{(2r)}\|_{L_p(D)}, \tag{9.7.7}$$

where φ relates to L_n according to Definition 9.3.1.

We now prove our result for $P_n(f, x)$.

Lemma 9.7.2. *For positive integer r we have*

$$\left\| x^{2r} \left(\frac{d}{dx}\right)^{2r} P_n(f,x) \right\|_{L_p(R_+)} \le M \|x^{2r} f^{(2r)}\|_{L_p(R_+)}. \tag{9.7.8}$$

PROOF. Expanding $f(u)$ by

$$f(u) = \sum_{i=0}^{2r-1} \frac{1}{i!}(u-x)^i f^{(i)}(x) + \frac{1}{(2r-1)!}\int_u^x (v-u)^{2r-1} f^{(2r-1)}(v)\,dv$$

and recalling that $P_n((\cdot - x)^i, x)$ is a polynomial in x of degree i, we have

$$\left(\frac{d}{dx}\right)^{2r} P_n(f,x) = \left(\frac{d}{dx}\right)^{2r} P_n(R_{2r}(f,\cdot,x), x).$$

Using Lemma 9.6.1, we have

$$|R_{2r}(f,u,x)| = \frac{1}{(2r-1)!}\left|\int_u^x (v-u)^{2r-1} f^{(2r)}(v)\,dv\right|$$

$$\le \frac{|x-u|^{2r-1}}{(2r-1)! x^{2r-1}}\left[\frac{1}{x}+\frac{1}{u}\right]\left|\int_u^x v^{2r}|f^{(2r)}(v)|\,dv\right|.$$

For $1 < p \le \infty$ we follow the result of the last section and (9.4.11) to get

$$\left(\frac{d}{dx}\right)^{2r} P_n(f,x) = \sum_{i=0}^{2r} Q_i(n,x) P_n((\cdot - x)^i R_{2r}(f,\cdot,x), x),$$

where $x^{2r}|Q_i(n,x)| \le Cn^{r+(i/2)}/x^i$. In other words, we have to estimate for $0 \le i \le 2r$

$$n^{r+(i/2)}\|x^{-i} P_n((\cdot - x)^i R_{2r}(f,\cdot,x), x)\|_{L_p(R_+)} \quad \text{for } 1 \le p \le \infty.$$

We set $g(v) = v^{2r} f^{(2r)}(v)$ and $M(g,x)$ its maximal function, and write for $1 < p \le \infty$

9.7. The Estimate $\|\varphi(x)^{2r}L_n^{(2r)}(f)\|_{L_p} \le M\|\varphi^{2r}f^{(2r)}\|_p$

$$n^{r+(i/2)}\|x^{-i}P_n((\cdot-x)^iR_{2r}(f,\cdot,x),x)\|_{L_p}$$
$$\le C_1 n^{r+(i/2)}\|x^{-2r-i}P_n(|\cdot-x|^{i+2r},x)\|_{L_\infty(R_+)}\|M(g,x)\|_{L_p}$$
$$\le C_1 n^{r+(i/2)}\left\|x^{-2r-i+1}P_n\left(|\cdot-x|^{i+2r}\frac{1}{\varphi(\cdot)},x\right)\right\|_{L_\infty(R_+)}\|M(g,x)\|_{L_p}.$$

Recalling Lemmas 9.4.4 and 9.6.3, we have

$$P_n(|\cdot-x|^{i+2r},x) \le P_n((\cdot-x)^{2i+4r},x)^{1/2} = O(xn^{-1/2})^{i+2r}$$

and

$$P_n\left(|\cdot-x|^{i+2r}\frac{1}{\varphi(\cdot)},x\right) \le P_n((\cdot-x)^{2i+4r},x)^{1/2}P_n(\varphi(\cdot)^{-2},x)^{1/2}$$
$$= O((xn^{-1/2})^{i+2r}/x),$$

and we obtain our result for $1 < p \le \infty$, using the inequality $\|M(g)\|_p \le C_p\|g\|_p$. For $p = 1$ we have to estimate for $0 \le i \le 2r$ the integrals

$$\int_0^\infty n^{r+(i/2)}x^{-i-2r+1}\left(\frac{n}{x}\right)^n \frac{1}{(n-1)!}\int_0^\infty e^{-nu/x}u^{n-1}|u-x|^{2r+i-1}$$
$$\times \left[\frac{1}{u}+\frac{1}{x}\right]\left|\int_u^x v^{2r}|f^{(2r)}(v)|dv\right|du\,dx.$$

Using the technique of Lemmas 9.6.5 and 9.6.6,

$$n^{r+(i/2)}\int_0^\infty x^{-2r-i+1}\left(\frac{n}{x}\right)^n \frac{1}{(n-1)!}\int_{x/2}^{2x} e^{-nx/u}u^{n-1}|u-x|^{2r+i-1}\left[\frac{1}{u}+\frac{1}{x}\right]$$
$$\times \left|\int_u^x v^{2r}|f^{(2r)}(v)|dv\right|du\,dx \le M\|v^{2r}f^{(2r)}(v)\|_{L_1}. \qquad \square$$

CHAPTER 10
WEIGHTED APPROXIMATIONS BY EXPONENTIAL-TYPE OPERATORS

Direct and converse results will be obtained for the rate of weighted approximation. The results for $w(x) = 1$ were proved earlier by V. Totik, and were extended in a different direction in the last chapter.

10.1. The Direct and Inverse Result

We first outline the operators, their definition, their related step-weight function $\varphi(x)$ and weight function $w(x)$, all of which will be given in the following table.

Definition 10.1.1. L_n, φ, w, p, and the domain of L_n are related by

$L_n(f, x)$	$B_n^*(f, x)$	$S_n^*(f, x)$	$V_n^*(f, x)$	$G_n(f, x)$
defined in	(9.2.1)	(9.2.2)	(9.2.3)	(9.1.11)
related $w(x)$	$x^{\gamma(0)}(1-x)^{\gamma(1)}$	$x^{\gamma(0)}(1+x)^{\gamma(\infty)}$	$x^{\gamma(0)}(1+x)^{\gamma(\infty)}$	$x^{\gamma(0)}(1+x)^{\gamma(\infty)}$
related $\varphi(x)^2$	$x(1-x)$	x	$x(1+x)$	x^2
domain of L_n	$wf \in L_p[0,1]$	$wf \in L_p(R_+)$	$wf \in L_p(R_+)$	$wf \in L_p(R_+)$

where $\gamma(\infty)$ is arbitrary and $\gamma(i)$ ($i = 0, 1$) satisfies $-1/p < \gamma(i) < 1 - (1/p)$ for $1 \leq p \leq \infty$. For $p = 1$ and $p = \infty$, $\gamma(0)$ or $\gamma(1)$ may also be equal to zero and for $G_n(f)$ $\gamma(0)$ is arbitrary.

We remark here that $\gamma(i) > -1/p$ is needed so that $wL_n f \in L_p$, and $\gamma(i) < 1 - 1/p$ is needed in the implication $wf \in L_p \Rightarrow f \in L_1[0, 1]$ which is used in the definition of $L_n = B_n^*$, S_n^*, or V_n^*.

10.1. The Direct and Inverse Result

We can now state the direct and converse theorems for the various operators in Definition 10.1.1.

Theorem 10.1.2. *Suppose* $wf \in L_p[0,1]$ *and either* $1 \le p \le \infty$ *and* $\alpha < 1$, *or* $1 < p < \infty$ *and* $\alpha \le 1$. *Then for w and φ related to B_n^* by Definition 10.1.1.*

$$\|w(B_n^* f - f)\|_{L_p[0,1]} = O(n^{-\alpha}) \Leftrightarrow \|w\Delta_{h\varphi}^2 f\|_{L_p[2h^2, 1-2h^2]} = O(h^{2\alpha}). \quad (10.1.1)$$

Theorem 10.1.3. *Suppose* $wf \in L_p(R_+)$ *and either* $1 \le p \le \infty$ *and* $\alpha < 1$ *or* $1 < p < \infty$ *and* $\alpha \le 1$, *then for* $L_n = S_n^*$ *or* $L_n = V_n^*$ *and both w and φ related to them by Definition 10.1.1 we have*

$$\|w(L_n f - f)\|_{L_p(R_+)} = O(n^{-\alpha}) \Leftrightarrow \|w\Delta_{h\varphi}^2 f\|_{L_p[2h^2, \infty]} = O(h^{2\alpha}). \quad (10.1.2)$$

Theorem 10.1.4. *Suppose* $wf \in L_p(R_+)$, $1 \le p \le \infty$, *and* $\alpha \le 1$, *then for w and φ related to G_n by Definition 10.1.1 we have*

$$\|w(G_n f - f)\|_{L_p(R_+)} = O(n^{-\alpha}) \Leftrightarrow \|w\Delta_{h\varphi}^2 f\|_{L_p(R_+)} = O(h^{2\alpha}). \quad (10.1.3)$$

Remark 10.1.5. (a) The saturation problem $\alpha = 1$ for $B = L_1$ remains open for $w(x) \ne 1$.

(b) Since the proof near 0, 1 or ∞ can be separated, we actually solved the case $\gamma(i) = 0$ in the last chapter. During the proof we will at times be faced with the need to deal with that case separately, and while not difficult, we will not do it, as for such a case more general theorems were already proved.

The proof of Theorems 10.1.2, 10.1.3, and 10.1.4 for $\alpha < 1$ will follow a similar line to the proof of Theorem 9.3.2; that is, we will prove the inequalities

$$\|wL_n f\|_{L_p(D)} \le A(p, w) \|wf\|_{L_p(D)}, \quad (10.1.4)$$

$$\left\| w(x)\varphi(x)^2 \left(\frac{d}{dx}\right)^2 L_n(f, x) \right\|_{L_p(D)} \le B(p, w)n \|wf\|_{L_p(D)}, \quad (10.1.5)$$

$$\left\| w(x)\varphi(x)^2 \left(\frac{d}{dx}\right)^2 L_n(f, x) \right\|_{L_p(D)} \quad (10.1.6)$$
$$\le C(p, w)(\|w\varphi^2 f''\|_{L_p(D)} + \|wf\|_{L_p(D)}),$$

and

$$\|w(x)(L_n f - f)\|_{L_p(D)} \le D(p, w)n^{-1}(\|w\varphi^2 f''\|_{L_p(D)} + \|wf\|_{L_p(D)}). \quad (10.1.7)$$

For $\gamma(1) = 0$, $p = \infty$ and the operators B_n^*, S_n^*, and V_n^* we will not prove (10.1.7) (which is probably not valid in that case) but because of Remark 10.1.5 (b) it would not be necessary for $\alpha < 1$. In this section we will show how our theorems for $\alpha < 1$ and the implication "$\Leftarrow$" for $\alpha = 1$ and $p < \infty$ follow from (10.1.4), (10.1.5), (10.1.6), and (10.1.7), and how (10.1.7) follows from

$$\|w(x)(L_n f - f)\|_{L_p(E_n)} \le D(p, w)n^{-1}(\|w\varphi^2 f''\|_{L_p(D)} + \|f\|_{L_p(D)}), \quad (10.1.8)$$

where $E_n = (A/n, 1 - (A/n))$ for $L_n = B_n^*$, $E_n = (A/n, \infty)$ for $L_n = S_n^*$, or $L_n = V_n^*$ and $E_n = R_+$ for $L_n = G_n$. In Sections 10.2, 10.3, 10.4, and 10.5 we will demonstrate (10.1.4), (10.1.5), (10.1.6), and (10.1.8) respectively, for the various operators (a highly uneven division of labor) and in Section 10.6 we will prove the saturation result ($\alpha = 1$). We will use many estimates from the last chapter. We do not deal here with combinations; however, we have to overcome the added difficulty of a weight function.

For the proof of Theorems 10.1.2 and 10.1.3 we observe that the right hand side of (10.1.1) and (10.1.2) imply $\|w\Delta_{h\varphi}^2 f\|_{L_p[2h^2, 1-2h^2]} = O(h^{2\alpha})$ and that all φ in question satisfy $\varphi(x) \geq \sqrt{x(1-x)}$ in $[0,1]$. Therefore, we may now use Theorem 6.2.1 and Corollary 8.2.5 applied to best weighted polynomial approximation in $[0,1]$ to obtain $K_{2,\varphi}(f, t^2)_{w, L_p[0, 3/4]} = O(t^{2\alpha})$. Similarly, we deal with the interval $[1/4, 1]$ and $L_n = B_n^*$ and obtain $K_{2,\varphi}(f, t^2)_{w, L_p[1/4, 1]} = O(t^{2\alpha})$ for $\varphi(x)^2 = x(1-x)$. In $[1/4, \infty]$ and for the operators S_n^* and V_n^* $\|w\Delta_{h\varphi}^2 f\|_{L_p(2h^2, \infty)} = O(h^{2\alpha})$ implies $K_{2,\varphi}(f, t^2)_{w, L_p[1/4, \infty)} = O(t^{2\alpha})$ via the main equivalence theorem. (See Theorem 6.2.1 (6.2.6) and observe that $K_{2,\varphi}(f, t^2)_{w, L_p(1/4, \infty)} \leq \mathscr{K}_{2,\varphi}(1, f, t^2)_{w, p}$.) Therefore, following (2.2.14) (the patching together technique), we have $K_{2,\varphi}(f, t^2)_{w, L_p(D)} = O(t^{2\alpha})$. Writing $f = f - g_t + g_t$ where g_t satisfies

$$\|f - g_t\|_{L_p(D)} \leq Mt^{2\alpha} \quad \text{and} \quad t^2\|\varphi^2 g_t^{(2)}\|_{L_p(D)} \leq Mt^{2\alpha},$$

setting $t = 1/\sqrt{n}$ and using the estimate (10.1.4) for $f - g_t$ and (10.1.8) for g_t we have

$$\|w(L_n f - f)\|_{L_p(E_n)} = O(n^{-\alpha}).$$

To show that

$$\|w(L_n f - f)\|_{L_p(D)} = O(n^{-\alpha})$$

we follow the same process as in Section 9.3 and observe that for B_n^* we simply use (8.1.4) with the weight that is certainly a Jacobi weight, with $g_{1/\sqrt{n}} = P_{[\sqrt{n}]}$ and the interval $[0, 1]$. For S_n^* and V_n^* we use the weight $x^{\gamma(0)}$ and best polynomial approximation in $[0, 2]$ and observe that $x^{\gamma(0)}(1 + x)^{\gamma(\infty)}$ would make only a slight change (and only if $\gamma(0) + \gamma(\infty) > 0$) in the proof used in Section 9.3. For G_n, (10.1.4) and (10.1.7) yield that

$$K_{2,\varphi}(f, t^2)_{w, p} \leq Mt^{2\alpha} \quad \text{implies} \quad \|w(G_n f - f)\|_{L_p(R_+)} = O(n^{-\alpha}).$$

As in this case $\varphi(x) = x$, Section 6.2 implies the equivalence of $K_{2,\varphi}(f, t^2)_{w, p} \leq Mt^{2\alpha}$ with the right hand side of (10.1.3) and therefore the implication "⇐" there follows. (Up to this point for $p < \infty$, $\alpha \leq 1$.)

To prove the other direction we use (10.1.5) and (10.1.6) to derive

$$K_{2,\varphi}(f, n^{-1})_{w, p} \leq \|w(L_k f - f)\|_p + M\frac{k}{n} K_{2,\varphi}(f, k^{-1})_{w, p} + Mn^{-1}\|wf\|_p \tag{10.1.9}$$

which will be used for $k < n$. This inequality and the obvious inequality $K_{2,\varphi}(f, t^2)_p \le \|wf\|_p$, lead via the Berens–Lorentz lemma [1] (Lemma 9.3.4) to $K_{2,\varphi}(f, t^2)_{w,p} = O(t^{2\alpha})$ and therefore $\Omega_\varphi^2(f, t)_{w,p} = O(t^{2\alpha})$. Following the idea that was used in the proof of Theorem 9.3.6, namely proving

$$\|w\varphi^4 L_n^{(4)}(f)\|_p \le Cn^2 \|wf\|_p$$

and

$$\|w\varphi^4 L_n^{(4)}(f)\|_p \le C(\|w\varphi^4 f^{(4)}\|_p + \|wf\|_p)$$

together with a suitable form of Marchaud's inequality, one can also get

$$\Omega_\varphi^2(f, n^{-1/2})_{w,p} \le C_1 K_{2,\varphi}(f, n^{-1})_{w,p} \tag{10.1.10}$$

$$\le C_2 n^{-1} \sum_{k=1}^n \|w(L_k f - f)\|_p + C_2 n^{-1} \|wf\|_p.$$

However, the corresponding direct estimate is missing; therefore, we omit the details.

10.2. The Boundedness of the Operators in Weighted Norm

In this section we will prove the inequality

$$\|wL_n f\|_p \le A(p, w)\|wf\|_p.$$

We will give the estimate regarding $G_n(f, x)$ at the end. We define $w_i(k, n)$ in relation to $w(x)$ by

$$w_1(k, n) = \left(\frac{k+1}{n}\right)^{\gamma(0)} \left(1 + \frac{k}{n}\right)^{\gamma(\infty)} \quad \text{if} \quad w(x) = x^{\gamma(0)}(1 + x)^{\gamma(\infty)}$$

$$w_2(k, n) = \left(\frac{k+1}{n}\right)^{\gamma(0)} \left(1 - \frac{k-1}{n}\right)^{\gamma(1)} \quad \text{if} \quad w(x) = x^{\gamma(0)}(1 - x)^{\gamma(\infty)}. \tag{10.2.1}$$

The following computational lemma will be useful.

Lemma 10.2.1. *Suppose* $w(x) = x^{\gamma(0)}(1 + x)^{\gamma(\infty)}$, *where* $-1/p < \gamma(0) < 1 - (1/p)$ *or* $\gamma(0) = 0$, $\gamma(\infty)$ *is arbitrary,* $g \ge 0$ *and* $wg \in L_p$. *Then* $g \in L_1[0, a]$ *for all a and*

$$\left| n \int_{k/n}^{k+1/n} g(u) \, du \right|^p \le Cw_1(k, n)^{-p} n \int_{k/n}^{k+1/n} |w(u) g(u)|^p \, du, \quad p < \infty$$

$$n \int_{k/n}^{k+1/n} g(u) \, du \le Cw_1(k, n)^{-1} \|wg\|_\infty, \quad p = \infty. \tag{10.2.2}$$

Suppose $w(x) = x^{\gamma(0)}(1 - x)^{\gamma(1)}$, *where* $-1/p < \gamma(i) < 1 - (1/p)$ *or* $\gamma(i) = 0$, $g(x) \ge 0$ *and* $wg \in L_p[0, 1]$. *Then* $g \in L_1[0, 1]$ *and*

$$\left|(n+1)\int_{k/n+1}^{k+1/n+1} g(u)\,du\right|^p \le Cw_2(k,n)^{-p}n\int_{k/n+1}^{k+1/n+1} |w(u)g(u)|^p\,du, \quad p<\infty$$
(10.2.3)
$$(n+1)\int_{k/n+1}^{k+1/n+1} g(u)\,du \le Cw_2(k,n)^{-1}\|wg\|_\infty, \quad p=\infty.$$

PROOF. For $k>0$, (10.2.2) is clear as $w_1(k,n) \sim w(u)$ in $(k/n, (k+1)/n)$. Similarly, for $0<k<n$, (10.2.3) is clear as $w_2(k,n) \sim w(u)$ in $(k/(n+1), (k+1)/(n+1))$. To prove (10.2.2) for $k=0$, we write for $p<\infty$

$$n\int_0^{1/n} g(u)\,du \le \left(n\int_0^{1/n} |w(u)g(u)|^p\,du\right)^{1/p}\left(n\int_0^{1/n} w(u)^{-q}\,du\right)^{1/q}$$

and

$$\left(n\int_0^{1/n} w(u)^{-q}\,du\right)^{1/q} \le cn^{\gamma(0)};$$

and for $p=\infty$

$$n\int_0^{1/n} g(u)\,du \le \|wg\|_\infty n\int_0^{1/n} w(u)^{-1}\,du \le cn^{\gamma(0)}\|wg\|_\infty.$$

Similarly, we prove the case $k=0$ or $k=n$ of the inequality (10.2.3). □

Theorem 10.2.2. *Suppose L_n is B_n^*, S_n^*, or V_n^*. Then we have (10.1.4).*

PROOF. Write $P_{n,k}(x)$ for $p_{n,k}(x)$, $s_{n,k}(x)$, or $v_{n,k}(x)$ (for B_n^*, S_n^*, and V_n^* respectively) and $w_i(k,n)$ where $i=2$ for B_n^* and $i=1$ for S_n^* and V_n^*. Lemma 10.2.1 implies for $p=\infty$

$$\|w(x)L_n(f,x)\|_\infty \le C\|\sum P_{n,k}(x)(w(x)/w_i(k,n))\|_\infty \|wf\|_\infty.$$

For $1 \le p < \infty$, using the Jensen inequality and $\sum P_{n,k}(x) = 1$, we write

$$\|w(x)L_n(f,x)\|_p \le C\left\|\left\{\sum P_{n,k}(x)(w(x)/w_i(k,n))^p n \int_{I(k,n)} |w(u)f(u)|^p\,du\right\}^{1/p}\right\|_p,$$

where $I(k,n) = [k/(n+1), (k+1)/(n+1)]$ for B_n^* and $I(k,n) = [k/n, (k+1)/n]$ for S_n^* and V_n^*.

To complete the proof we have to show

$$\sum P_{n,k}(x)(w(x)/w_i(k,n)) \le C \quad \text{and} \quad w_i(k,n)^{-p}\int_D P_{n,k}(x)w(x)^p\,dx \le C/n$$

for $p=\infty$ and $p<\infty$ respectively. For $p=\infty$ we note that $0 \le \gamma(0) < 1$ and use $\sum P_{n,k}(x) = 1$ and Hölder's inequality to show for $w_1(k,n)$, for instance,

$$\sum P_{n,k}(x)\left(\frac{n}{k+1}\right)^{\gamma(0)}\left(\frac{n}{n+k}\right)^{\gamma(\infty)} \le \left\{\sum P_{n,k}(x)\left(\frac{n}{k+1}\right)^l\right\}^\mu$$

$$\times \left\{\sum P_{n,k}(x)\left(\frac{n}{n+k}\right)^m\right\}^\nu,$$

10.2. The Boundedness of the Operators in Weighted Norm

where if $\gamma(0) > 0$ and $\gamma(\infty) \neq 0$, we set $l \in N$, $l \geq \gamma(0)$; $\mu = \gamma(0)/l$, $m \in Z$, $m/\gamma(\infty) \geq 1/(l - \gamma(0))$, and $\nu = \gamma(\infty)/m$; if $\gamma(0) = 0$ and $\gamma(\infty) \neq 0$, we write $l = 0$, $m \in Z$, and $\nu^{-1} = m/\gamma(\infty) \geq 1$; if $\gamma(\infty) = 0$ and $\gamma(0) > 0$, we write $m = 0$ and $\mu^{-1} = l/\gamma(0) \geq 1$; and if $\gamma(\infty) = \gamma(0) = 0$ we write $l = m = 0$. Similarly, we deal with $w_2(k, n)$. This implies that it is sufficient to show

$$\sum P_{n,k}(x)\left(\frac{n}{k+1}\right)^l \leq Cx^{-l},$$

$$\sum_{k=0}^{n} p_{n,k}(x)\left(\frac{n}{n-k+1}\right)^l \leq C(1-x)^{-l} \tag{10.2.4}$$

for $l \in N$ ($l = 0, 1, 2, \ldots$) and

$$\sum s_{n,k}(x)\left(1 + \frac{n}{k}\right)^m \leq C(1+x)^m, \quad \sum v_{n,k}(x)\left(1 + \frac{k}{n}\right)^m \leq C(1+x)^m$$

for $m \in Z$. These inequalities are known in part (see Lemma 9.6.2 and Ditzian [1, p. 280]), and the other cases follow similar computation. We now prove the result needed for $1 \leq p < \infty$, i.e.,

$$w_i(k, n)^{-p} \int_D P_{n,k}(x) w(x)^p \, dx \leq C/n. \tag{10.2.5}$$

For $L_n = V_n^*$ or $L_n = S_n^*$ and both $\gamma(0) \neq 0$ and $\gamma(\infty) \neq 0$ we choose $\eta > -1$ such that $0 < p\gamma(0)/\eta < 1$ and $m \in Z$ such that $0 < \gamma(\infty)p/m(1 - \gamma(0)p/\eta) \leq 1$, and write, using Hölder's inequality and $\int P_{n,k}(x) \, dx = n^{-1}(1 + o(1))$,

$$\int P_{n,k}(x) w_1(x)^p \, dx$$

$$= \int P_{n,k}(x) x^{\gamma(0)p}(1 + x)^{\gamma(\infty)p} \, dx$$

$$\leq \left\{\int P_{n,k}(x) x^\eta \, dx\right\}^{p\gamma(0)/\eta} \left\{\int P_{n,k}(x)(1 + x)^{\gamma(\infty)p\eta/(\eta - \gamma(0)p)} \, dx\right\}^{1 - (p\gamma(0)/\eta)}$$

$$\leq \left\{\int P_{n,k}(x) x^\eta \, dx\right\}^{p\gamma(0)/\eta} \left\{\int P_{n,k}(x)(1 + x)^m \, dx\right\}^{p\gamma(\infty)/m}$$

$$\times n^{-1 + (p\gamma(0)/\eta) + (\gamma(\infty)p/m)}(1 + o(1)).$$

For $\gamma(0) = 0$ we choose m satisfying $\operatorname{sgn} m = \operatorname{sgn} \gamma(\infty)$ and $|m| \geq |\gamma(\infty)|_p$ to obtain

$$\int P_{n,k}(x) w(x)^p \, dx \leq \left\{\int P_{n,k}(x)(1 + x)^m \, dx\right\}^{\gamma(\infty)/m} n^{-1 + (\gamma(\infty)/m)}(1 + o(1)),$$

and for $\gamma(\infty) = 0$ we have

$$\int P_{n,k}(x) w(x)^p \, dx = \int P_{n,k}(x) x^{\gamma(0)p} \, dx.$$

For $L_n = B_n^*$ in case $\gamma(0) \neq 0$ and $\gamma(1) \neq 0$, we have for η satisfying $\eta > -1$ and $0 < p\gamma(0)/\eta < 1$ and $m \in Z$ satisfying $0 < \gamma(1)p/m(1 - \gamma(0)p/\eta) \leq 1$

$$\int_0^{1/2} P_{n,k}(x) w_2(x)^p \, dx \leq \left\{ \int_0^1 P_{n,k}(x) x^\eta \, dx \right\}^{p\gamma(0)/\eta} \left\{ \int_0^{1/2} P_{n,k}(x)(1-x)^m \, dx \right\}^{p\gamma(1)/m}$$
$$\times n^{-1+(\gamma(0)p/\eta)+(\gamma(1)p/m)}(1 + o(1)).$$

For negative m and $(\gamma(1))$ we use

$$\int_0^{1/2} P_{n,k}(x)(1-x)^m \, dx \leq 2^{|m|} \int_0^1 P_{n,k}(x) \, dx \leq Bn^{-1}$$

and for $m \geq 0$ we use

$$\int_0^{1/2} P_{n,k}(x)(1-x)^m \, dx \leq \int_0^1 P_{n,k}(x)(1-x)^m \, dx \leq n^{-1}.$$

Similarly, we deal with the integral on $[1/2, 1]$ where the roles of $\gamma(0)$ and $\gamma(1)$ are interchanged. The cases $\gamma(0) = 0$ and $\gamma(1) = 0$ for which we may refer to Remark 10.1.5 (b) (since the behavior near 0 and 1 can be separated) can also be treated in this way.

From these considerations it is clear that it is enough to show

$$\int P_{n,k}(x) x^\eta \, dx \leq C(\eta) \left(\frac{k+1}{n} \right)^\eta \frac{1}{n} \quad \text{for } \eta > -1, \tag{10.2.6}$$

$$\int_0^\infty s_{n,k}(x)(1+x)^m \, dx \leq C(m) \frac{1}{n} \left(1 + \frac{k}{n} \right)^m,$$
$$\int_0^\infty v_{n,k}(x)(1+x)^m \, dx \leq C(m) \frac{1}{n} \left(1 + \frac{k}{n} \right)^m \quad \text{for } m \in Z, \tag{10.2.7}$$

and for $p_{n,k}(x)$ we need also

$$\int_0^1 p_{n,k}(x)(1-x)^\zeta \, dx \leq C(\zeta) \left(\frac{n-k+1}{n} \right)^\zeta \frac{1}{n} \quad \text{for } \zeta > -1, \tag{10.2.8}$$

where the constants $C(\eta)$, $C(m)$, and $C(\zeta)$ are independent of k and n.

These inequalities are computational and are essentially a straightforward itemizing of cases. Actually $0 > \eta > -1$ for (10.2.6) ($-1 < \zeta < 0$ for (10.2.8)) should be treated only for $k = 0$ ($k = n$), otherwise it is easier to use again Hölder's inequality and prove (10.2.6) with $\eta = -1$ (or (10.2.8) with $\zeta = -1$). Combining all these inequalities, we obtain

$$w_i(k, n)^{-p} \int_D P_{n,k}(x) w(x)^p \, dx \leq C/n$$

and complete the proof of Theorem 10.2.2. □

We now prove (10.1.4) for G_n. Using for $\gamma(\infty) \geq 0$

$$w(x)/w(nx/\tau) \leq (\tau/n)^{\gamma(0)}(1 + (\tau/n))^{\gamma(\infty)},$$

and following Totik [11], we obtain for $1 \leq p < \infty$ via Jensen's inequality

$$\|w(x)G_n(f,x)\|_p = \left\| w(x) \frac{1}{n!} \int_0^\infty e^{-\tau}\tau^n f\left(\frac{nx}{\tau}\right) d\tau \right\|_p$$

$$\leq \left\{ \int_0^\infty \frac{1}{n!} \int_0^\infty e^{-\tau}\tau^n (\tau/n)^{\gamma(0)p}(1 + (\tau/n))^{\gamma(\infty)p} |w(nx/\tau)f(nx/\tau)|^p \, d\tau \, dx \right\}^{1/p}$$

$$= \|wf\|_p \left\{ \frac{1}{n!} \int_0^\infty e^{-\tau}\tau^{n+1/p} n^{-1/p} (\tau/n)^{\gamma(0)p}(1 + (\tau/n))^{\gamma(\infty)p} \, d\tau \right\}^{1/p}.$$

Inequality (10.1.4) for $1 \leq p < \infty$ now follows from

$$\int_0^\infty e^{-\tau}\tau^{n+\eta}(n! n^\eta)^{-1} \, d\tau \leq C(\eta).$$

For $p = \infty$ we use

$$\|wG_n(f)\|_\infty \leq \|wf\|_\infty \left\| \frac{1}{n} w(x) \int_0^\infty e^{-\tau}\tau^n w\left(\frac{nx}{\tau}\right)^{-1} d\tau \right\|_\infty$$

which, using

$$\frac{1}{n!} \int_0^\infty e^{-\tau}\tau^n \left(\frac{\tau}{n}\right)^{\gamma(0)} \left(1 + \frac{\tau}{n}\right)^{\gamma(\infty)} d\tau \leq \frac{2^{|\gamma(\infty)|}}{n!} \int_0^\infty e^{-\tau}\tau^n \left[\left(\frac{\tau}{n}\right)^{\gamma(0)} + \left(\frac{\tau}{n}\right)^{\gamma(0)+\gamma(\infty)} \right] dt$$

$$\leq C,$$

implies $\|wG_n(f)\|_\infty \leq A\|wf\|_\infty$. When $\gamma(\infty) < 0$ the proof is similar if we use $w(x)/w(nx/\tau) \leq (\tau/n)^{\gamma(0)}(1 + (n/\tau))^{-\gamma(\infty)}$. □

10.3. Bernstein-Type Inequality

The Bernstein-type inequality

$$\left\| w(x)\varphi(x)^2 \left(\frac{d}{dx}\right)^2 L_n(f,x) \right\|_p \leq B(p,w)n \|wf\|_p$$

could be proved on the whole domain in one step here, as we deal here only with the second derivative. We will use, however, two estimates, one on E_n $((A/n, 1 - (A/n))$ or $(A/n, \infty))$ and the other on E_n^c. This will be done following Section 9.4 as this will allow cruder and easier calculations. On E_n we use for B_n^*, S_n^*, and V_n^*

$$\varphi(x)^2 \left(\frac{d}{dx}\right)^2 L_n(f,x) = \frac{n^2}{\varphi(x)^2} \sum P_{n,k}(x) \left(\frac{k}{n} - x\right)^2 b_k(n) - nL_n(f,x)$$

$$- \frac{d}{dx}((\varphi(x)^2) \frac{n}{\varphi(x)^2} \sum P_{n,k}(x) \left(\frac{k}{n} - x\right) b_k(n) \quad (10.3.1)$$

$$\equiv I_1(n,x) + I_2(n,x) + I_3(n,x),$$

where for S_n^* and V_n^* $b_k(n) = a_k(n) = n \int_{k/n}^{(k+1)/n} f(u)\,du$ and for B_n^* $b_k(n) = a_k(n+1)$. Using the result of Section 10.2, we have

$$\|wI_2\|_{L_p(D)} \leq nA(p,w)\|wf\|_p.$$

Following the method of Section 10.2 for $p = \infty$, we have

$$|n^{-1}w(x)I_1(n,x)| \leq \frac{n}{\varphi(x)^2} \sum P_{n,k}(x)\left(\frac{k}{n} - x\right)^2 (w(x)/w_i(k,n))\|wf\|_\infty$$

and

$$|n^{-1}w(x)I_3(n,x)| \leq \left|\frac{d}{dx}(\varphi(x))^2\right| \varphi(x)^2 \sum P_{n,k}(x)\left|\frac{k}{n} - x\right|(w(x)/w_i(k,n))\|wf\|_\infty.$$

To estimate $\|n^{-1}w(x)I_1(n,x)\|_{L_\infty(E_n)}$ we recall that $0 \leq \gamma(0) < 1$ (and for B_n^* also $0 \leq \gamma(1) < 1$) and choose q' satisfying $\gamma(0) < 1/q'$ (and for B_n^* also $\gamma(1) < 1/q'$). We now recall Lemma 9.4.4 and the estimates of Section 10.2, and write

$$\frac{n}{\varphi(x)^2} \sum P_{n,k}(x)\left(\frac{k}{n} - x\right)^2 (w(x)/w_i(k,n)) \leq \frac{n}{\varphi(x)^2}\left\{\sum P_{n,k}(x)\left|\frac{k}{n} - x\right|^{2p'}\right\}^{1/p'}$$

$$\times \left\{\sum P_{n,k}(x)((w(x)/w_i(k,n))^{q'}\right\}^{1/q'}$$

$$\leq C,$$

for $x \in E_n$. (We note that $\sum P_{n,k}(x)(w(x)/w_i(k,n))^{q'} \leq C_1$ was proved for the weight $w(x)^{q'}$ provided that $0 \leq q'\gamma(0) < 1$.) To estimate $n^{-1}w(x)I_3(n,x)$ we use the same technique and recall also that $|(d/dx)(\varphi(x)^2)/\varphi(x)n^{1/2}| \leq C$ in E_n.

For $p < \infty$ we use

$$n\varphi(x)^{-2} \sum P_{n,k}(x)\left(\frac{k}{n} - x\right)^2 = 1$$

and Jensen's inequality to write

$$|n^{-1}w(x)I_1(n,x)|^p \leq (1 + o(1))n\varphi(x)^{-2} \sum P_{n,k}(x)\left(\frac{k}{n} - x\right)^2$$

$$\times \left(\frac{w(x)}{w_i(k,n)}\right)^p n \int_{I(k,n)} |w(u)f(u)|^p\,du,$$

where $I(k,n) = [(k/n), (k+1)/n]$ for S_n^* and V_n^* and $I(k,n) = [k/(n+1), (k+1)/(n+1)]$ for B_n^*. To complete the estimate of this term in $L_p(E_n)$ it will be enough to show

$$n^2 \int_{E_n} P_{n,k}(x)\left(\frac{k}{n} - x\right)^2 (w(x)/w_i(k,n))^p \varphi(x)^{-2}\,dx \leq C.$$

For this estimate we use Lemma 9.4.5 (10.2.5), and Hölder's inequality with $(p')^{-1} + (q')^{-1} = 1$ and $p' > 1$ such that $w(x)$ satisfies the condition of Definition 10.1.1 with pp' as well, (we may choose p' as close to 1 as we like and disregard the case $\gamma(i) = 0$) and write

$$n^2 \int_{E_n} P_{n,k}(x) \left(\frac{k}{n} - x\right)^2 \left(\frac{w(x)}{w_i(k,n)}\right)^p \varphi(x)^{-2} \, dx$$

$$\leq n^2 \left\{ \int_{E_n} P_{n,k}(x) \left(\frac{k}{n} - x\right)^{2q'} \varphi(x)^{-2q'} \, dx \right\}^{1/q'} \left\{ \int_{E_n} P_{n,k}(x) \left(\frac{w(x)}{w_i(k,n)}\right)^{pp'} dx \right\}^{1/p'}$$

$$\leq Cn^2 (n^{-q'-1})^{1/q'} n^{-1/p'}$$

$$\leq C.$$

Similarly, as Lemma 9.4.5 implies

$$\left|\frac{d}{dx}(\varphi(x)^2)\right| \varphi(x)^{-2} \sum P_{n,k}(x) \left|\frac{k}{n} - x\right| \leq \left|\frac{d}{dx}(\varphi(x))^2\right| \Big/ n^{1/2} \varphi(x) \leq C,$$

for all $x \in E_n$, we have, using Jensen's inequality for the sum,

$$|n^{-1} w(x) I_3(n,x)|^p \leq C_1 \left|\frac{d}{dx}(\varphi(x)^2)\right| \varphi(x)^{-2} \sum P_{n,k}(x) \left|\frac{k}{n} - x\right|$$

$$\times \left(\frac{w(x)}{w_i(k,n)}\right)^p n \int_{I(k,n)} |w(u) f(u)|^p \, du.$$

To complete the proof we have to show

$$J(n,k) \equiv n \int_{E_n} \left|\frac{d}{dx}(\varphi(x)^2)\right| \varphi(x)^{-2} P_{n,k}(x) \left|\frac{k}{n} - x\right| \left(\frac{w(x)}{w_i(k,n)}\right)^p dx \leq C_2.$$

To show this we follow the above, again with p' such that $w(x)$ satisfies the conditions of Definition 10.1.1 with pp' as well as with p, to obtain

$$J(n,k) \leq n \left\{ \int_{E_n} \left|\frac{d}{dx}(\varphi(x)^2)\right|^{q'} \varphi(x)^{-2q'} P_{n,k}(x) \left|\frac{k}{n} - x\right|^{q'} dx \right\}^{1/q'}$$

$$\times \left\{ \int_{E_n} P_{n,k}(x) \left(\frac{w(x)}{w_i(k,n)}\right)^{pp'} dx \right\}^{1/p'}.$$

Using $|(d/dx)(\varphi(x)^2)| \leq M_1$ and $|n^{1/2}\varphi(x)|^{-1} \leq M_2$ and on $E_n \cap [0,1]$ and (only for S_n^* and V_n^*) $|(d/dx)(\varphi(x)^2)/\varphi(x)| \leq M_3$, on $[1, \infty)$, we have

$$n \int_{E_n} \left|\frac{d}{dx}\varphi(x)^2\right|^{q'} \varphi(x)^{-2q'} P_{n,k}(x) \left|\frac{k}{n} - x\right|^{q'} dx$$

$$\leq M_1 M_2 n^{1+q'/2} \int_{E_n \cap [0,1]} P_{n,k}(x) \left|\frac{k}{n} - x\right|^{q'} \varphi(x)^{-q'} dx$$

$$+ M_3 n \int_1^\infty P_{n,k}(x) \left|\frac{k}{n} - x\right|^{q'} \varphi(x)^{-q'} dx$$

$$\leq M_4 n^{1+q'/2} \int_{E_n} P_{n,k}(x) \left|\frac{k}{n} - x\right|^{q'} \varphi(x)^{-q'} dx.$$

Since all we required of p' is that it be close to 1, we may assume $q' \geq 2$, and therefore, using Lemma 9.4.5, we complete the estimate of $\|n^{-1}wI_3(n,x)\|_{L_p(E_n)}$ and hence of $\|n^{-1}w\varphi^2 L_n^{(2)}(f)\|_{L_p(E_n)}$.

The estimate of $\|w(x)\varphi(x)^2(d/dx)^2 L_n(f,x)\|_{L_p(E_n^c)}$ is actually that of Section 10.2 when we recall $n\varphi(x)^2 \leq A$ on E_n^c and utilize (9.4.4), which means that estimates of Section 10.2 will be applied to $p_{n-2,k}(x)$, $s_{n,k}(x)$ and $v_{n+2,k}(x)$ instead of $p_{n,k}(x)$, $s_{n,k}(x)$, and $v_{n,k}(x)$.

We will not give details of the estimate (10.1.5) for $G_n(f,x)$ which essentially can follow Totik [11] and remarks at the end of Section 10.2.

10.4. The Estimate $\|w\varphi^2 L_n^{(2)}(f)\| \leq C(\|w\varphi^2 f^{(2)}\| + \|f\|)$

For all operators except G_n we use (9.7.4), (9.7.5), and (9.7.6) to obtain

$$\left|\varphi(x)^2 \left(\frac{d}{dx}\right)^2 L_n(f,x)\right| \leq Cn \sum P_{n,k+1}(x) \int_{I(k,n)} \varphi(x)^2 |f''(u)| \, du, \quad (10.4.1)$$

where $I(k,n) = [k/(n+1), (k+3)/(n+1)]$ (and the sum on $k = 0, \ldots, n-2$) for B_n^* and $I(k,n) = [k/n, (k+3)/n]$ (and the sum on $k \in N$) for S_n^* and V_n^*. The technique of Section 10.2 and the fact that we treat the same $P_{n,k}$ (but with $k+1$) and w yield the result for these operators when we observe that for $k > 0$ ($k < n-2$ too for B_n^*) $w_i(k,n) \sim w_i(k+2,n) \sim w(k/n)$ in $I(k,n)$, and near 0 (and 1 too for B_n^*) the treatment is again the same as that of Section 10.2.

For $G_n(f,x)$ we use $G_n(f,x) = \frac{1}{n!} \int_0^\infty e^{-\tau} \tau^n f(nx/\tau) \, d\tau$ to derive from (10.1.4)

$$\|w(x) x^2 G_n''(f,x)\|_p = \left\| w(x) \frac{1}{n!} \int_0^\infty e^{-\tau} \tau^n \left(\frac{nx}{\tau}\right)^2 f''\left(\frac{nx}{\tau}\right) d\tau \right\|_p$$

$$= \|w(x) G_n((\cdot)^2 f''(\cdot), x)\|_p$$

$$\leq C \|w\varphi^2 f''\|_p.$$

10.5. The Estimate of $L_n f - f$ for Smooth Functions

In this section we will estimate $w(L_n f - f)$ to obtain (10.1.7). We use the Taylor formula

$$f(u) = f(x) + f'(x)(u-x) + \int_x^u (u-v)f''(v)\, dv$$

and obtain

$$w(x)(L_n(f,x) - f(x)) = w(x)f'(x)L_n((\cdot - x), x) + w(x)L_n(R_2(f, \cdot, x), x),$$

10.5. The Estimate of $L_n f - f$ for Smooth Functions

where
$$R_2(f, u, x) = \int_x^u (u - v) f''(v) \, dv.$$

We will deal with the operator $G_n(f, x)$ at the end. As in Sections 9.5 and 9.6, the proof is divided into two parts estimating the two terms. First we estimate the first term. Recalling that for B_n^*, S_n^*, and V_n^* $|L_n((\cdot - x), x)| \leq C/n$, it is enough to show:

Lemma 10.5.1. *For $\varphi(x)^2 = x(1-x)$ and $\varphi(x)^2 = x$*

$$\|w(x)f'\|_p \leq C(\|w(x)\varphi(x)^2 f''(x)\|_p + \|wf\|_p), \quad 1 \leq p \leq \infty, \quad (10.5.1)$$

and for $\varphi(x)^2 = x(1+x)$

$$\|w(x)(1+x)f'\|_p \leq C(\|w(x)\varphi(x)^2 f''(x)\|_p + \|wf\|_p), \quad 1 \leq p \leq \infty \quad (10.5.2)$$

both excluding $\gamma(i) = 0$ (where $i = 0$ or $i = 1$) in case $p = \infty$.

Since $\gamma(i) = 0$ is a case which was settled for all p in the last chapter, and in case $\gamma(0) = 0$, $\gamma(1) \neq 0$ or vice versa, we can separate the results into two domains, the exclusion will not cause difficulty.

PROOF. On $[1, \infty)$ the result follows from the estimate

$$\|f'\|_{L_p[2^k, 2^{k+1}]} \leq M\{h\|f''\|_{L_p[2^k, 2^{k+1}]} + h^{-1}\|f\|_{L_p[2^k, 2^{k+1}]}\},$$

where M is independent of k and h and where $h \leq 2^k$ (see (2.2.14)). Using $w(2^k) \sim w(u)$ and $\varphi(2^k) \sim \varphi(u)$ for $u \in [2^k, 2^{k+1}]$, and choosing $h = c\varphi(2^k)$, we have

$$\|w\varphi f'\|_{L_p[2^k, 2^{k+1}]} \leq M_1\{\|w\varphi^2 f''\|_{L_p[2^k, 2^{k+1}]} + \|wf\|_{L_p[2^k, 2^{k+1}]}\}$$

which implies for $p < \infty$

$$\|w\varphi f'\|_{L_p[1, \infty)} = \left(\sum_k \|w\varphi f'\|_{L_p[2^k, 2^{k+1}]}^p\right)^{1/p}$$

$$\leq M_2 \left(\sum_{k=0}^\infty \|w\varphi^2 f''\|_{L_p[2^k, 2^{k+1}]}^p + \sum_{k=0}^\infty \|wf\|_{L_p[2^k, 2^{k+1}]}^p\right)^{1/p}$$

$$\leq M_3(\|w\varphi^2 f''\|_{L_p[1, \infty)} + \|wf\|_{L_p[1, \infty)}).$$

The treatment for $p = \infty$ is somewhat easier, and the proof of (10.5.1) and (10.5.2) in $[1, \infty)$ follows again from the inequality (2.2.14) and the fact that $1 \leq \sqrt{x}$ for S_n^* and $(1 + x) \leq 2\sqrt{x(1+x)}$ for V_n^*. The proof for B_n^* on its domain and for V_n^* and S_n^* in $[0, 1]$ follows exactly the argument used in the proof of Theorem 9.5.3, that is, using Hardy's inequality for $1 \leq p < \infty$ and integration by parts and a straightforward estimate for $p = \infty$. □

To deal with the remainder term (again just for B_n^*, S_n^*, and V_n^* first) which we will do separately for $p = 1$ and $p > 1$, we will need the following lemma.

170   10. Weighted Approximations by Exponential-Type Operators

Lemma 10.5.2. *Let $g(v) = w(v)\varphi^2(v)|f''(v)|$ and k^* be k or $k+1$ in a way maximizing the expression in which it is written. Then we have the following estimates for $x \in E_n$:*

(a) *For $\gamma(i)_- \equiv \min(\gamma(i), 0)$, $\gamma(i)_+ \equiv \gamma(i) - \gamma(i)_-$, $w_-(x) = x^{\gamma(0)_-}(1-x)^{\gamma(1)_-}$ and $\varphi(x)^2 = x(1-x)$ we have*

$$(n+1)\int_{k/n+1}^{k+1/n+1} \left|R_2(f, u, x)\right| du$$

$$\leq C\left(\left|\frac{k}{n} - x\right| + \frac{1}{n}\right)(\varphi(x)^2 w_-(x))^{-1}$$

$$\times \left[\left(\frac{n}{k+1}\right)^{\gamma(0)_+} + \left(\frac{1}{x}\right)^{\gamma(0)_+}\right] \qquad (10.5.3)$$

$$\times \left[\left(\frac{n}{n-k+1}\right)^{\gamma(1)_+} + \left(\frac{1}{1-x}\right)^{\gamma(1)_+}\right]$$

$$\times \left|\int_x^{k^*/n+1} g(v)\, dv\right|;$$

(b) *for $\gamma(0)_+$ as above, $w_-(x) = x^{\gamma(0)_-}(1+x)^{-\gamma(0)_-}$ or $w_-(x) = x^{\gamma(0)_-}(1+x)^{-\gamma(0)_- - 1}$, $\eta = \gamma(\infty) + \gamma(0)$ or $\eta = \gamma(\infty) + \gamma(0) + 1$ for $\varphi(x)^2 = x$ and $\varphi(x)^2 = x(1+x)$ respectively, we have*

$$n\int_{k/n}^{k+1/n}\left|R_2(f, u, x)\right| du \leq C\left(\left|\frac{k}{n} - x\right| + \frac{1}{n}\right)(\varphi(x)^2 w_-(x))^{-1}$$

$$\times \left[\left(\frac{n}{k+1}\right)^{\gamma(0)_+} + \left(\frac{1}{x}\right)^{\gamma(0)_+}\right] \qquad (10.5.4)$$

$$\times \left[\left(\frac{n}{k+n}\right)^{\eta} + \left(\frac{1}{1+x}\right)^{\eta}\right]$$

$$\times \left|\int_x^{k^*/n} g(v)\, dv\right|;$$

where C is independent of k, n, g, and x in both cases.

PROOF. Following Lemma 9.6.1 we first observe that

$$\left|\int_x^u |u-v||f''(v)|\, dv\right| \leq \frac{|x-u|}{\varphi(x)^2 w_-(x)}\left|\int_x^u \varphi(v)^2 w_-(v)|f''(v)|\, dv\right|. \quad (10.5.5)$$

For $k > 0$ in case of S_n^* and V_n^* or $0 < k < n$ in case of B_n^* we use the fact that $x^{\gamma(0)_+}$, $(1+x)^\eta$ and $(1-x)^{\gamma(1)_+}$ are monotonic and achieve maximum at an endpoint of $[k/n, (k+1)/n]$ or $[k/(n+1), (k+1)/(n+1)]$; and also observe that $|u-x| \leq |(k/n) - x| + (1/n)$ in those intervals to obtain (10.5.3) and (10.5.4). For $k = 0$ (for instance) we write for $x > n^{-1}$ ($x \in E_n$)

10.5. The Estimate of $L_n f - f$ for Smooth Functions

$$n \int_0^{1/n} \frac{|u-x|}{\varphi(x)^2 w_-(x)} \left\{ \left| \int_{1/n}^x \varphi(v)^2 w_-(v) |f''(v)| \, dv \right| \right.$$

$$\left. + \int_u^{1/n} \varphi(v)^2 w_-(v) |f''(v)| \, dv \right\} du \equiv I_1 + I_2.$$

The terms I_1 and I_2 can be estimated by

$$I_1 \leq \frac{(x+n^{-1})}{\varphi(x)^2 w_-(x)} \left[\left(\frac{n}{1}\right)^{\gamma(0)_+} + \left(\frac{1}{x}\right)^{\gamma(0)_+} \right] \left[\left(\frac{n}{n+1}\right)^\eta + \left(\frac{1}{1+x}\right)^\eta \right]$$

$$\times \left| \int_x^{k^*/n} \varphi(v)^2 w(v) |f''(v)| \, dv \right|,$$

and

$$I_2 \leq \frac{(x+n^{-1})}{\varphi(x)^2 w_-(x)} n \int_0^{1/n} \int_u^{1/n} \varphi(v)^2 w_-(v) |f''(v)| \, dv$$

$$\leq \frac{(x+n^{-1})}{\varphi(x)^2 w_-(x)} n \int_0^{1/n} \varphi(v)^2 w_-(v) |f''(v)| \left\{ \int_0^v du \right\} dv$$

(as $v/w_+(v) \leq n^{-1+\gamma(0)_+}$ for $v \in [0, 1/n]$)

$$\leq \frac{(x+n^{-1})}{\varphi(x)^2 w_-(x)} n^{\gamma(0)_+} \int_0^{1/n} \varphi(v)^2 w(v) |f''(v)| \, dv$$

$$\leq \frac{(x+n^{-1})}{\varphi(x)^2 w_-(x)} n^{\gamma(0)_+} \int_0^x \varphi(v)^2 w(v) |f''(v)| \, dv,$$

which yields (10.5.4). Similarly, we estimate the terms $k = 0$ and $k = n$ for B_n^*. □

We can now state and prove our result for $1 \leq p \leq \infty$.

Theorem 10.5.3. Suppose $1 \leq p \leq \infty$, $w(x)\varphi(x)^2 f''(x) \in L_p$ and $L_n(f, x)$ is B_n^*, S_n^*, or V_n^*. Then

$$\|w(x) L_n(R_2(f, \cdot, x), x)\|_{L_p(E_n)} \leq M n^{-1} \|w \varphi^2 f''\|_{L_p(D)}.$$

PROOF FOR $p > 1$. We denote $g(x) = w(x) \varphi(x)^2 |f''(x)|$ and the maximal function of g by $M(g, x)$ and recall that

$$\|\Phi_n(x) M(g, x)\|_{L_p(E_n)} \leq \|\Phi_n(x)\|_{L_\infty(E_n)} \|M(g, x)\|_p$$

$$\leq C \|\Phi_n(x)\|_{L_\infty(E_n)} \|g\|_p$$

for $1 < p \leq \infty$. We also recall

$$\left| \int_x^{k^*/n} \varphi(v)^2 w(v) |f''(v)| \, dv \right| \leq \left|\frac{k^*}{n} - x\right| M(g, x) \leq \left(\left|\frac{k}{n} - x\right| + \frac{1}{n}\right) M(g, x)$$

and

$$\left(\left|\frac{k}{n}-x\right|+\frac{1}{n}\right)^2 \le 2\left(\left|\frac{k}{n}-x\right|^2+n^{-2}\right).$$

These estimates, together with Lemma 10.5.2, imply that the expression $\|\Phi_n(x)\|_{L_\infty(E_n)}$, which we have to estimate, is given in case $L_n = B_n^*$ by

$$\Phi_n(x) = C_1 \sum p_{n,k}(x)\left(\left|\frac{k}{n}-x\right|^2+n^{-2}\right)\varphi(x)^{-2}\left[\left(\frac{xn}{k+1}\right)^{\gamma(0)_+}+1\right]$$

$$\times \left[\left(\frac{(1-x)n}{n-k+1}\right)^{\gamma(1)_+}+1\right]$$

and in case $L_n = S_n^*$ or $L_n = V_n^*$ by

$$\Phi_n(x) = C_1 \sum P_{n,k}(x)\left(\left|\frac{k}{n}-x\right|^2+n^{-2}\right)\varphi(x)^{-2}\left[\left(\frac{xn}{k+1}\right)^{\gamma(0)_+}+1\right]$$

$$\times \left[\left(\frac{(1+x)n}{n+k}\right)^n+1\right].$$

In both of these expressions the necessary estimates for all terms were already made. We use Lemma 9.4.4, Hölder's inequality as well as (10.2.4), (10.2.5) and $(n\varphi(x)^2)^{-1} \ge A$ to obtain $\|\Phi_n(x)\|_{L_\infty(E_n)} \le Cn^{-1}$. □

PROOF OF THEOREM 10.5.3 FOR $p = 1$. We use Lemma 10.5.2 and observe that for $p = 1$, $\gamma(0)_+ = \gamma(1)_+ = 0$, and therefore,

$$\|w(x)B_n^*(R_2(f,\cdot,x),x)\|_{L_1(E_n)}$$

$$\le C \int_{E_n} \sum p_{n,k}(x)\left(\left|\frac{k}{n}-x\right|+\frac{1}{n}\right)\varphi(x)^{-2}\left|\int_x^{k^*/n+1} g(v)\,dv\right|dx, \quad (10.5.6)$$

the right hand side of which is the same expression as the right hand side of (9.6.8) with $r = 1$ and $g(v) = w(v)\varphi(v)^2|f''(v)|$ replacing $g(v) = \varphi(v)^2|f''(v)|$. Consequently, we obtain the same estimate with the new expression for g, that is

$$\|w(x)B_n^*(R_2(f,\cdot,x),x)\|_{L_1(E_n)} \le C_1 n^{-1}\|w\varphi^2 f''\|_{L_1[0,1]}.$$

Somewhat more complicated are the S_n^* and V_n^* cases. We still have $\gamma(0)_+ = 0$ and also

$$\|w(x)L_n(R_2(f,\cdot,x),x)\|_{L_1(E_n)} \le C\int_{E_n}\sum P_{n,k}(x)\left(\left|\frac{k}{n}-x\right|+\frac{1}{n}\right)\varphi(x)^{-2}$$

$$\times \left[\left(\frac{(1+x)n}{k+n}\right)^n+1\right]\left|\int_x^{k^*/n} g(v)\,dv\right|dx,$$

10.5. The Estimate of $L_n f - f$ for Smooth Functions

where for $L_n = S_n^*$ or $L_n = V_n^*$, $P_{n,k}(x)$ is $s_{n,k}(x)$ or $v_{n,k}$; $\varphi(x)^2$ is x or $x(1+x)$; and η is $\gamma(\infty) + \gamma(0)$ or $\gamma(\infty) + \gamma(0) + 1$ respectively. The number η can be any real number. If $\eta = 0$, we have the result using the estimate of Section 9.6. For other η we choose $m \in Z$ such that $m\eta > 0$ and $|m| > \eta$, and therefore, $((1+x)n/(k+n))^\eta \leq C(((1+x)n/(k+n))^m + 1)$. Using the results of Section 9.6 for $r = 1$, we have

$$\int_{E_n} \sum P_{n,k}(x) \left(\left|\frac{k}{n} - x\right| + \frac{1}{n}\right) \varphi(x)^{-2} \left|\int_x^{k^*/n} g(v)\,dv\right| dx \leq Cn^{-1}\|g\|_1. \quad (10.5.7)$$

Therefore, we only have to show for $P_{n,k}(x) = s_{n,k}(x)$ or $P_{n,k}(x) = v_{n,k}(x)$ and $m \in Z$ that

$$\int_{E_n} \sum P_{n,k}(x) \left(\left|\frac{k}{n} - x\right| + \frac{1}{n}\right) \varphi(x)^{-2} \left(\frac{(1+x)n}{n+k}\right)^m \left|\int_x^{k^*/n} g(v)\,dv\right| dx \quad (10.5.8)$$
$$\leq Cn^{-1}\|g\|_1.$$

For $P_{n,k}(x) = v_{n,k}(x)$ we have $v_{n,k}(x)((1+x)n/(n+k))^m \leq C(m)v_{n-m,k}(x)$ and therefore, for $n > 2m$ and k^{**} equal to k or $k+1$ in a way maximizing the expression in which it appears, we have

$$\int_{E_n} \sum_{k=0}^{\infty} v_{n,k}(x) \left(\left|\frac{k}{n} - x\right| + \frac{1}{n}\right) \varphi(x)^{-2} \left(\frac{(1+x)n}{n+k}\right)^m \left|\int_x^{k^*/n} g(v)\,dv\right| dx$$

$$\leq C \int_{E_n} \sum_{k=0}^{\infty} v_{n-m,k}(x) \left(\left|\frac{k}{n-m} - x\right| + \frac{km}{n(n-m)}\right) \varphi(x)^{-2} \left|\int_x^{k^*/n} g(v)\,dv\right| dx$$

$$\leq C_1 \int_{E_n} \sum_{k=0}^{\infty} v_{n-m,k}(x) \left(\left|\frac{k}{n-m} - x\right| + \frac{x}{n}\right) \varphi(x)^{-2} \left|\int_x^{k^*/n} g(v)\,dv\right| dx$$

$$\equiv I_1 + I_2.$$

We write now

$$I_1 \leq C \int_{E_n} \sum_{k=0}^{\infty} v_{n-m,k}(x) \left|\frac{k}{n-m} - x\right| \varphi(x)^{-2} \left\{\left|\int_x^{k^{**}/n-m} g(v)\,dv\right|\right.$$
$$\left. + \left|\int_{k^*/n}^{k^{**}/n-m} g(v)\,dv\right|\right\} dx \equiv J_1 + J_2.$$

Recall that m is fixed but n tends to infinity. The term J_1 is estimated like (10.5.7) with $n - m$ replacing n. To estimate J_2 we observe that the number of intervals $(k^*/n, k^{**}/(n-m))$ or $(k^{**}/(n-m), k^*/n)$ which contain v is bounded by $C_m(1+v)$, and since such a v satisfies $k/n < v < (k+1)/(n-m) < 2k/n$ for $n > n_0$, this number is also bounded by $2C_m(1 + (k/n))$ (for any k which satisfies $k/n < v < 2k/n$). We now use Lemma 9.4.5 and the estimate

$$v_{n-m,k}(x)\varphi(x)^{-2} \leq C(1)\frac{n}{k}\left(1 + \frac{k}{n}\right)^{-1} v_{n-m+2,k-1}(x)$$

to obtain for $k \neq 0$

174   10. Weighted Approximations by Exponential-Type Operators

$$J_3(k) \equiv \int_{E_n} v_{n-m,k}(x) \left| \frac{k}{n-m} - x \right| \varphi(x)^{-2} \left(1 + \frac{k}{n}\right) dx$$

$$\leq C \left\{ \int_{E_n} v_{n-m,k}(x) \left| \frac{k}{n-m} - x \right|^2 \varphi(x)^{-2} dx \right\}^{1/2}$$

$$\times \left\{ \int_{E_n} v_{n-m,k}(x) \frac{(1 + (k/n))^2}{\varphi(x)^2} dx \right\}^{1/2}$$

$$\leq C n^{-1} k^{-1},$$

and, as we can easily get $J_3(0) = O(1/n)$ directly, we conclude the estimate of J_2. To estimate I_2 we follow the proof of (9.6.11), and instead of (9.6.13) we estimate

$$I_2 \leq C n^{-1} \sum_{l=0}^{\infty} \frac{1}{(l+1)^4} \int_{E_n} \frac{1}{\varphi(x)} \int_{F(l,x)} g(v) \, dv.$$

The rest follows the estimate (9.6.13) in Section 9.6 very closely.

To prove (10.5.8) for $s_{n,k}(x)$ we estimate it first for $m > 0$ using for $j < m < n$ $s_{n,k+j}(x)((1+x)n/n+k) \leq s_{n,k+j}(x) + s_{n,k+j+1}(x)$, and therefore,

$$s_{n,k}(x)((1+x)n/n+k)^m \leq \sum_{j=0}^{m} \binom{m}{j} s_{n,k+j}(x).$$

We can now write

$$\int_{E_n} \sum s_{n,k}(x) \left(\left| \frac{k}{n} - x \right| + \frac{1}{n} \right) \varphi(x)^{-2} \left(\frac{(1+x)n}{n+k} \right)^m \left| \int_x^{k^*/n} g(v) \, dv \right| dx$$

$$\leq \sum_{j=0}^{m} \binom{m}{j} \int_{E_n} \sum_{k=0}^{\infty} s_{n,k+j}(x) \left(\left| \frac{k}{n} - x \right| + \frac{1}{n} \right) \varphi(x)^{-2} \left| \int_x^{k^*/n} g(v) \, dv \right| dx$$

$$\equiv \sum_{j=0}^{m} \binom{m}{j} I(j).$$

To estimate $I(j)$ we write

$$I(j) \leq C \int_{E_n} s_{n,k+j}(x) \left(\left| \frac{k+j}{n} - x \right| + \frac{1}{n} \right) \varphi(x)^{-2} \left| \int_x^{(k+j)^*/n} g(v) \, dv \right| dx$$

$$+ C \int_{E_n} s_{n,k+j}(x) \left(\left| \frac{k+j}{n} - x \right| + \frac{1}{n} \right) \varphi(x)^{-2} \left| \int_{k/n}^{(k+j+1)/n} g(v) \, dv \right| dx$$

$$\equiv I_1(j) + I_2(j).$$

$(I_1(j)$ is of the form (10.5.7). The fact that the intervals $(k/n, (k+j+1)/n)$ overlap at most $j+1$ times and

$$\int_{E_n} s_{n,k+j}(x) \left(\left| \frac{k+j}{n} - x \right| + \frac{1}{n} \right) \varphi(x)^{-2} dx \leq M/n$$

will yield $I_2(j) \leq n^{-1} \|g\|_1$. To prove the result for $m < 0$ we recall for $k > n$ $n > 2|m|$ and for $j < |m|$

$$S_{n,k-j}(x)\left(\frac{n+k}{(1+x)n}\right) \le 4s_{n,k-j-1}(x),$$

and therefore,

$$\int_{E_n} \sum_{k=0}^{\infty} s_{n,k}(x)\left(\left|\frac{k}{n}-x\right|+\frac{1}{n}\right)\varphi(x)^{-2}\left(\frac{n+k}{(1+x)n}\right)^{-m}\left|\int_x^{k^*/n} g(v)\,dv\right|dx$$

$$\le \int_{E_n} \sum_{k=0}^{n} s_{n,k}(x)\left(\left|\frac{k}{n}-x\right|+\frac{1}{n}\right)\varphi(x)^{-2}2^{-m}\left|\int_x^{k^*/n} g(v)\,dv\right|dx$$

$$+ 4^{-m}\int_{E_n} \sum_{k=n}^{\infty} s_{n,k+m}(x)\left(\left|\frac{k}{n}-x\right|+\frac{1}{n}\right)\varphi(x)^{-2}\left|\int_x^{k^*/n} g(v)\,dv\right|dx.$$

The first expression is bounded because of (10.5.7) and the second can be split as we did for S_n^* in case $m > 0$ to complete the proof of our theorem.

To prove the result for $G_n(f, x)$ we observe that $G_n(\cdot - x, x) = 0$, and therefore, there is no need for an analogue of Lemma 10.5.1. Following Lemma 10.5.2., we may write

$$I(n,x) = w(x)\frac{1}{n!}x^{n+1}\int_0^\infty e^{-xu}u^n\left|R_2\left(f,\frac{n}{u},x\right)\right|du$$

$$\le w(x)\frac{x^{n+1}}{n!}\int_0^\infty e^{-xu}u^n\left(\frac{n}{u}-x\right)x^{-1}\left(\frac{1}{x}+\frac{u}{n}\right)\left|\int_x^{n/u} v^2|f''(v)|\,dv\right|du$$

$$\le \frac{Cx^{n+1}}{n!}\int_0^\infty e^{-xu}\frac{((n/u)-x)}{x}\left(\frac{1}{x}+\frac{u}{n}\right)\left(1+\left(\frac{n}{u}x\right)^{\gamma(0)}\right)$$

$$\times\left(1+\left(\frac{1+x}{1+(n/u)}\right)^{\gamma(\infty)}\right)\left|\int_x^{n/u} v^2 w(v)|f''(v)|\,dv\right|du.$$

For $1 < p \le \infty$ we use $|\int_x^{n/u} g(v)\,dv| \le |x - (n/u)|\,M(g,x)$ and have to show only that

$$\sup_x\left|\frac{x^{n+1}}{n!}\int_0^\infty e^{-xu}u^n\left(\frac{n}{u}-x\right)^2 x^{-2}(1+(ux/n))\left(1+\left(\frac{u}{n}x\right)^{\gamma(0)}\right)\right.$$

$$\left.\times\left(1+\left((1+x)\Big/\left(1+\frac{n}{u}\right)\right)^{\gamma(\infty)}\right)du\right| \le Cn^{-1} \tag{10.5.9}$$

which can be verified directly or following the method used by V. Totik [11]. The L_1 case is a little more complicated. We denote $g(v) = w(v)v^2|f''(v)|$ and follow calculations in Section 9.6 for $r = 1$ and the Post-Widder operator to obtain our result. □

10.6. The Saturation Result

The following theorem constitutes the saturation result.

Theorem 10.6.1. *For $L_n f$, φ and w given in Definition 10.1.1 and for $1 < p < \infty$ the following statements are equivalent:*

(a) $\|w(L_n f - f)\|_p = O(n^{-1})$,
(b) f is differentiable, $f' \in \text{A.C.}_{\text{loc}}$ and $\|w\varphi^2 f''\|_p < \infty$,
(c) $\Omega_\varphi^2(f, t)_{w,p} = O(t^2)$,
(d) $\mathscr{K}_{2,\varphi}(f, t^2)_{w,p} = O(t^2)$,
(e) $K_{2,\varphi}(f, t^2)_{w,p} = O(t^2)$.

The characterization of (a) when $p = 1$ remains open (see for $w(x) = 1$ Maier [1] and Totik [10]). For $L_n = G_n$, however, the result holds for $p = 1$ and $p = \infty$ as well. This was proved for $w = 1$ (see Totik [11]) and for this operator $w \sim 1$ does not add significant difficulty here.

PROOF. By the definition of the main-part K-functional $\mathscr{K}$ (see Section 6.2) (b) implies (d). The implications (d) $\Leftrightarrow$ (c) and (b) $\Rightarrow$ (a) were proved in Theorem 6.2.1 and Section 10.5 respectively. Obviously (e) implies (d). Corollary 8.2.5 yields the implication (d) $\Rightarrow$ (e) (splitting the interval in case $D = R_+$). Thus it remains to show that (a) implies (b).

Consider first $L_n = B_n^*$, in which case we have

$$w(x) = x^{\gamma(0)}(1 - x)^{\gamma(1)}, \quad -1/p < \gamma(0), \quad \gamma(1) < 1 - 1/p, \tag{10.6.1}$$

$$\varphi(x) = \sqrt{x(1-x)}.$$

For $g \in C^2(0, 1)$ with compact support in $(0, 1)$ we have, following Maier [1] (see also Ditzian and May [1, Lemma 5.3]),

$$\left| n \int_0^1 w(x)(B_n^*(f, x) - f(x))g(x) \, dx \right| \leq C_g \|wf\|_p \tag{10.6.2}$$

uniformly in n and $wf \in L_p(0, 1)$. (To be more precise the considerations in Maier [1] imply only (10.6.2) with $w = 1$, but taking into account the strong localization of B_n^*, (10.6.2) follows without much difficulty.) For twice continuously differentiable functions the identities

$$B_{n-1}^*(t - x, \cdot) = \frac{1 - 2x}{2n} \equiv \frac{\psi(x)}{2n} \quad \text{and}$$

$$B_{n-1}^*((t - x)^2, x) = \frac{x(1 - x)}{n} + O(1/n^2)$$

and Taylor's formula imply

$$\lim_{n \to \infty} 2n \int_0^1 w(x)(B_{n-1}^*(f, x) - f(x))g(x) \, dx$$

$$= \int_0^1 (\psi(x)f'(x) + \varphi(x)^2 f''(x))w(x)g(x) \, dx$$

$$= \int_0^1 f(x)(-(w(x)\psi(x)g(x))' + (w(x)\varphi(x)^2 g(x))'') \, dx.$$

Hence, by (10.6.2)

10.6. The Saturation Result

$$\lim_{n\to\infty} 2n \int_0^1 w(x)(B_n^*(f,x) - f(x))g(x)\,dx$$

$$= \int_0^1 f(x)(-(w(x)\psi(x)g(x))' + (w(x)\varphi(x)^2 g(x))'')\,dx$$

holds for every $wf \in L_p(0,1)$. Using weak* compactness and the separability of L_p, (a) implies that there exists a subsequence $\{2n_k w(B_{n_k}^* f - f)\}$ converging to a function $h \in L_p(0,1)$ in the weak* topology on $L_p(0,1)$. Hence

$$\int_0^1 f(x)(-(w(x)\psi(x)g(x))' + (w(x)\varphi^2(x)g(x))'')\,dx = \int_0^1 h(x)g(x)\,dx$$

$$= \int_0^1 (h(x)/w(x))w(x)g(x)\,dx,$$

and since this is valid for every $g \in C^2(0,1)$ with compact support in $(0,1)$, elementary results on distributions (cf. Maier [1], Totik [10]) yield that f is differentiable, f' is locally absolutely continuous, and

$$\psi f' + \varphi^2 f'' = h/w \qquad \text{a.e.} \qquad (10.6.3)$$

Solving this differential equation, we obtain

$$f(x) = \int_{1/2}^x \frac{1}{u(1-u)} \int_0^u \frac{h(\tau)}{w(\tau)}\,d\tau\,du + c + d\left(\log\frac{x}{1-x}\right)$$

$$\equiv f_1(x) + c + d\left(\log\frac{x}{1-x}\right)$$

with some constants c and d. Our next aim is to prove that here $d = 0$ and $w\varphi^2 f'' \in L_p(0, 1/2)$. In fact,

$$(w\varphi^2 f_1'')(x) = h(x) - \frac{(1-2x)w(x)}{x(1-x)}\int_0^x (h(u)/w(u))\,du$$

$$\equiv h(x) - \frac{(1-2x)w(x)}{x(1-x)}h_1(x).$$

Making use of Hardy's inequality (see Stein [1, p. 272])

$$\left(\int_0^\infty \left(\int_0^x g(y)\,dy\right)^p x^{-r-1}\,dx\right)^{1/p} \leq \frac{p}{r}\left(\int_0^\infty (yg(y))^p y^{-r-1}\,dy\right)^{1/p},$$

$$g \geq 0, \qquad p \geq 1, \qquad r > 0$$

with $g(y) = h(y)/w(y)$ and $r = p - p\gamma(0) - 1$ (which is bigger than 0 by (10.6.1)), we obtain

$$\left\|\frac{w(x)(1-2x)}{x(1-x)}h_1(x)\right\|_{L_p(0,1/2)} \leq 2\|w(x)h_1(x)/x\|_{L_p(0,1/2)} \leq C\|h\|_{L_p(0,1/2)}.$$

Hence $w\varphi^2 f_1'' \in L_p(0, 1/2)$, and by the results of Section 6.2 this implies

$$\Omega_\varphi^2(f_1, t)_{w, L_p(0, 1/2)} = O(t^2). \tag{10.6.4}$$

Since Theorem 10.1.2 has already been proved for $0 < \alpha < 1$, we can deduce from (a) that

$$\Omega_\varphi^2(f, t)_{w, L_p(0, 1/2)} = O(t^{2-\varepsilon}) \tag{10.6.5}$$

for every $\varepsilon > 0$. Now (10.6.5) and (10.6.4) imply that if in the expression of f, $d \neq 0$, then

$$\Omega_\varphi^2(\log x, t)_{w, L_p(0, 1/2)} = O(t^{2-\varepsilon})$$

holds for any $\varepsilon > 0$. This leads to a contradiction since

$$\Omega_\varphi^2(\log x, t) \geq Ct^2 \left\{ \int_{4t^2}^{1/4} |w(x)\varphi^2(x)(\log x)''|^p \, dx \right\}^{1/p}$$

$$\geq C_1 t^2 \left(\int_{4t^2}^{1/4} x^{(\gamma(0)-1)p} \, dx \right)^{1/p}$$

$$\geq C_2 t^{2(\gamma(0)) + 1/p},$$

and here, by (10.6.1), $\gamma(0) + 1/p < 1$. Therefore, we conclude that $d = 0$ and

$$w\varphi^2 f'' = w\varphi^2 f_1'' \in L_p(0, 1/2).$$

Similarly we can show $w\varphi^2 f'' \in L_p(1/2, 1)$ and this completes the proof for the Kantorovich operators.

Now let $L_n = S_n^*$ or V_n^*. The analogue of (10.6.2) can be verified using the considerations of Totik [3], [10] instead of those in Maier [1], from which (10.6.3) (with $\varphi(x) = \sqrt{x}$ for $L_n = S_n^*$ and $\varphi(x) = \sqrt{x(1 + x)}$ for $L_n = V_n^*$ and $\psi(x) = 1$ in both cases) follows exactly as above. The solution of (10.6.3) for $\varphi^2(x) = x$ and $\psi(x) = 1$ ($L_n = S_n^*$) is

$$f(x) = \int_{1/2}^x \frac{1}{u} \int_0^u \frac{h(\tau)}{w(\tau)} \, d\tau \, du + c + d \log x$$

and for $\varphi^2(x) = x(1 + x)$ and $\psi(x) = 1$, ($L_n = V_n^*$) is

$$f(x) = \int_{1/2}^x \frac{1+u}{u} \int_0^u \frac{h(\tau)}{w(\tau)(1+\tau)^2} \, d\tau + c + d(\log x + x).$$

Using an argument analogous to that applied above for $L_n = B_n^*$, we obtain $w\varphi^2 f'' \in L_p(0, 1)$.

To show that $w\varphi^2 f'' \in L_p(1, \infty)$ is also satisfied we first recall that

$$2n \|w(L_n f - f)\|_p \leq C. \tag{10.6.6}$$

Since $f' \in \text{A.C.}_{\text{loc}}$, simple deduction based on Taylor's formula shows that

$$\lim_{n \to \infty} 2n(L_n(f, x) - f(x)) = f'(x) + \varphi(x)^2 f''(x)$$

is satisfied almost everywhere. Hence (10.6.6) and Fatou's lemma imply

10.6. The Saturation Result

$w(f' + \varphi^2 f'') \in L_p(0, \infty)$. One can derive as an easy consequence of the next lemma that $w\varphi^2 f'' \in L_p[1, \infty)$ follows from $w(f' + \varphi^2 f'') \in L_p(0, \infty)$ which will conclude the proof of Theorem 10.6.1. □

Lemma 10.6.2. *Suppose* $1 \le p < \infty$, $r \in R$, f *is differentiable,* $f' \in A.C._{\text{loc}}$, $x^{r-1}f(x) \in L_p(1, \infty)$ *and* $x^r(f'(x) + \varphi^2(x)f''(x)) \in L_p(1, \infty)$, *where* $\varphi^2(x) \ge x$. *Then* $x^r\varphi^2(x)f''(x) \in L_p(1, \infty)$.

PROOF. We will use the inequality

$$\|f'\|_{L_p(2^k, 2^{k+1})} \le A(h^{-1}\|f\|_{L_p(2^k, 2^{k+1})} + h\|f''\|_{L_p(2^k, 2^{k+1})}) \quad (10.6.7)$$

which is valid for every k and $0 < h \le 2^k$ (see Ditzian [3] or (2.2.14)). Multiplying (10.6.7) by 2^{kr} and setting $h = C2^k$ with some sufficiently small fixed $C > 0$, we obtain

$$\|x^r f'(x)\|_{L_p(2^k, 2^{k+1})} \le A_1 \|x^{r-1}f(x)\|_{L_p(2^k, 2^{k+1})} + \tfrac{1}{2}\|x^{r+1}f''(x)\|_{L_p(2^k, 2^{k+1})}$$

which implies

$$\|x^r\varphi^2(x)f''(x)\|_{L_p(2^k, 2^{k+1})} \le \|x^r(f'(x) + \varphi^2(x)f''(x))\|_{L_p(2^k, 2^{k+1})}$$
$$+ A_1\|x^{r-1}f(x)\|_{L_p(2^k, 2^{k+1})}$$
$$+ \tfrac{1}{2}\|x^r x \cdot f''(x)\|_{L_p(2^k, 2^{k+1})}$$

and therefore

$$\|x^r\varphi^2(x)f''(x)\|_{L_p(2^k, 2^{k+1})} \le 2\|x^r(f'(x) + \varphi^2(x)f''(x))\|_{L_p(2^k, 2^{k+1})}$$
$$+ 2A_1\|x^{r-1}f(x)\|_{L_p(2^k, 2^{k+1})}.$$

Raising these inequalities to the pth power and summing on $k = 0, 1, 2, \ldots$, we conclude the proof of the lemma. □

CHAPTER 11
WEIGHTED POLYNOMIAL APPROXIMATION IN $L_p(R)$

For a weight function $W(x) = \exp(-Q(x))$ the behavior of the rate of best weighted polynomial approximation in $L_p(R)$ will be related to a new modulus of continuity. This will be done via characterization of certain K-functionals which solve a problem of G. Freud and H.N. Mhaskar.

11.1. Introduction

The investigation of weighted polynomial approximation on the real line has a long history. Like many other branches of approximation theory it was introduced by S.N. Bernstein who posed (and solved in many special cases) the problem: determine those weight functions W for which the polynomials are dense in $C_0(W)$. Several outstanding mathematicians contributed to the solution of this problem which is strongly related to that of the moment problem. A solution was finally given by N.I. Akhiezer [1] who showed, roughly speaking, that the condition

$$\int_{-\infty}^{\infty} \frac{\log(W(x))}{1+x^2} dx = -\infty$$

must be satisfied.

The quantitative investigation of the problem for weights of the type $W(x) = \exp(-Q(x))$ began with the works of M.M. Dzrbasyan (see Dzrbasyan [1], [2] and Dzrbasyan and Tavadyan [1]). However, since analogy with trigonometric approximation was not expected, the progress was rather slow until the seventies when G. Freud developed a new tool, the use of special orthogonal polynomials now commonly known as Freud polynomials. As a result of Freud's work, the strong similarity with the trigonometric case became visible. He proved the correct analogues of the Markov–Bernstein and Jackson–Favard inequalities (see Section 11.3) which are fundamental in finding the results on best weighted polynomial approximation on R that correspond to those on best trigonometric polynomial approximation. These led to the characterization of best weighted approximation through the behavior of the K-functional given by

$$K_r(f,t^r)_{W,p} = \inf_{g^{r-1} \in A.C._{\text{loc}}} [\|W(f-g)\|_p + t^r \|Wg^{(r)}\|_p]. \tag{11.1.1}$$

However, the characterization of this K-functional using structural properties of f, which would complete the analogy with the trigonometric case, has remained open (except for $r = 1, 2$, cf. Section 11.5). In this chapter we give this characterization. Although (11.1.1) is a special case of (6.1.1), we cannot use the moduli developed in Part I, as the weight in (11.1.1) will not satisfy the conditions imposed in Chapter 6 (since its behavior is exponential). However, we will use a technique similar to that which we used in the first part of our book.

We will define our new moduli that solve the above problem and state the equivalence theorem with the K-functional in Section 11.2 The proof will be given in Section 11.4. Section 11.3 is devoted to the characterization of best weighted polynomial approximation. Finally, in Section 11.5 a comparison will be made with earlier moduli together with a few remarks generalizing the results further.

11.2. The Equivalence Result

We first describe the class of weights $W(Q)$ for which the equivalence result is proved.

Definition 11.2.1. A weight function W belongs to the class of weights $W(Q)$ if

$$W(x) = W_Q(x) = \exp(-Q(x)), \tag{11.2.1}$$

where the function $Q(x)$ satisfies the conditions:

(a) $Q(x)$ is a continuous even function in $C^1(R_+)$,
(b) $Q'(x) \nearrow \infty$ as $x \nearrow \infty$,
(c) $Q'(x+1) \leq AQ'(x)$ for $x > 0$ and some A independent of x.

(See also the remarks in Section 11.5 about slightly less restrictive conditions on $Q(x)$.)

Typical examples of W above are $W(x) = \exp(-c|x|^\lambda)$, where $\lambda > 1$ and $c > 0$.

Let $L_{p,W}(a,b)$ be the Banach space of measurable functions f with finite norm $\|Wf\|_{L_p(a,b)}$. In this chapter when we write $\|f\|_p$ we will mean $\|f\|_{L_p(R)}$.

We will need a function $t \to t^*(t)$ for defining the modulus of smoothness given in this chapter.

Definition 11.2.2. For $Q(x)$ given in Definition 11.2.1, $t^* \equiv t^*(t)$ is given by $tQ'(t^*) = 1$ if this condition defines t^* uniquely; otherwise we set $t^* = 0$.

With the aid of t^* we now define our modulus as

$$\omega_r^*(f,t)_{W,p} \equiv \omega_r^*(f,t) = \sup_{0 < h \le t} \|W\Delta_h^r f\|_{L_p[-h^*,h^*]} + \|W(f - P_{r,t}(f))\|_{L_p[t^*,\infty)}$$
$$+ \|W(f - \tilde{P}_{r,t}(f))\|_{L_p(-\infty,-t^*]}, \qquad (11.2.2)$$

where $P_{r,t}(f)$ and $\tilde{P}_{r,t}(f)$ are the orthogonal projections of f onto Π_{r-1} in $L_{2,W}(t^*,\infty)$ and $L_{2,W}(-\infty,-t^*)$ respectively. We note that for the weight function $W(x)$ of Definition 11.2.1 and $f \in L_{p,W}(R)$ we have

$$\int_{-\infty}^{\infty} |f(x)||x|^i W(x)^2 \, dx < \infty$$

and therefore the projections $P_{r,t}(f)$ given as the unique polynomials satisfying

$$\int_{t^*}^{\infty} x^i P_{r,t}(f)(x) \, dx = \int_{t^*}^{\infty} x^i f(x) \, dx, \qquad 0 \le i < r$$

and $\tilde{P}_{r,t}$ given similarly are well defined.

The K-functional is given by

$$K_r(f,t^r)_{W,p} = \inf_{g^{(r-1)} \in \text{A.C.}_{\text{loc}}} \{\|W(f-g)\|_p + t^r\|Wg^{(r)}\|_p\}. \qquad (11.2.3)$$

We are now able to state our characterization result which we will prove in Section 11.4.

Theorem 11.2.3. Suppose $W_Q = W$ is as given in Definition 11.2.1, ω_r^* and K_r are given by (11.2.2) and (11.2.3) respectively, r is any positive integer, $1 \le p \le \infty$ and $0 < t \le 1$. Then

$$C^{-1}\omega_r^*(f,t)_{W,p} \le K_r(f,t^r)_{W,p} \le C\omega_r^*(f,t)_{W,p}, \qquad (11.2.4)$$

where C is independent of $f \in L_{p,W}(R)$ and t.

The reason for the choice of $P_{r,t}(f)$ and $\tilde{P}_{r,t}(f)$ in the definition of ω_r^* in (11.2.2) is that these expressions are directly computable. A choice like

$$\inf_{P \in \Pi_{r-1}} \|W(f - P)\|_{L_p[t^*,\infty)} \quad \text{and} \quad \inf_{P \in \Pi_{r-1}} \|W(f - P)\|_{L_p(-\infty,-t^*]}$$

could replace

$$\|W(f - P_{r,t}(f))\|_{L_p[t^*,\infty)} \quad \text{and} \quad \|W(f - \tilde{P}_{r,t}(f))\|_{L_p(-\infty,-t^*]}$$

in (11.2.2) and still maintain (11.2.4). But while these projections are directly computable they certainly are not simple to compute, and simpler moduli $\bar{\omega}_r$

and Ω_r, which will yield almost all the strength of Theorem 11.2.3, will now be introduced. We define $\bar{\omega}_r(f,t)_{w,p}$ by

$$\bar{\omega}_r(f,t)_{w,p} = \sup_{0<h\leq t} \|W\Delta_h^r f\|_{L_p[-h^*,h^*]} + \|Wf\|_{L_p[t^*,\infty)} + \|Wf\|_{L_p(-\infty,-t^*]},$$
(11.2.5)

and the main-part modulus $\Omega_r(f,t)_{w,p}$ by

$$\Omega_r(f,t)_{w,p} = \sup_{0<h\leq t} \|W\Delta_h^r f\|_{L_p[-h^*,h^*]}.$$
(11.2.6)

For these moduli one can deduce the following theorems as corollaries of Theorem 11.2.3.

Theorem 11.2.4. *Under the assumptions of Theorem 11.2.3 we have for $\bar{\omega}_r$ given in (11.2.5)*

$$C^{-1}K_r(f,t^r)_{W,p} \leq \bar{\omega}_r(f,t)_{W,p} \leq CK_r(f,t^r)_{W,p} + Ce^{-c/t}\|Wf\|_p, \quad (11.2.7)$$

where C and $c > 0$ are independent of $f \in L_{p,W}(R)$ and $0 < t < 1$.

Since $\bar{\omega}_r(f,t)$ and $K_r(f,t^r)$ are greater than $Bt^r > 0$ (except when f is a polynomial of degree at most $r-1$), the second, exponentially small term on the right of (11.2.7) is negligible.

Theorem 11.2.5. *Under the assumptions of Theorem 11.2.3 we have for $\Omega_r(f,t)_{W,p}$ given in (11.2.6)*

$$C^{-1}\Omega_r(f,t)_{W,p} \leq K_r(f,t^r)_{W,p} \leq C\int_0^t (\Omega_r(f,\tau)_{W,p}/\tau)\,d\tau + Ce^{-c/t}\|Wf\|_p,$$
(11.2.8)

where C and $c > 0$ are independent of $f \in L_{p,W}(R)$ and $0 < t < 1$. In particular, for $0 < \alpha < r$

$$K_r(f,t^r)_{W,p} = O(t^\alpha) \Leftrightarrow \Omega_r(f,t)_{W,p} = O(t^\alpha) \Leftrightarrow \|W\Delta_h^r f\|_{L_p[-h^*,h^*]} = O(h^\alpha).$$

In fact it is $\|W\Delta_h^r f\|_{L_p[-h^*,h^*]}$ that is the most accessible expression from the computational point of view. For example, if

$$W_\lambda(x) = \exp(-|x|^\lambda), \qquad \lambda > 1,$$

then $t^* \sim t^{1/1-\lambda}$ and we can show for f given by

$$f(x) = (\exp|x|^\lambda)|x|^\gamma, \qquad \gamma < -1/p, \qquad |x| > 1 \quad \text{and} \quad f \in C^\infty(R)$$

that

$$\omega_r^*(f,t)_{W_\lambda,p} \sim \Omega_r(f,t)_{W_\lambda,p} \sim \begin{cases} t^r & \text{if } (\lambda-1)r + \gamma < -1/p \\ t^r \log 1/t & \text{if } (\lambda-1)r + \gamma = -1/p \\ t^{-(\gamma+1/p)/(\lambda-1)} & \text{if } (\lambda-1)r + \gamma > -1/p, \end{cases}$$
(11.2.9)

and, of course, the K-functional from (11.2.3) has also the same order.

As a second example we write
$$W(x) = \exp(-\exp|x|),$$
$f \in C^{\infty}(R)$ and
$$f(x) = \exp(\exp|x|)(\exp \gamma|x|)|x|^{\delta} \qquad (\gamma < 0) \quad \text{for } |x| > 1.$$
Then $t^* = \log 1/t$ and for large r
$$\omega_r^*(f,t)_{W,p} \sim \Omega_r(f,t)_{W,p} \sim t^{-\gamma}(\log 1/t)^{\delta}.$$

11.3. The Direct and Converse Results

In this section we apply the results of Section 11.2 to weighted polynomial approximation. For a weight function $W(x)$ the best weighted polynomial approximation is given by
$$E_n(f)_{W,p} = \inf_{P_n \in \Pi_n} \|W(f - P_n)\|_p \tag{11.3.1}$$
(where Π_n is the set of polynomials of degree at most n). Our aim is to characterize $E_n(f)_{W,p}$ using the moduli of smoothness defined in Section 11.2. For this characterization we will make somewhat stronger assumptions on the weight $W(x) = \exp(-Q(x))$ than in (11.2.1) which will be given in the following definitions.

Definition 11.3.1. A weight function W belongs to the Freud class of weights $W^*(Q)$ if W is given by (11.2.1) and $Q(x)$ satisfies:

(a) Q is a positive even function in $C^2(0, \infty)$,
(b) $\lim_{x \to \infty} xQ''(x)/Q'(x) = \gamma > 0$, and
(c) if $\gamma = 1$, or 3, then Q'' is nondecreasing.

These conditions are a sort of mixture of conditions used by G. Freud and H.N. Mhaskar [1] and A.L. Levin and D.S. Lubinsky [2].

The results below are valid for somewhat more general weights than those given in Definition 11.3.1 (those weights are dealt with for convenience). At any rate, the weights
$$W_\lambda(x) = \exp(-|x|^\lambda), \qquad \lambda > 1$$
satisfy these conditions.

For every n we define $q_n > 0$ by
$$q_n Q'(q_n) = n. \tag{11.3.2}$$
This is well-defined for large n and it is easy to see that
$$1 < d_1 \leq q_{2n}/q_n \leq d_2 < 2 \qquad (n \geq n_0). \tag{11.3.3}$$

11.3. The Direct and Converse Results

With these q_n the analogue of the Markov–Bernstein inequality is

$$\|WP_n'\|_p \le C(n/q_n)\|WP_n\|_p, \qquad P_n \in \Pi_n \tag{11.3.4}$$

and that of the Jackson–Favard inequality

$$E_n(g)_{W,p} \le C(q_n/n) E_{n-1}(g')_{W,p}, \qquad g \in \text{A.C.}_{\cdot\text{loc}}. \tag{11.3.5}$$

In the parameter range $\gamma \ge 1$ (see Definition 11.3.1) (11.3.4) was proved by Freud [3], [4] and his result was recently extended to $0 < \gamma < 1$ by Levin and Lubinsky [1], [2]. The inequality (11.3.5) was established by Freud [1], [2], [4], again only for $\gamma \ge 1$, but his method is applicable to $0 < \gamma < 1$ as well, since the necessary estimate for the (generalized) Christoffel functions was proved by Levin and Lubinsky [1], [2].

Theorem 11.3.2 (Freud and Mhaskar). *The inequalities (11.3.4) and (11.3.5) together with (11.3.3) imply*

$$E_n(f)_{W,p} \le CK_r(f,(q_n/n)^r)_{W,p} \tag{11.3.6}$$

and

$$K_r(f,(q_n/n)^r)_{W,p} \le C(q_n/n)^r \sum_{k=1}^n k^{r-1} q_k^{-r} E_k(f)_{W,p}. \tag{11.3.7}$$

The proof follows easily from (11.3.2)–(11.3.5) (see also Freud and Mhaskar [1] and [2]).

We can now use Theorem 11.2.3 and substitute here $\omega_r^*(f,t)_{W,p}$ for $K_r(f,t^r)_{W,p}$ by which we obtain:

Theorem 11.3.3. *Suppose $W \in W^*(Q)$ (see Definition 11.3.1). Then for every $1 \le p \le \infty$ and $r \ge 1$ we have*

$$E_n(f)_{W,p} \le M\omega_r^*\left(f, \frac{q_n}{n}\right)_{W,p} \tag{11.3.8}$$

and

$$\omega_r^*\left(f, \frac{q_n}{n}\right)_{W,p} \le M\left(\frac{q_n}{n}\right)^r \sum_{k=0}^n k^{r-1} q_k^{-r} E_k(f)_{W,p}, \tag{11.3.9}$$

where M is independent of $f \in L_{p,W}(R)$ and q_n is given by (11.3.2). In particular, if $0 < \alpha < r$, then

$$E_n(f)_{W,p} = O\left(\left(\frac{q_n}{n}\right)^\alpha\right) \qquad (n = 1, 2, \ldots)$$

if and only if

$$\omega_r^*(f,t)_{W,p} = O(t^\alpha) \qquad (t \to 0),$$

or, equivalently,

$$\|\Delta_h^r f\|_{L_p(-h^*, h^*)} = O(h^\alpha) \qquad (h \to 0).$$

We remind the reader that $E_n(f)_{W,p}$, $\omega_r^*(f,t)$, and h^* were defined in (11.3.1), (11.2.2), and Definition 11.2.2 respectively. For the proof of Theorem 11.3.3 it is enough to remark that $W \in W^*(Q)$ (Definition 11.3.1) implies $W \in W(Q)$ (Definition 11.2.1), and therefore, Theorems 11.2.3 and 11.2.5 are applicable.

The following corollary is worth mentioning separately.

Corollary 11.3.4. *Let* $W_\lambda(x) \equiv \exp(-|x|^\lambda)$ *with* $\lambda > 1$. *Then for any* $1 \le p \le \infty$, $r \ge 1$ *and* $0 < \alpha < r$,

$$E_n(f)_{W_\lambda,p} = O(n^{-\alpha(1-(1/\lambda))}) \Leftrightarrow \|W_\lambda \Delta_h^r f\|_{L_p(-h^{1/1-\lambda}, h^{1/1-\lambda})} = O(h^\alpha).$$

PROOF. For W_λ the function $t^*(t) \equiv t^*$ (cf. Definition 11.2.2) is given by

$$t^* = t^{1/1-\lambda} \lambda^{1/1-\lambda} < t^{1/1-\lambda}$$

and the sequence q_n (see (11.3.2)) by $q_n = (n/\lambda)^{1/\lambda}$. Therefore, the sufficiency ($\Leftarrow$) follows from Theorem 11.3.3. By Theorems 11.3.3 (11.3.9), 11.2.3, and 11.2.4

$$E_n(f)_{W_\lambda,p} = O(n^{-\alpha(1-(1/\lambda))})$$

implies (cf. (11.2.5))

$$\bar{\omega}_r(f,t)_{W_\lambda,p} = O(t^\alpha)$$

from which the necessity ($\Rightarrow$) is easy to deduce. $\square$

As an illustration let us consider the functions f from the end of Section 11.2, i.e.,

$$f(x) = (\exp|x|^\lambda)|x|^\gamma, \qquad \gamma < -1/p \quad \text{for } |x| \ge 1 \quad \text{and } f \in C^\infty(R).$$

By (11.2.9) and Theorem 11.3.3 we have for this function ($\lambda > 1$)

$$E_n(f)_{W_\lambda,p} \sim n^{-(\gamma+1/p)/(\lambda-1)}.$$

11.4. Proof of the Equivalence Result

In this section we shall prove the equivalence results between the K-functional and the various moduli; that is, we shall prove the theorems stated in Section 11.2. To proceed with the proof, we will follow a sequence of lemmas that will be needed for proving

$$\|W(f - P_{r,t}(f))\|_{L_p(t^*,\infty)} \le MK_r(f,t^r)_{W,p}$$

and for proving Theorem 11.2.5.

Lemma 11.4.1. *Suppose* $W = W_Q$ *and* t^* *are given by Definitions 11.2.1 and 11.2.2 respectively. Then for* $1 \le p \le \infty$, $g(u) \ge 0$, *and* $\xi \ge t^*$ *we have*

11.4. Proof of the Equivalence Result

$$\left\| W(x) \int_\xi^x g(u)\, du \right\|_{L_p(\xi,\infty)} \le Mt \|Wg\|_{L_p(\xi,\infty)}, \tag{11.4.1}$$

where M is independent of g, t, and ξ.

PROOF. We may assume t so small that in (t^*, ∞) the functions $W(x)$ and $Q'(x)$ are decreasing and increasing respectively. We can now write

$$W(x) \int_\xi^x g(u)\, du \le W(x)^{1/2} \int_\xi^x W(u)^{1/2} g(u)\, du$$

$$\le W(x)^{1/2} t \int_\xi^x W(u)^{-1/2} Q'(t^*) [W(u) g(u)]\, du$$

$$\le W(x)^{1/2} t \int_\xi^x W(u)^{-1/2} Q'(u) [W(u) g(u)]\, du.$$

The simple estimate

$$W(x)^{1/2} \int_\xi^x W(u)^{-1/2} Q'(u)\, du \le 2$$

(which we obtain using integration by substitution) will imply (11.4.1) for $p = \infty$. Moreover, the above estimates, Jensen's inequality, and Fubini's theorem imply for $1 \le p < \infty$

$$\left\{ \int_\xi^\infty \left(W(x) \int_\xi^x g(u)\, du \right)^p dx \right\}^{1/p}$$

$$\le 2t \left\{ \int_\xi^\infty W(x)^{1/2} \int_\xi^x W(u)^{-1/2} Q'(u) |W(u)g(u)|^p\, du\, dx \right\}^{1/p}$$

$$\le 2t \left\{ \int_\xi^\infty W(x)^{1/2} Q'(x) \int_\xi^x W(u)^{-1/2} |W(u)g(u)|^p\, du\, dx \right\}^{1/p}$$

$$\le 2t \left\{ \int_\xi^\infty |W(u)g(u)|^p W(u)^{-1/2} \int_u^\infty W(x)^{1/2} Q'(x)\, dx\, du \right\}^{1/p}$$

$$= 2^{1+(1/p)} t \left\{ \int_\xi^\infty |W(u)g(u)|^p\, du \right\}^{1/p}. \qquad \square$$

Lemma 11.4.2. *There exists a constant $M \equiv M(r, p)$ independent of t and $P \in \Pi_r$ such that*

$$\|PW\|_{L_p[t^*,\infty)} \le M \|PW\|_{L_p[t^*, t^*+t]}, \qquad P \in \Pi_r. \tag{11.4.2}$$

PROOF. We may assume t as small as we wish. In particular t can be chosen so small that W and Q are decreasing and increasing in (t^*, ∞) respectively. We use induction on r. For $r = 0$ we deal only with $1 \le p < \infty$, as for $p = \infty$

and $r = 0$ the result is obvious. We follow the proof of the last lemma and obtain

$$\left\{\int_{t^*}^{\infty} W(u)^p \, du\right\} \leq \left\{t \int_{t^*}^{\infty} W(u)^p Q'(u) \, du\right\}^{1/p} = p^{-1/p} t^{1/p} W(t^*) \leq t^{1/p} W(t^*).$$

We observe that

$$\|W\|_{L_p[t^*, t^*+t]} \geq t^{1/p} W(t^* + t),$$

and as $W(t^* + t) \geq M_1 W(t^*)$, which follows from

$$Q(t^* + t) = Q(t^*) + tQ'(t^* + \zeta) \leq Q(t^*) + tQ'(t^* + 1)$$
$$\leq Q(t^*) + tAQ'(t^*) \leq Q(t^*) + A$$

(see condition (c) of Definition 11.2.1), we have (11.4.2) for $r = 0$. Assuming (11.4.2) for $r - 1$, we write

$$\|PW\|_{L_p[t^*, t^*+t]} = |P(\xi)| \|W\|_{L_p[t^*, t^*+t]} \geq |P(\xi)| t^{1/p} W(t + t^*)$$
$$\geq M|P(\xi)| t^{1/p} W(t^*).$$

With this ξ we have $P(x) = P(\xi) + \int_{\xi}^{x} P'(u) \, du$ and hence by Lemma 11.4.1 and the induction hypothesis we obtain

$$\|PW\|_{L_p[t^*, \infty)} \leq \|PW\|_{L_p[t^*, \xi]} + |P(\xi)| \|W\|_{L_p[t^*, \infty)}$$
$$+ \left\|W(x) \int_{\xi}^{x} P'(u) \, du\right\|_{L_p(\xi, \infty)}$$
$$\leq \|PW\|_{L_p[t^*, t^*+t]} + |P(\xi)| t^{1/p} W(t^*) + M_2 t \|WP'\|_{L_p[t^*, \infty)}$$
$$\leq M_3(\|PW\|_{L_p[t^*, t^*+t]} + t\|WP'\|_{L_p[t^*, t^*+t]}).$$

For a polynomial q of fixed degree r the Markov–Bernstein inequality implies

$$\|q'(x)\|_{L_p[0,1]} \leq B \|q(x)\|_{L_p[0,1]},$$

where B depends only on r, and therefore

$$t\|WP'\|_{L_p[t^*, t^*+t]} \leq tW(t^*)\|P'\|_{L_p[t^*, t^*+t]} \leq BW(t^*)\|P\|_{L_p[t^*, t^*+t]}$$
$$\leq M_4 \|WP\|_{L_p[t^*, t^*+t]},$$

which completes the proof. $\square$

Lemma 11.4.3. *For $1 \leq p \leq \infty$ and $r \geq 0$ there exists a constant $M = M(r, p)$ independent of $P \in \Pi_r$ and $0 < t \leq 1$ such that*

$$\|PW\|_{L_p[t^*, \infty)} \leq Mt^{1/p - 1/2} \|PW\|_{L_2[t^*, \infty)}, \qquad P \in \Pi_r. \qquad (11.4.3)$$

PROOF. For any $1 \leq p \leq \infty$

$$\|P\|_{L_p[0,1]} \leq M_1(r, p) \|P\|_{L_2[0,1]} \quad \text{if } P \in \Pi_r,$$

and therefore, following Lemma 11.4.2,

$$\|PW\|_{L_p[t^*,\infty)} \leq A\|PW\|_{L_p[t^*,t^*+t]} \leq A_1 W(t^*)\|P\|_{L_p[t^*,t^*+t]}$$
$$\leq A_1 M_1(r,p)W(t^*)t^{1/p-1/2}\|P\|_{L_2[t^*,t^*+t]} \leq Mt^{1/p-1/2}\|WP\|_{L_2[t^*,t^*+t]}$$
$$\leq Mt^{1/p-1/2}\|WP\|_{L_2[t^*,\infty)}. \qquad \square$$

Lemma 11.4.4. *For W, $P_{r,t}(f)$, and t^* as in Theorem 11.2.3 and $Wf \in L_p$ we have*

$$\|WP_{r,t}(f)\|_{L_p[t^*,\infty)} \leq M\|Wf\|_{L_p[t^*,\infty)}, \qquad (11.4.4)$$

where M is independent of $f \in L_{p,W}$ and $0 < t < 1$.

PROOF. For $P_0, \ldots, P_{r-1}$, the orthonormal sequence of polynomials with respect to W^2 on (t^*, ∞), we may write

$$P_{r,t}(f) = \sum_{i=0}^{r-1} c_i P_i,$$

where for q satisfying $q^{-1} + p^{-1} = 1$ we have

$$|c_i| = \left|\int_{t^*}^{\infty} f(u)P_i(u)W(u)^2\, du\right| \leq \|fW\|_{L_p[t^*,\infty)}\|P_i W\|_{L_q[t^*,\infty)}.$$

Using Lemma 11.4.3, we write

$$\|WP_{r,t}(f)\|_{L_p[t^*,\infty)} \leq \sum_{i=0}^{r-1} |c_i|\|WP_i\|_{L_p[t^*,\infty)}$$
$$\leq M\sum_{i=0}^{r-1}\|Wf\|_{L_p[t^*,\infty)}t^{1/q-1/2}\|WP_i\|_{L_2[t^*,\infty)}t^{1/p-1/2}\|WP_i\|_{L_2[t^*,\infty)}$$
$$= rM\|Wf\|_{L_p[t^*,\infty)}. \qquad \square$$

PROOF OF THEOREM 11.2.3. First we estimate the K-functional from below; that is, we show

$$\omega_r^*(f,t)_{W,p} \leq MK_r(f,t^r)_{W,p}.$$

To do that, we have to prove

$$\|W\Delta_h^r f\|_{L_p[-h^*,h^*]} \leq M_1 K_r(f,t^r)_{W,p} \quad \text{for all } 0 < h \leq t, \qquad (11.4.5)$$
$$\|W(f - P_{r,t}(f))\|_{L_p[t^*,\infty)} \leq M_2 K_r(f,t^r)_{W,p}, \qquad (11.4.6)$$

and an analogue of (11.4.6) for $\tilde{P}_{r,t}(f)$ and $L_p(-\infty, -t^*]$ (the proof of which is identical to the proof of (11.4.6) and will be omitted). To prove (11.4.6), we recall that for any $P \in \Pi_{r-1}$, $\omega_r(f - P, t) = \omega_r(f,t)$, $K_r(f - P, t^r)_{W,p} = K_r(f,t^r)_{W,p}$, and $P_{r,t}(P) = P$. We choose g such that

$$\|W(f-g)\|_p + t^r\|Wg^{(r)}\|_p \leq 2K_r(f,t^r)_{W,p},$$

and therefore for any $P \in \Pi_{r-1}$ $g - P = g_1$ satisfies

$$\|W(f - P - g_1)\|_p + t^r\|Wg_1^{(r)}\|_p \leq 2K_r(f - P, t^r)_{W,p} = 2K_r(f,t^r)_{W,p}.$$

We now choose P such that $g_1^{(i)}(t^*) = 0$, for $0 \leq i < r$, and write

$$\|W(f - P_{r,t}(f))\|_{L_p[t^*, \infty)} = \|W(f - P) - P_{r,t}(f - P)\|_{L_p[t^*, \infty)}$$
$$\leq \|W(f - P - g_1)\|_{L_p[t^*, \infty)}$$
$$+ \|WP_{r,t}(f - P - g_1)\|_{L_p[t^*, \infty)} + \|Wg_1\|_{L_p[t^*, \infty)}$$
$$+ \|WP_{r,t}(g_1)\|_{L_p[t^*, \infty)}.$$

Applying Lemma 11.4.4, we have

$$\|WP_{r,t}(f - P - g_1)\|_{L_p[t^*, \infty)} \leq M\|W(f - P - g_1)\|_{L_p[t^*, \infty)}$$

and

$$\|WP_{r,t}(g_1)\|_{L_p[t^*, \infty)} \leq M\|Wg_1\|_{L_p[t^*, \infty)}.$$

Using the choice of g and g_1, we have

$$\|W(f - P - g_1)\|_{L_p[t^*, \infty)} \leq 2K_r(f, t^r)_{W, p},$$

and therefore, to complete the proof of (11.4.6), we need only to show the estimate

$$\|Wg_1\|_{L_p[t^*, \infty)} \leq MK_r(f, t^r)_{W, p}.$$

We use Lemma 11.4.1, with t^* and $g_1^{(r)}$ playing the roles of ξ and g there respectively, to obtain

$$\|Wg_1^{(r-1)}\|_{L_p[t^*, \infty)} \leq Mt\|Wg_1^{(r)}\|_{L_p[t^*, \infty)}.$$

Repeating the above process inductively with $g_1^{(r-i)}$ taking the place of g in Lemma 11.4.1, and observing $g_1^{(i)}(t^*) = 0$ for $0 \leq i < r$ we derive

$$\|Wg_1\|_{L_p[t^*, \infty)} \leq M^r t^r \|Wg_1^{(r)}\|_{L_p[t^*, \infty)} \leq 2M^r K_r(f, t^r)_{W, p}$$

which completes the proof of (11.4.6).

To prove (11.4.5) we again choose g such that

$$\|W(f - g)\|_p + t^r\|Wg^{(r)}\|_p \leq 2K_r(f, t^r)_{W, p}$$

and write

$$\|W\Delta_h^r f\|_{L_p[-h^*, h^*]} \leq \|W\Delta_h^r(f - g)\|_{L_p[-h^*, h^*]} + \|W\Delta_h^r g\|_{L_p[-h^*, h^*]}.$$

Using conditions on the weight function W and following

$$Q(x + \eta h) = Q(x) + \eta h Q'(\xi) \leq Q(x) + \eta h O(Q'(h^*)) = Q(x) + O(1),$$

for $|\eta| \leq r$ and $|x| \leq h^* + 1$, we have

$$C^{-1}W(x) \leq W(x + \eta h) \leq CW(x) \quad \text{for } |\eta| < \frac{r}{2} \quad \text{and}$$
$$x \in [-h^* - rh/2, h^* + rh/2]. \tag{11.4.7}$$

11.4. Proof of the Equivalence Result

Therefore,

$$\|W\Delta_h^r(f-g)\|_{L_p[-h^*,h^*]} \le L\|W(f-g)\|_{L_p[-h^*-1,h^*+1]} \le L_1 K_r(f,t^r)_{W,p}.$$

Using standard evaluation, we have

$$|\Delta_h^r g(x)| = \left|\int_{-h/2}^{h/2} \cdots \int_{-h/2}^{h/2} g^{(r)}(x+u_1+\cdots+u_r)\,du_1\cdots du_r\right|$$

$$\le Ah^{r-1}\int_{-rh/2}^{rh/2}|g^{(r)}(x+u)|\,du.$$

The above, (11.4.7), and $0 < h \le t$ imply now

$$\|W\Delta_h^r g\|_{L_p[-h^*,h^*]} \le A_1 h^{r-1}\left\|\int_{-rh/2}^{rh/2} W(\cdot+u)|g^{(r)}(\cdot+u)|\,du\right\|_{L_p[-h^*,h^*]}$$

$$\le A_2 h^r \|Wg^{(r)}\|_p \le A_3 K_r(f,t^r)_{W,p}.$$

The estimate of the K-functional from above is more routine. We construct g_t in the following way:

$$g_t(x) = \psi(1+(t^*-x)/t)\psi(1+(t^*+x)/t)f_\tau(x)$$
$$+ (1-\psi(1+(t^*-x)/t))P_{r,t}(x) \qquad (11.4.8)$$
$$+ (1-\psi(1+(x+t^*)/t))\tilde{P}_{r,t}(x),$$

where $\psi(x) = 0$ for $x \le 0$, $\psi(x) = 1$ for $x \ge 1$, and $\psi(x) \in C^\infty$; and where

$$f_\tau(x) \equiv r^r \int_{-1/2r}^{1/2r}\cdots\int_{-1/2r}^{1/2r} \sum_{k=1}^r (-1)^{k+1}\binom{r}{k} f(t+k\tau(u_1+\cdots+u_r))\,du_1\cdots du_r$$

with $0 < \tau \le t$ to be chosen so that $\tau^* > t^* + t$ but $t > \tau > At$ for some $A > 0$ independent of t. (Definition 11.2.1 (c) guarantees that such a choice of τ is possible.) Obviously, for $-t^* \le x \le t^*$, $g_t = f_\tau$, for $x \ge t^* + t$, $g_t = P_{r,t}(x)$, for $x < -t^* - t$, $g_t = \tilde{P}_{r,t}$, for $t^* \le x \le t^* + t$,

$$g_t(x) = P_{r,t}(x) + \psi(1-(t^*-x)/t)(f_\tau(x) - P_{r,t}(x)),$$

and for $-t^* - t \le x < -t^*$,

$$g_t(x) = \tilde{P}_{r,t}(x) + \psi(1+(x+t^*)/t)(f_\tau(x) - \tilde{P}_{r,t}(x)).$$

With f_τ defined as above, we have

$$f(x) - f_\tau(x)$$
$$= (-1)^r r^r \int_{-1/2r}^{1/2r}\cdots\int_{-1/2r}^{1/2r} \Delta_{\tau(u_1+\cdots+u_r)}^r f(x+k\tau(u_1+\cdots+u_r)/2)\,du_1\cdots du_r$$
$$\qquad\qquad(11.4.9)$$

and

$$\left(\frac{d}{dx}\right)^r f_\tau(x) = r^r \sum_{k=1}^r (-1)^{k+1}\binom{r}{k}(k\tau)^{-r}\Delta_{k\tau/r}^r f(x). \qquad (11.4.10)$$

Using (11.4.7), (11.4.9), (11.4.10), $\tau > At$, and the definition of $\omega_r^*(f,t)_{W,p}$ (see (11.2.2)), one can see that for $D = (-t^*, t^*)$, $D = (t^* + t, \infty)$, or $D = (-\infty, -t^* - t)$,

$$\|W(f - g_t)\|_{L_p(D)} + t^r \|Wg_t^{(r)}\|_{L_p(D)} \leq M\omega_r^*(f,t)_{W,p}.$$

The identities (11.4.9) and (11.4.10) together with $\tau^* > t + t^*$ yield

$$\|W(f - f_\tau)\|_{L_p[-t-t^*, t+t^*]} + t^r \|Wf_\tau^{(r)}\|_{L_p[-t-t^*, t+t^*]} \\ \leq M_1 \omega_r^*(f,\tau)_{W,p} \leq M_1 \omega_r^*(f,t)_{W,p}. \tag{11.4.11}$$

Since we also have

$$\|W(f - P_{r,t}(f))\|_{L_p(t^*, t^*+t)} + t^r \|W(P_{r,t}(f))^{(r)}\|_{L_p(t^*, t^*+t)} \\ \leq \|W(f - P_{r,t}(f))\|_{L_p(t^*, t^*+t)} \leq \omega_r^*(f,t)_{W,p} \tag{11.4.12}$$

and a similar inequality for $\tilde{P}_{r,t}(f)$ on $[-t^* - t, -t^*]$, the rest of the proof follows the "patching together" technique of Sections 2.2 and 2.3. More precisely, in the remaining intervals (i.e., $(-t^* - t, -t^*)$ and $(t^*, t^* + t)$) we have to recall only that $\psi + (1 - \psi) = 1$ for the estimate of $\|W(f - g_t)\|$. For the estimate of $\|Wg_t^{(r)}(r)\|$ in $D = (t^*, t^* + t)$, for example, we use

$$|\psi^{(j)}(1 + (t^* - x)/t)| \leq t^{-j} M(j)$$

to obtain

$$t^r \|Wg_t^{(r)}\|_{L_p[t^*, t^*+t]} \leq t^r \|W(P_{r,t}(f))^{(r)}\|_{L_p[t^*, t^*+t]} \\ + M_1 \sum_{i=0}^{r} \|W(f_\tau - P_{r,t}(f))^{(i)}\|_{L_p[t^*, t^*+t]} t^i$$

and for $G = f_\tau - P_{r,t}(f)$, $a = t^*$, and $b = t^* + t$ we use the inequality (2.2.14), that is

$$\|G^{(i)}\|_{L_p[a,b]} \leq M((b-a)^{-i} \|G\|_{L_p[a,b]} + (b-a)^{r-i} \|G^{(r)}\|_{L_p[a,b]}).$$

We need also the estimates

$$\|f_\tau - P_{r,t}(f)\|_{L_p[t^*, t^*+t]} \leq \|f - f_\tau\|_{L_p[t^*, t^*+t]} + \|f - P_{r,t}(f)\|_{L_p[t^*, t^*+t]}, \\ \leq M\omega_r^*(f,t)_{W,p},$$

and

$$t^r \|W(f_\tau - P_{r,t}(f))^{(r)}\|_{L_p[t^*, t^*+t]} = t^r \|Wf_\tau^{(r)}\|_{L_p[t^*, t^*+t]} \leq M\omega_r^*(f,t)_{W,p},$$

which follow from the inequalities (11.4.11) and (11.4.12). These estimates and the corresponding ones on $(-t^* - t, -t^*)$ complete the proof of the upper estimate for the K-functionals and therefore the proof of Theorem 11.2.3. $\square$

PROOF OF THEOREM 11.2.4. As $K\bar{\omega}_r(f,t)_{W,p} \geq \omega_r^*(f,t)_{W,p}$ (see Lemma 11.4.4), the inequality $C^{-1}K_r(f,t^r)_{W,p} \leq \bar{\omega}_r(f,t)_{W,p}$ is obvious. To obtain the estimate

$$\bar{\omega}_r(f,t)_{W,p} \leq CK_r(f,t^r)_{W,p} + Ce^{-c/t} \|Wf\|_p$$

11.4. Proof of the Equivalence Result

we first choose m so large that $(t/m)^* > t^* + 2$ for all t which is possible because of Definition 11.2.1(c). It is enough to show

$$\bar{\omega}_r(f, t/m^2)_{W,p} \leq C K_r(f, t^r)_{W,p} + C e^{-c/t} \|Wf\|_p,$$

and therefore it suffices to prove

$$\|Wf\|_{L_p[(t/m^2)^*, \infty)} \leq C_1 (K_r(f, t^r)_{W,p} + e^{-c/t} \|Wf\|_p)$$

(and the analogous estimate on $(-\infty, -(t/m^2)^*)$. We now write, using (11.4.6) and Lemma 11.4.2,

$$\|Wf\|_{L_p[(t/m^2)^*, \infty)} \leq \|W(f - P_{r,t}(f))\|_{L_p[(t/m^2)^*, \infty)} + \|WP_{r,t}(f)\|_{L_p[(t/m^2)^*, \infty)}$$

$$\leq M K_r(f, t^r)_{W,p} + M \|WP_{r,t}(f)\|_{L_p[(t/m^2)^*, (t/m^2)^* + t/m^2]},$$

and therefore we have to estimate only

$$\|WP_{r,t}(f)\|_{L_p[(t/m^2)^*, (t/m^2)^* + t/m^2]} \leq W((t/m^2)^*) \|P_{r,t}(f)\|_{L_p[(t/m^2)^*, (t/m^2)^* + t/m^2]}.$$

For $\lambda \geq 1$ and $P \in \Pi_{r-1}$ one has the simple estimate

$$\|P\|_{L_p[-\lambda, \lambda]} \leq C(r,p) \lambda^r \|P\|_{L_p[-1, 1]},$$

which, using translation by a, yields

$$\|P\|_{L_p[-\lambda + a, \lambda + a]} \leq C(r,p) \lambda^r \|P\|_{L_p[-1 + a, 1 + a]}$$

(also for $\lambda \geq 1$ and $P \in \Pi_{r-1}$), where $C(r,p)$ is independent of λ and a. We choose $a = (t/m)^* - 1$ and $\lambda = (t/m^2)^* + 2 - (t/m)^*$ to obtain for small t

$$W((t/m^2)^*) \|P_{r,t}(f)\|_{L_p[(t/m^2)^*, (t/m^2)^* + 1]}$$

$$\leq C(r,p) W((t/m^2)^*)((t/m^2)^* + 2 - (t/m)^*)^r \|P_{r,t}(f)\|_{L_p[(t/m)^* - 2, (t/m)^*]}$$

$$\leq 2^r C(r,p) W((t/m^2)^*) W((t/m)^*)^{-1} ((t/m^2)^* - (t/m)^*)^r \|WP_{r,t}(f)\|_{L_p[t^*, (t/m)^*]}.$$

Recalling Lemma 11.4.4, we have

$$\|WP_{r,t}(f)\|_{L_p[t^*, (t/m)^*]} \leq \|WP_{r,t}(f)\|_{L_p[t^*, \infty)} \leq M \|Wf\|_{L_p[t^*, \infty)}.$$

We now complete the proof by estimating the coefficient of $\|WP_{r,t}(f)\|_{L_p[t^*, (t/m)^*]}$; that is

$$W((t/m^2)^*) W((t/m)^*)^{-1} ((t/m^2)^* - (t/m)^*)^r$$

$$= \exp\left(-\int_{(t/m)^*}^{(t/m^2)^*} Q'(v) \, dv\right) ((t/m^2)^* - (t/m)^*)^r$$

$$\leq \exp\left(-\frac{m}{t} \int_{(t/m)^*}^{(t/m^2)^*} dv\right) ((t/m^2)^* - (t/m)^*)^r \leq M e^{-c/t},$$

where at the last step we used

$$\int_{(t/m)^*}^{(t/m^2)^*} dv = (t/m^2)^* - (t/m)^* \geq 2.$$

This completes the proof and for future reference we mention that in the course of the proof we actually verified the inequality

$$\|WP\|_{L_p((t/m^2)^*, \infty)} \leq M e^{-c/t} \|WP\|_{L_p(t^*, (t/m)^*)}, \tag{11.4.13}$$

where $M = M(r, p)$ is independent of $P \in \Pi_r$ and $0 < t < 1$. □

PROOF OF THEOREM 11.2.5. Using Theorems 11.2.3 and 11.2.4, we have to show only

$$\|Wf\|_{L_p[(t/m^2)^*, \infty)} \leq C \int_0^{t/m^2} (\Omega_r(f, \tau)_{W, p}/\tau) \, d\tau + Ce^{-c/t} \|Wf\|_p$$

(and an analogous estimate on $(-\infty, -t^*]$). Using the process of the proof of Theorem 11.2.3, we can construct G_t such that

$$\|W(f - G_t)\|_{L_p(-(t/m^3)^*, (t/m^3)^*)} + t^r \|WG_t^{(r)}\|_{L_p[-(t/m^3)^*, (t/m^3)^*]} \leq M\Omega_r(f, t/m^3)_{W, p}.$$

We can now choose P such that $P^{(i)}(t^*) - G_t^{(i)}(t^*) = 0$ for $i = 0, 1, \ldots, r - 1$, and write $g_t = G_t - P$ which implies

$$\|W(f - P - g_t)\|_{L_p[-(t/m^3)^*, (t/m^3)^*]} + t^r \|Wg_t^{(r)}\|_{L_p[-(t/m^3)^*, (t/m^3)^*]}$$
$$\leq M\Omega_r(f, t/m^3)_{W, p}.$$

Since repeated use of Lemma 11.4.1 implies

$$\|Wg_t\|_{L_p[t^*, (t/m)^*)} \leq M_1 t^r \|Wg_t^{(r)}\|_{L_p[t^*, (t/m)^*)} \leq M_2 \Omega_r(f, t/m^3)_{W, p},$$

we have

$$\|W(f - P)\|_{L_p[t^*, (t/m^3)^*]} \leq M_3 \Omega_r(f, t/m^3)_{W, p}.$$

We now use (11.4.13) obtained in the proof of the preceding theorem and write

$$\|Wf\|_{L_p[(t/m^2)^*, (t/m^3)^*]} \leq \|W(f - P)\|_{L_p[t^*, (t/m^3)^*]} + \|WP\|_{L_p[(t/m^2)^*, (t/m^3)^*]}$$
$$\leq M_3(\Omega_r(f, t/m^3)_{W, p} + \|WP\|_{L_p[(t/m^2)^*, \infty)})$$
$$\leq M_4(\Omega_r(f, t/m^3)_{W, p} + e^{-c/t}\|WP\|_{L_p[t^*, (t/m)^*]}).$$

The inequality

$$\|WP\|_{L_p[t^*, (t/m)^*]} \leq \|W(f - P)\|_{L_p[t^*, (t/m)^*]} + \|Wf\|_{L_p[t^*, (t/m)^*]}$$

now implies

$$\|Wf\|_{L_p[(t/m^2)^*, (t/m^3)^*]} \leq M_5(\Omega_r(f, t/m^3)_{W, p} + e^{-c/t}\|Wf\|_{L_p[t^*, (t/m)^*]},$$

and therefore,

$$\|Wf\|_{L_p[(t/m^2)^*, \infty)} \leq \sum_{i=0}^{\infty} \|Wf\|_{L_p[(t/m^{2+i})^*, (t/m^{3+i})^*]}$$
$$\leq A \left(\sum_{i=0}^{\infty} \Omega_r(f, t/m^{3+i})_{W, p} + \sum_{i=0}^{\infty} e^{-cm^i/t} \|Wf\|_p \right)$$
$$\leq A_1 \left\{ \int_0^{t/m^2} (\Omega_r(f, v)_{W, p}/v) \, dv + e^{-c/t} \|Wf\|_p \right\}. \quad □$$

11.5. Comparisons and Generalizations

We first note that the condition that $Q(x)$ is even, that everyone seems to impose, is not necessary. Each side, however, would have to satisfy separately the same type of conditions. The result would look a bit messier (though somewhat more general) but the proof would be the same as we treated each side separately anyway.

The equivalence theorem 11.2.3 is valid for the more general class of weights $W_Q(x) = \exp(-Q(x))$, where Q satisfies:

(a) $Q(x)$ is even continuous and increasing in R_+,
(b) $Q(x + \eta) - Q(x)$ is increasing and tends to infinity for all $\eta > 0$, and
(c) $t^* = \sup\{x; Q(x + rt) - Q(x) \le 1\}$.

Theorem 11.2.3 can be proved using the above with minor changes in the main body of the proof and substantial changes in the proof of Lemmas 11.4.1 and 11.4.2. The above conditions, however, do not seem to be sufficient for the proof of Theorems 11.2.4 and 11.2.5.

Freud and Mhaskar had for $r = 1$ and 2 a different definition of the modulus of smoothness. They denote

$$\omega_1(p, Q, f, \delta) = \sup_{|t| \le \delta} \|\Delta_t(W_Q f)\|_p + \delta \|Q'_\delta W_Q f\|_p \tag{11.5.1}$$

and

$$\omega_2(p, Q, f, \delta) = \sup_{|t| \le \delta} \|\Delta_t^2(W_Q f)\|_p + \delta \sup_{|t| \le \delta} \|Q'_\delta \Delta_t(W_Q f)\|_p$$
$$+ \delta^2 \|Q'^2_\delta W_Q f\|_p, \tag{11.5.2}$$

where $Q'_\delta = \min(\delta^{-1}, (1 + Q'(x)^2)^{1/2})$ and their moduli are given by

$$\Omega_1(p, Q, f, \delta) = \inf_{a \in R} \omega_1(p, Q, f - a, \delta) \tag{11.5.3}$$

and

$$\Omega_2(p, Q, f, \delta) = \inf_{a, b \in R} \omega_2(p, Q, f - a - bx, \delta). \tag{11.5.4}$$

Because of the equivalence theorem, we have

$$\Omega_1(p, Q, f, \delta) \sim \omega_1^*(f, \delta)_{W, p} \quad \text{and} \quad \Omega_2(p, Q, f, \delta) \sim \omega_2^*(f, \delta)_{W, p}$$

for W as restricted by Freud and Mhaskar. Our modulus $\omega_r^*(f, t)_{W, p}$ is simpler, and available (and equivalent to the K-functional) for all r. Moreover, if the behavior of $K_r(f, h^r)_{W, p}$ or $\omega_r^*(f, h)_{W, p}$ is $O(h^\alpha)$, it is sufficient to examine the expression $\sup_{0 < h \le t} \|W\Delta_h^r f\|_{L_p[-h^*, h^*]}$, and that expression is really substantially simpler.

Note also that, as we mentioned, $\inf_{P \in \Pi_{r-1}} \|W(f - P)\|_{L_p(t^*, \infty)}$ can replace the $L_{2, W}$ projection on Π_{r-1} that we used, with almost no change in the proof.

Discussions and results in this chapter raise the natural question of how to characterize $E_n(f)_{W_\lambda,p} = O(n^{-\alpha})$ when $W_\lambda(x) = \exp(-|x|^\lambda)$ for $0 < \lambda \le 1$. For $\lambda < 1$ this is not possible as the polynomials are not dense in $L_{p,W}$ (see Akhiezer [2, Ch. 2, pr. 14] and Nevai–Totik [1]). For $\lambda = 1$ and $p = 1$ the problem was recently solved (see Ditzian, Lubinsky, Nevai, and Totik [1]). For $\lambda = 1$ and $1 < p$ the problem remains open.

CHAPTER 12
POLYNOMIAL APPROXIMATION IN SEVERAL VARIABLES

In this chapter we investigate polynomial approximation on polytopes in R^m, $m \geq 2$. A modulus of smoothness will be introduced and its behavior will be related to the rate of polynomial approximation. It will turn out that in many cases the problem of polynomial approximation in several variables can be reduced to that of ordinary polynomial approximation on a finite interval.

12.1. Approximation on Cubes

As the fundamental problem of best polynomial approximation on $[-1, 1]$ has been resolved only very recently (see Chapters 7 and 13), it is natural that the more complicated problem of polynomial approximation in several variables has barely been touched upon. In this direction the literature is very scarce and is mostly restricted to L_∞ approximation in R^2. Furthermore, the results announced are formulated in a rather complicated form (see Fuksman [1], Ivanov [8], and Konovalov [1]).

In this chapter we will show how our results from Chapter 7 can be utilized to solve the higher dimensional problem for certain polytopes that include simplices, cubes, and (in R^2) every convex polygon (two-dimensional polytope). The most significant consequence of our results is that in these cases results on polynomial approximation in several variables can be reduced to those on ordinary polynomial approximation on a finite interval.

Let $Q^m = [-1, 1]^m = \{(x_1, \ldots, x_m) | |x_i| \leq 1\}$ and $f \in L_p(Q^m)$, $1 \leq p \leq \infty$. The problem we consider is the characterization (in the sense of Chapter 7) of the best polynomial approximation

$$E_n(f)_{L_p(Q^m)} = \inf_{P, \deg P \leq n} \|f - P\|_{L_p(Q^m)}$$

of f by polynomials of several variables of degree at most n (in each variable). As a measure of smoothness we use for $p < \infty$

$$\omega_Q^r(f,t)_p = \sup_{1 \leq i \leq m, 0 < h \leq t} \left\{ \int_{Q^m} |\Delta_{i,h\varphi(x_i)}^r f(x)|^p \, dx \right\}^{1/p}, \qquad (12.1.1)$$

where $\varphi(\tau) = \sqrt{1 - \tau^2}$, $x = (x_1, \ldots, x_m)$ and $\Delta_{i,h}$ means that the difference is formed in the ith coordinate with the understanding that $\Delta_{i,h}^r f(x) = 0$ if $(x_i - rh/2, x_i + rh/2) \not\subset [-1, 1]$. For $p = \infty$ we write instead of (12.1.1)

$$\omega_Q^r(f,t)_\infty = \sup_{1 \leq i \leq m, 0 < h \leq t, x \in Q^m} |\Delta_{i,h\varphi(x_i)}^r f(x)|. \qquad (12.1.2)$$

Using the results of Chapters 2, 3, and 7, we can prove:

Theorem 12.1.1. *Let* $1 \leq p \leq \infty$, $m \geq 1$, $r \geq 1$, *and* $f \in L_p(Q^m)$. *Then*

$$E_n(f)_{L_p(Q^m)} \leq M \omega_Q^r \left(f, \frac{1}{n} \right)_p \qquad (12.1.3)$$

and

$$\omega_Q^r(f,t)_p \leq M t^r \sum_{0 \leq k \leq 1/t} (k+1)^{r-1} E_k(f)_{L_p(Q^m)}, \qquad (12.1.4)$$

where M is independent of f, $n \geq r$, and $0 < t < 1$.

PROOF. The main idea of the proof can be summarized as follows. While proving the results of Chapter 7 on best polynomial approximation, we actually constructed a linear operator that was close to the operator of best polynomial approximation and served the role of that operator. Now we can apply this operator separately in each variable yielding a polynomial of m variables that approximates f well in the above sense. The details are somewhat more elaborate because we need the commutativity and uniform boundedness of the operators in question.

We begin our proof by the construction of an operator for functions of a single variable. Let

$$S_n^1(g, x) = \int_{-\pi}^{\pi} g(\cos(\arccos x - t)) T_n(t) \, dt \qquad (x \in [-1, 1])$$

be the polynomials from the proof of Theorem 7.2.1, where T_n are the trigonometric polynomials introduced in Lemma 7.2.2 (with some large l). We define inductively

$$S_n^{j+1}(g, x) = S_n^j \left(\left(g(t) - \int_0^t S_{n-1}^j(g', \tau) \, d\tau \right), x \right) + \int_0^x S_{n-1}^j(g', \tau) \, d\tau$$

provided $g^{(j)}$ exists and is in $L_\infty[-1, 1]$. Observe that $S_n^{j+1}(g) \in \Pi_n$, i.e., it is a polynomial of degree at most n.

12.1. Approximation on Cubes

Let $\varphi(x) = \sqrt{1-x^2}$, $\delta_n(x) = \sqrt{1-x^2} + 1/n$, and $\|\ \|_p = \|\ \|_{L_p[-1,1]}$. In Lemma 7.2.3 we proved that (with appropriate choice of $\{T_n\}$)

$$\|\delta_n^\gamma(g - S_n^1(g))\|_p \leq \frac{M_0}{n}\|\delta_n^{\gamma+1}g'\|_p, \quad g \in \text{A.C.}, \quad 0 \leq \gamma \leq r.$$

Hence

$$\|g - S_n^r(g)\|_p \leq \frac{M_0}{n}\|\delta_n(x)(g(x) - \int_0^x S_{n-1}^{r-1}(g', \tau)d\tau)'\|_p$$

$$\leq \frac{M_0}{n}\|\delta_n(g' - S_{n-1}^{r-1}(g'))\|_p \leq \cdots$$

$$\leq \frac{M_0^{r-1}}{n(n-1)\cdots(n-r+2)}\|\delta_n^{r-1}(g^{(r-1)} - S_{n-r+1}^1(g^{(r-1)}))\|_p$$

$$\leq \frac{M_0^r}{n\cdots(n-r+1)}\|\delta_n^r g^{(r)}\|_p \leq M_1 n^{-r}(\|\varphi^r g^{(r)}\|_p + n^{-r}\|g^{(r)}\|_p)$$

for $n \geq r$ and $g^{(r-1)} \in \text{A.C.}_{\text{loc}}$.

In Sections 2.2, 2.3, and 3.1 we proved that for every small t, say $0 < t \leq t_0$ there exists a linear operator V_t constructed with the aid of integration, change of variable, addition, and multiplication by fixed functions such that $(V_t f)^{(r-1)} \in \text{A.C.}[-1, 1]$ and

$$\|f - V_t f\|_p + t^r(\|\varphi^r(V_t f)^{(r)}\|_p + t^{2r}\|(V_t f)^{(r)}\|_p) \leq M_2 \omega_\varphi^{*r}(f, t)_p,$$

where M_2 is the independent of $f \in L_p[-1, 1]$ and $0 < t < t_0$. We recall from Section 2.2 that

$$\omega_\varphi^{*r}(f, t)_p = \left\{\frac{1}{t}\int_0^t \int_{-1}^1 |\Delta_{\tau\varphi(x)}^r f(x)|^p dx\, d\tau\right\}^{1/p}, \quad 1 \leq p < \infty;$$

and

$$\omega_\varphi^{*r}(f, t)_\infty = \sup_{0 < \tau \leq t, x \in [-1,1]} |\Delta_{\tau\varphi(x)}^r f(x)|.$$

We now define

$$T_n = S_n^r \circ V_{1/n}.$$

For these T_n we have demonstrated above that

$$T_n f \in \Pi_n \tag{12.1.5}$$

and

$$\|f - T_n f\|_p \leq \|f - V_{1/n} f\|_p + \|V_{1/n} f - S_n^r(V_{1/n} f)\|_p \leq M_1 M_2 \omega_\varphi^{*r}\left(f, \frac{1}{n}\right)_p. \tag{12.1.6}$$

Since here

$$\omega_\varphi^{*r}(f,t)_p \leq \omega_\varphi^r(f,t)_p \leq C\|f\|_p$$

(cf. Section 2.2), we can see that

$$T_n: L_p[-1,1] \to \Pi_n$$

are uniformly bounded linear operators (in n) on $L_p[-1,1]$; that is,

$$\|T_n\|_p = \|T_n\|_{L_p \to L_p} \leq K. \tag{12.1.7}$$

After these preliminaries let $f \in L_p(Q^m)$ and $x = (x_1, \ldots, x_m) \in Q^m$. For $1 \leq i \leq m$ we set $x_i^* = (x_1, \ldots, x_{i-1}, x_{i+1}, \ldots, x_m) \in Q^{m-1}$,

$$f_{i,x_i^*}(\tau) = f(x_1, \ldots, x_{i-1}, \tau, x_{i+1}, \ldots, x_m)$$

and

$$T_{n,i}f(x) = T_n f_{i,x_i^*}(x_i), \quad x \in Q^m,$$

i.e., $T_{n,i}$ is the operator T_n applied in the ith variable while the rest of the variables are kept fixed. For $1 \leq p < \infty$ we get from (12.1.7)

$$\|T_{n,i}f\|_{L_p(Q^m)} = \left(\int_{Q^{m-1}} \int_{-1}^{1} |T_n f_{i,x_i^*}(x_i)|^p \, dx_i \, dx_i^* \right)^{1/p}$$
$$\leq \left\{ \int_{Q^{m-1}} K^p \int_{-1}^{1} |f_{i,x_i^*}(x_i)|^p \, dx_i \, dx_i^* \right\}^{1/p} = K\|f\|_{L_p(Q^m)}, \tag{12.1.8}$$

and a similar reasoning yields that the operators $T_{n,i}$, $1 \leq i \leq m$, $n = 1, \ldots$ are uniformly bounded on $L_\infty(Q^m)$. A second very important property of these operators is their commutativity, namely every $T_{n,i}$ commutes with every $T_{n,j}$ (with the same n). This is an immediate consequence of the facts that $T_{n,i}$ is the operator T_n applied in the ith variable and both S_n^r and $V_{1/n}$ are constructed with the aid of differentiation, integration, addition, multiplication by fixed functions, and change of variable, and if we perform these operations in different variables, they commute.

Define

$$T_n^* = T_{n,1} \circ \cdots \circ T_{n,m}.$$

We claim that T_n^*f is a good polynomial approximation to f, that is, one for which

$$\|T_n^*f - f\|_{L_p(Q)} \leq M\omega_Q^r\left(f, \frac{1}{n}\right)_p.$$

In fact, by commutativity and (12.1.5)

$$T_n^*f = T_{n,i}(T_{n,1} \cdots T_{n,i-1} T_{n,i+1} \cdots T_{n,m} f)$$

is a polynomial of degree at most n in the ith variable for every $1 \leq i \leq m$, and hence T_n^*f if a polynomial (of m variables) of degree at most n. From (12.1.6) and (12.1.8) we derive

$$\|f - T_n^* f\|_{L_p(Q^m)} \le \sum_{i=0}^{m-1} \|T_{n,1} \cdots T_{n,i} f - T_{n,1} \cdots T_{n,i+1} f\|_{L_p(Q^m)}$$

$$\le \sum_{i=0}^{m-1} K^i \|f - T_{n,i+1} f\|_p$$

$$\le M_1 M_2 \sum_{i=1}^{m} K^{i-1} \left\| \omega_\varphi^{*r}\left(f_{i,x_i^*}, \frac{1}{n}\right)_p \right\|_{L_p(Q^{m-1})},$$

where the last norm is taken with respect to $x_i^* \in Q^{m-1}$. Examining the definition of $\omega_Q^r(f, t)_p$ ((12.1.1) and (12.1.2)), we can see that, e.g., for $1 \le p < \infty$,

$$\left\| \omega_\varphi^{*r}\left(f_{i,x_i^*}, \frac{1}{n}\right)_p \right\|_{L_p(Q^{m-1})}^p$$

$$= \int_{-1}^{1} \cdots \int_{-1}^{1} n \int_{0}^{1/n} \int_{-1}^{1} |\Delta_{i,\tau\varphi(x_i)}^r f(x_1, \ldots, x_m)|^p \, dx_i \, d\tau \, dx_1 \cdots dx_{i-1} \, dx_{i+1} \cdots dx_m$$

$$= n \int_{0}^{1/n} \|\Delta_{i,\tau\varphi(x_i)}^r f(x)\|_{L_p(Q^m)}^p \, d\tau \le \omega_Q^r\left(f, \frac{1}{n}\right)_p^p, \tag{12.1.9}$$

and this completes the proof of (12.1.3).

We omit the proof of (12.1.4) here as a stronger result will proved in the next section. $\square$

12.2. Approximation on Polytopes

Following Section 12.1 and Chapter 7, one would like to characterize the rate of best polynomial approximation on a body $S \subset R^m$ given by

$$E_n(f)_{L_p(S)} \equiv \inf\{\|f - P\|_{L_p(S)}; \deg P \le n\}$$

by some measure of smoothness of f on S. Recall that the degree is the maximal one-dimensional degree in each variable. This makes $E_n(f)_{L_p(S)}$ dependent on the coordinates chosen for R^m but does not cause difficulty because if $\bar{E}_n(f)_{L_p(S)}$ is the same expression for another coordinate system, we have

$$E_{nm}(f)_{L_p(S)} \le \bar{E}_n(f)_{L_p(S)} \quad \text{and} \quad \bar{E}_{nm}(f)_{L_p(S)} \le E_n(f)_{L_p(S)}.$$

In this section we show how to characterize $E_n(f)_{L_p(S)}$ when S is a simple polytope (cf. below). The characterizing modulus can be defined as follows. For a vector $e \in R^m$ and a point $x \in R^m$ we write

$$\Delta_{e,h}^1 f(x) = \Delta_{e,h} f(x) = f(x + he/2) - f(x - he/2), \quad \Delta_{e,h}^r f(x) = \Delta_{e,h} \Delta_{e,h}^{r-1} f(x)$$

with the understanding that $\Delta_{e,h}^r f(x) = 0$ if either of $x \pm rhe/2$ does not belong to S. With the Euclidean distance $d(x, y)$ we set

$$\tilde{d}_S(e, x) = \left(\min_{x + \lambda e \notin S} d(x, x + \lambda e)\right)\left(\max_{x + \lambda_i e \in S} d(x + \lambda_1 e, x + \lambda_2 e)\right),$$

the square root of which will replace $\varphi(x) = \sqrt{1-x^2}$ in the higher dimensional analogue of ω_φ^r. The modulus of smoothness is now given for $p = \infty$ by

$$\tilde{\omega}_S^r(f,t)_\infty = \sup_{\substack{x \in S, e \in V^{m-1} \\ 0 < h \le t}} |\Delta^r_{e,h\tilde{d}_S(e,x)^{1/2}} f(x)| \tag{12.2.1}$$

and for $1 \le p < \infty$ by

$$\tilde{\omega}_S^r(f,t)_p = \sup_{\substack{e \in V^{m-1} \\ 0 < h \le t}} \left\{ \int_{x \in e^\perp} \int_{-\infty}^\infty |\Delta^r_{e,h\tilde{d}_S(e,x+\lambda e)^{1/2}} f(x + \lambda e)|^p \, d\lambda \, dm_e(x) \right\}^{1/p}, \tag{12.2.2}$$

where V^{m-1} denotes the set of unit vectors in R^m and m_e is the $(m-1)$-dimensional Lebesgue measure on $e^\perp$ (the orthogonal complement of e). For a polytope S, i.e., the convex hull of a finite set of points in R_m, it is worthwhile introducing a somewhat different modulus $\omega_S^r(f,t)_p$, which can be defined by replacing V^{m-1} in (12.2.1) and (12.2.2) with V_S, the set of unit vectors pointing in the direction of the edges of S.

Remarks 12.2.1. (1) In the integrals in (12.2.2) the real domains of integrations are bounded sets (outside which the integrands are 0).

(2) For $S = Q^m$ the definitions of $\omega_S^r(f,t)_p$ given above and in the preceding section essentially coincide.

(3) We may have (e.g., for $S = Q^m$) $\tilde{\omega}_S^r(f,h)_\infty \sim h^r \log h$ while $\omega_S^r(f,h)_\infty \sim h^r$ (see Ditzian [6]), and therefore $\tilde{\omega}_S^r$ and ω_S^r are not equivalent. (This is also shown by the function $f(x_1, \ldots, x_m) = x_1 \cdots x_m$, for which $\omega_{Q^m}^m(f,h)_p \equiv 0$ but $\tilde{\omega}_{Q^m}^m(f,h)_p \sim h^m$.) However, the difference between ω_S^r and $\tilde{\omega}_S^r$ cannot be "too big" as Theorem 12.2.3 below implies that for $0 < \alpha < r$ $\omega_S^r(f,t)_p = O(t^\alpha)$ and $\tilde{\omega}_S^r(f,t)_p = O(t^\alpha)$ are equivalent.

Definition 12.2.2. A polytope $S \subset R^m$, with an interior point, is called simple if each vertex is joined to other vertices by exactly m edges.

Cubes and simplices in R^m and every polygon in R^2 are simple. Moreover, we can get from any polytope a simple one by "chopping off" small neighborhoods of its vertices.

The analogue of Theorem 12.1.1 for simple polytopes is as follows.

Theorem 12.2.3. Let $S \subseteq R^m$ be a simple polytope, $r \ge 1$, $1 \le p \le \infty$ and $f \in L_p(S)$. Then

$$E_n(f)_{L_p(S)} \le M \left[\omega_S^r\left(f, \frac{1}{n}\right)_p + n^{-r} \|f\|_{L_p(S)} \right]$$

$$\le M \left[\tilde{\omega}_S^r\left(f, \frac{1}{n}\right)_p + n^{-r} \|f\|_{L_p(S)} \right] \tag{12.2.3}$$

and

$$\omega_S^r(f,t)_p \le \tilde{\omega}_S^r(f,t)_p \le Mt^r \sum_{0 \le k \le 1/t} (k+1)^{r-1} E_k(f)_{L_p(S)}, \tag{12.2.4}$$

where M is independent of f, $n \ge r$, and $0 < t < 1$.

12.2 Approximation on Polytopes

We were not able to settle the question whether Theorem 12.2.3 is valid for any polytope in R^m. In the proof we need the following technical lemma.

Lemma 12.2.4. *If $\psi \in C^\infty(R^m)$ with bounded derivatives, Then*

$$\tilde{\omega}_S^r(f\psi, t)_p \leq M(\tilde{\omega}_S^r(f, t)_p + t^r \|f\|_{L_p(S)})$$

with an M independent of f and small t. A similar statement also holds for $\omega_S^r(f, t)_p$.

PROOF OF LEMMA 12.2.4. First let $1 \leq p < \infty$. It is enough to show that for $0 < h \leq t$

$$\int_{-\infty}^{\infty} |\Delta_{e, h\tilde{d}_S(e, x+\lambda e)^{1/2}}^r (f\psi)(x+\lambda e)|^p d\lambda$$

$$\leq M \left(\frac{1}{t} \int_0^t \int_{-\infty}^{\infty} |\Delta_{e, \tau\tilde{d}_S(e, x+\lambda e)^{1/2}}^r f(x+\lambda e)|^p d\lambda\, d\tau + t^r \int_{-\infty}^{\infty} |f(x+\lambda e)|^p d\lambda \right) \quad (12.2.5)$$

(let $f(x) \equiv 0$ if $x \notin S$), because we can then integrate this inequality with respect to $x \in e^\perp$ and on the right perform operations similar to those in (12.1.9); note also that

$$\int_{x \in e^\perp} \int_{-\infty}^{\infty} |f(x+\lambda e)|^p d\lambda\, dm_e(x) = \|f\|_{L_p(S)}^p.$$

Let L be the line $\{x + \lambda e | \lambda \in R\}$ and $I_L = S \cap L = \{x + \lambda e | x + \lambda e \in S\}$. Choose a linear transformation onto I_L

$$T: [-1, 1] \to I_L$$

and set

$$F(u) = f(T(u)), \quad \Psi(u) = \psi(T(u)), \quad u \in [-1, 1].$$

Clearly T and Ψ depend on L, but since ψ had bounded derivatives and since the length of I_L denoted by $|I_L|$ is bounded, we have for $u \in [-1, 1]$

$$|\Psi^{(k)}(u)| \leq \|\psi^{(k)}\|_\infty (|I_L|/2)^k \leq M_0, \quad k = 0, \ldots, r \quad (12.2.6)$$

independently of $u \in [-1, 1]$ and L. Using the transformation T, (12.2.5) can replaced by

$$J(h) \equiv \int_{-1}^{1} |\Delta_{h\min(\sqrt{1-u},\sqrt{1+u})/\sqrt{2}}^r (F\Psi)(u)|^p du$$

$$\leq M \left\{ \frac{1}{t} \int_0^t \int_{-1}^{1} |\Delta_{\tau\min(\sqrt{1-u},\sqrt{1+u})/\sqrt{2}}^r F(u)|^p du\, d\tau \right. \quad (12.2.7)$$

$$\left. + t^r \int_{-1}^{1} |F(u)|^p du \right\} = MJ^*(t)$$

for $0 < h < t$ (with the same M) which we prove below using the technique of Section 2.2. Let $\varphi_1(x) = \min(\sqrt{1-x}, \sqrt{1+x})\sqrt{2}$ and from now on until the end of the proof of Lemma 12.2.4 every norm, K-functional etc. is taken on $[-1, 1]$. By (2.3.2) and (2.3.3) (more precisely by their $[-1, 1]$ variant)

$$K_{r,\varphi_1}(F, t^r)_p \leq M_1 J^*(t) \qquad (0 < t < t_0 < 1),$$

and by Theorem 2.1.1

$$J(h) \leq M_2 K_{r,\varphi_1}(F\Psi, h^r)_p \qquad (h < t_0).$$

Hence it suffices to show that for small t, say $0 < t < t_0 < 1$,

$$K_{r,\varphi_1}(F\Psi, t^r)_p \leq M_3(K_{r,\varphi_1}(F, t^r)_p + t^r \|F\|_p)$$

with a constant M_3 which depends only on r, p, and M_0 (cf. (12.2.6)). By the definition of the K-functional (2.1.1), this is a consequence of the inequality

$$\|\varphi_1^r(g\Psi)^{(r)}\|_p \leq M_4(\|\varphi_1^r g^{(r)}\|_p + \|g\|_p) \qquad (g^{r-1} \in \text{A.C.}_{\text{loc}})$$

which in turn follows from

$$\|\varphi_1^r g^{(i)} \Psi^{(r-i)}\|_p \leq M_5(\|\varphi_1^r g^{(r)}\|_p + \|g\|_p) \qquad (0 \leq i \leq r),$$

or from the stronger inequality (see (12.2.6))

$$\|\varphi_1^i g^{(i)}\|_p \leq (2^{-r} M_5/M_0)(\|\varphi_1^r g^{(r)}\|_p + \|g\|_p) \qquad (0 \leq i \leq r). \quad (12.2.8)$$

Obviously we have to prove (12.2.8) with some M_5 only for $i = r - 1$. Using Fatou's lemma we have

$$\|\varphi_1^{r-1} g^{(r-1)}\|_p \leq \liminf_{h \to 0} \|\Delta_{h\varphi_1}^{r-1} g\|_p / h^{r-1} \leq \liminf_{h \to 0} \omega_{\varphi_1}^{r-1}(g, h)_p / h^{r-1},$$

and, by Marchaud's inequality (4.3.3), here

$$\omega_{\varphi_1}^{r-1}(g, h)_p / h^{r-1} \leq M_6 \left(\int_h^{t_0} (\omega_{\varphi_1}^r(g, u)_p / u^r) \, du + \|g\|_p \right)$$

$$\leq M_7 \left(\int_t^{t_0} (K_{r,\varphi_1}(g, u)_p / u^r) \, du + \|g\|_p \right)$$

$$\leq M_7 \left(\int_t^{t_0} \|\varphi_1^r g^{(r)}\|_p \, du + \|g\|_p \right)$$

$$\leq M_7 (\|\varphi_1^r g^{(r)}\|_p + \|g\|_p),$$

which proves (12.2.8) and hence the proof of Lemma 12.2.4 is complete for $1 \leq p < \infty$. When $p = \infty$, the proof follows along the same lines where the necessary modifications are obvious. □

PROOF OF THEOREM 12.2.3. Consider first (12.2.3). Since $\omega_S^r(f, t)_p \leq \tilde{\omega}_S^r(f, t)_p$, we have to prove only the first inequality in (12.2.3). Let $v_1, \ldots, v_\nu$ be the vertices of S and let $F_1^i, \ldots, F_m^i$ be those $(m - 1)$-dimensional faces of S that contain v_i (since S is simple, there are exactly m of them). Choose bounded open sets

12.2. Approximation on Polytopes

$U_i \subset R^m$ such that $v_i \in U_i$, $S \subseteq \bigcup_{i=1}^{\nu} U_i$, but the closure $\bar{U}_i$ of U_i is disjoint from every face of S not containing v_i, i.e.,

$$\bar{U}_i \cap F_k^j = \emptyset, \qquad j \neq i, \quad k = 1, \ldots, m, \quad v_i \notin F_k^j.$$

Choose a C^∞ partition of unity $\{\psi_i\}_{i=1}^{\nu}$ on S subordinated to $\{U_i\}_{i=1}^{\nu}$, i.e., $\psi_i \in C^\infty(R^m)$, $\operatorname{supp}\psi_i \subset U_i$, $0 \le \psi_i \le 1$, and $\sum_{i=1}^{\nu}\psi_i(x) = 1$ if $x \in S$; and set $f_i = f\psi_i$. Let S_i (cf. the illustration below) be a parallelepiped containing S such that v_i is a vertex of S_i and F_k^i, $k = 1, \ldots, m$ are all on the boundary of S_i. We claim that

$$\omega_{S_i}^r(f_i, t)_p \le C\omega_S^r(f_i, t)_p \tag{12.2.9}$$

with a C independent of f and small t. In fact, we can choose an open set V_i such that $\bar{U}_i \subset V_i$ and

$$\bar{V}_i \cap F_k^j = \emptyset, \qquad j \neq i, \quad k = 1, \ldots, m, \quad v_i \notin F_k^j.$$

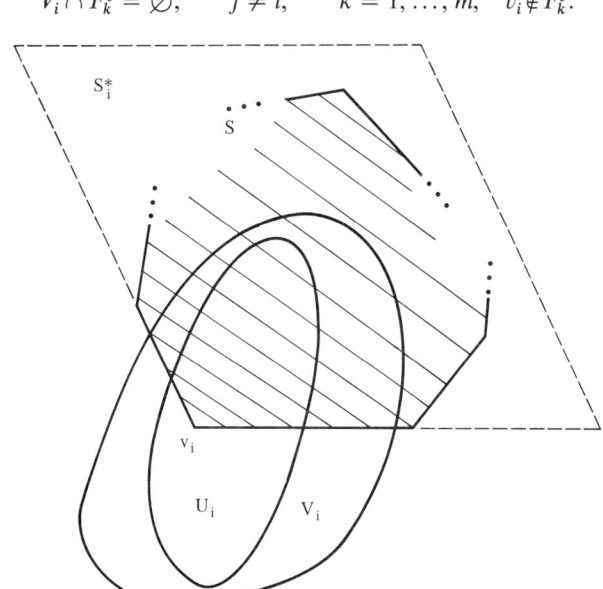

If for some $e \in V^{m-1}$, $x \in S$ and small h the difference

$$\Delta_{e,h\tilde{d}_{S_i}(e,x)^{1/2}}^r f_i(x)$$

is not zero, then $x \in V_i$ (note that $\operatorname{supp} f_i \subset U_i$), therefore

$$\tilde{d}_{S_i}(e, x) \le C_1 \tilde{d}_S(e, x). \tag{12.2.10}$$

Hence it is enough to show for $1 \le p < \infty$, $h \le t$, and small t, that

$$\int_{-\infty}^{\infty} |\Delta_{e,h\tilde{d}_{S_i}(e,x+\lambda e)^{1/2}}^r f_i(x+\lambda e)|^p \, d\lambda$$

$$\le C\frac{1}{t}\int_0^t \int_{-\infty}^{\infty} |\Delta_{e,\tau\tilde{d}_S(e,x+\lambda e)^{1/2}}^r f_i(x+\lambda e)|^p \, d\lambda \, d\tau$$

which, in view of (12.2.10), can be accomplished as in the proof of Lemma 12.2.4 via the K-functional.

Making use of (12.2.9) and Lemma 12.2.4, we have

$$\omega_{S_i}^r(f_i, t)_p \leq M(\omega_S^r(f, t)_p + t^r \|f\|_{L_p(S)}).$$

Let T_i be a linear mapping between Q^m and S_i and set $g_i = f_i \circ T_i$.

When we examine the definition of our moduli of smoothness it immediately becomes clear that these moduli are changed only slightly by linear transformations, that is

$$\omega_{Q^m}^r(g_i, t)_p \leq C_i \omega_{S_i}^r(f_i, t)_p$$

with a C_i independent of f and t. Hence

$$\omega_{Q^m}^r(g_i, t)_p \leq B(\omega_S^r(f, t)_p + t^r \|f\|_{L_p(S)}), \qquad i = 1, \ldots, v.$$

On applying Theorem 12.1.1 to $g_i(u)$ and making use of the change of variable $x = Tu$, we get a polynomial $P_{n,i}(x)$ of degree at most mn (linear transformations can increase the degree of polynomials as we counted the degree in each variable separately) such that

$$\|f_i - P_{n,i}\|_{L_p(S_i)} \leq B_1 \left[\omega_S^r\left(f, \frac{1}{n}\right)_p + n^{-r}\|f\|_{L_p(S)} \right].$$

Since

$$\left\| f - \sum_{i=1}^v P_{n,i} \right\|_{L_p(S)} \leq \sum_{i=1}^v \|f_i - P_{n,i}\|_{L_p(S)} \leq \sum_{i=1}^v \|f_i - P_{n,i}\|_{L_p(S_i)}$$

$$\leq vB_1\left(\omega_S^r\left(f, \frac{1}{n}\right)_p + n^{-r}\|f\|_{L_p(S)}\right),$$

we have only to show

$$\omega_S^r\left(f, \frac{1}{n}\right)_p \leq B_2 \omega_S^r\left(f, \frac{1}{mn}\right)_p,$$

to complete the proof and this can be verified by the K-functional argument applied in Lemma 12.2.4.

The proof of (12.2.4) is much easier. Let, e.g., $1 \leq p < \infty$ and let P_k be the best approximating polynomial to f in $L_p(S)$ of degree at most k. We write ($P_{1/2} \equiv 0$)

$$f = f - P_{2^l} + \sum_{k=-1}^{l-1} (P_{2^{k+1}} - P_{2^k})$$

with $2^l \leq 1/t < 2^{l+1}$, and by the standard technique (cf., e.g., Theorem 7.2.4) it is enough to show that

$$\tilde{\omega}_S^r(P_{2^{k+1}} - P_{2^k}, t)_p \leq Mt^r 2^{kr} E_{2^k}(f)_{L_p(S)}. \tag{12.2.11}$$

12.2. Approximation on Polytopes

Using Theorem 2.1.1, the definition of the K-functional in (2.1.1) and change of variable, we have

$$\int_{-\infty}^{\infty} |\Delta_{e,h\tilde{d}_S(e,x+\lambda e)^{1/2}}^r (P_{2^{k+1}} - P_{2^k})(x + \lambda e)|^p \, d\lambda$$

$$\leq M_1 h^r \int_{x+\lambda e \in S} \left| \tilde{d}_S(e, x + \lambda e)^{r/2} \frac{\partial^r (P_{2^{k+1}} - P_{2^k})(x + \lambda e)}{\partial \lambda^r} \right|^p d\lambda \qquad (12.2.12)$$

$$\leq M_2 h^r 2^{(k+1)r} \int_{x+\lambda e \in S} |(P_{2^{k+1}} - P_{2^k})(x + \lambda e)|^p \, d\lambda,$$

where, in the last step, we also used the inequality (7.2.7) and the fact that on any line $P_{2^{k+1}} - P_{2^k}$ is a polynomial of degree at most $m2^{k+1}$ in any coordinate system. The point is that by a simple change of variable, (12.2.12) can be transformed into an analogous expression on $[-1, 1]$, and hence M_2 is independent of x and e. Integrating (12.2.12) on $x \in e^\perp$, then taking the pth root and the supremum for $0 < h \leq t$ and $e \in V^{m-1}$, we arrive at

$$\tilde{\omega}_S^r(P_{2^{k+1}} - P_{2^k}, t)_p \leq M_2 t^r 2^{(k+1)r} \|P_{2^{k+1}} - P_{2^k}\|_{L_p(S)} \leq 2M_2 t^r 2^{(k+1)r} E_{2^k}(f)_{L_p(S)}$$

which proves (12.2.11), and the proof of Theorem 12.2.3 is complete. □

We formulate two consequences of Theorem 12.2.3 which show that in many cases the problem of polynomial approximation in several variables can be reduced to ordinary polynomial approximation on a finite interval. Let $x \in e^\perp$, $e \in V^{m-1}$, and $I_{e,x} = \{x + \lambda e \mid x + \lambda e \in S\}$. The quantity

$$E_n(f|_{I_{e,x}})_{L_p(I_{e,x})}$$

is the best approximation of f by (ordinary) polynomials in L_p on the segment $I_{e,x}$ of S (the measure is the linear Lebesgue measure on $I_{e,x}$). We define

$$E_n^*(f)_p = \sup_{e \in V^{m-1}} \left\{ \int_{x \in e^\perp} E_n(f|_{I_{e,x}})_{L_p(I_{e,x})}^p \, dm_e(x) \right\}^{1/p}, \qquad 1 \leq p < \infty$$

and

$$E_n^*(f)_\infty = \sup_{e \in V^{m-1}, x \in e^\perp} E_n(f|_{I_{e,x}})_{L_\infty(I_{e,x})}.$$

$E_n^*(f)_p$ measures how well f can be approximated by ordinary polynomials on segments of S. Obviously

$$E_n^*(f)_p \leq E_n(f)_{L_p(S)}$$

but to our knowledge the following converse theorem has not been observed up to now.

Theorem 12.2.5. *Let S be a simple polytope, $1 \leq p \leq \infty$ and $f \in L_p(S)$. Then for every r there exists a constant $M = M_{r,p,S}$ such that*

$$E_n(f)_{L_p(S)} \leq M\left(n^{-r} \sum_{k=0}^{n} (k+1)^{r-1} E_n^*(f)_p + n^{-r} \|f\|_{L_p(S)}\right). \quad (12.2.13)$$

In particular, for $\alpha > 0$

$$E_n(f)_{L_p(S)} = O(n^{-\alpha})$$

and

$$E_n^*(f)_p = O(n^{-\alpha})$$

are equivalent.

PROOF. All we have to mention is that while proving (12.2.4) we actually verified

$$\tilde{\omega}_S^r(f, t)_p \leq K t^r \sum_{0 \leq k \leq 1/t} (k+1)^{r-1} E_k^*(f)_p.$$

(In (12.2.12) choose $P_{2^k} = P_{2^k, I_{e,x}}$ such that

$$E_{2^k}(f|_{I_{e,x}})_{L_p(I_{e,x})} = \|f - P_{2^k}\|_{L_p(I_{e,x})}.)$$

Therefore, (12.2.13) is a consequence of Theorem 12.2.3. □

The following corollary is worth mentioning separately.

Corollary 12.2.6. *Let S be a simple polytope and $f \in C(S)$. If f can be approximated along every segment of S (as a function of a single variable) by (ordinary) polynomials of degree n with error smaller than $n^{-\alpha}$, $n = 1, 2, \ldots$, then, on the whole of S, the error of (m-variable) polynomial approximation to f is at most $Kn^{-\alpha}$ ($n = 1, \ldots$), where K depends only on S.*

This corollary is about as far as one can go using lines, as is shown by the following proposition.

Proposition 12.2.7. *Let $S \subseteq R^m$ be a convex body with piecewise C^2 boundary (i.e., ∂S is the union of finitely many C^2 surfaces). If Corollary 12.2.6 is true on S for some $\alpha > 0$, then S is a polytope.*

From the proof below it is equally obvious how one can step forward to include other domains; namely instead of lines one has to consider higher order algebraic curves along which the smoothness (or polynomial approximation) must be taken into account.

PROOF OF PROPOSITION 12.2.7. It is enough to prove the conclusion for the intersection of S with every plane; hence we may assume without loss of generality (cf. the construction below) that $m = 2$ and $S \subset R^2$ is convex. Suppose S is not a polygon. Then there is a point $z \in \partial S$ such that at z the curve

12.2 Approximation on Polytopes

describing the boundary of S has continuous and positive curvature, i.e., around z it is similar to a parabola, and the consideration below easily implies that Corollary 12.2.6 fails to hold for S. For the sake of simplicity we carry out the construction only for $2/3 < \alpha < 1$ and for the parabolic domain $S = \{(x, y) | -2 \le x \le 2, x^2 \le y \le 4\}$.

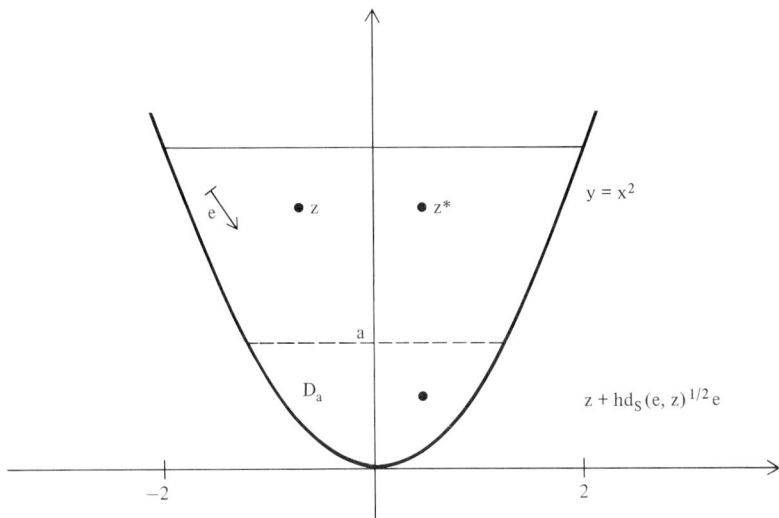

By Theorem 7.2.1 it is enough to show that $\tilde{\omega}_S^1(f, t)_C = O(t^\alpha)$ does not imply $E_n(f)_{C(S)} = O(n^{-\alpha})$ (notice that Theorem 7.2.1 yields

$$\tilde{\omega}_S^1(f, t)_C = O(t^\alpha) \Rightarrow E_n^*(f)_C = O(n^{-\alpha}).)$$

If it did, then a uniform boundedness principle (see the proof of the Banach Steinhaus theorem in Lorentz [3] for example) would imply that there is a constant C independent of f such that

$$\tilde{\omega}_S^1(f, t)_C \le Mt^\alpha, \qquad 0 < t \le 1 \qquad (12.2.14)$$

implies

$$E_n(f)_{C(S)} \le CMn^{-\alpha}, \qquad n = 1, 2, \ldots,$$

and this is what we will disprove below. Indeed for $0 < a < 1/2$ let

$$f_a(x, y) = \begin{cases} \sin\left(\dfrac{\pi}{2} x/a^{3/2}\right)(a - y) & \text{if } -\sqrt{a} \le x \le \sqrt{a},\, x^2 < y < a \\ 0 & \text{otherwise.} \end{cases}$$

We claim that (12.2.14) holds for $f = f_a$ with an M independent of a. Taking this for granted, the nonexistence of the above C can be shown by contradiction. In fact, if C existed, then we would get

$$E_n(f_a)_{C(S)} \le CMn^{-\alpha}, \qquad n = 1, 2, \ldots$$

with CM independent of $0 < a < 1/2$, but then for the function

$$g_a(x) = f_a\left(x, \frac{a}{2} + x^2\right), \qquad x \in [-1, 1],$$

belonging to $C[-1, 1]$ we would have

$$E_{3n}(g_a)_{C[-1,1]} \leq E_n(f_a)_{C(S)} \leq CMn^{-\alpha}, \qquad n = 1, 2, \ldots$$

which in turn implies (see Theorem 7.2.4) that

$$|g_a(0) - g_a(h)| \leq M_1 h^\alpha, \qquad 0 < h < 1/2$$

with a constant M_1 depending only on CM and $2/3 < \alpha < 1$. However, with $h = a^{3/2}$ we would get

$$a - \left(\frac{a}{2} + (a^{3/2})^2\right) = g_a(h) - g_a(0) \leq M_1(a^{3/2})^\alpha$$

which is false if a is small enough (note that $\alpha > 2/3$).

Thus, it remains to show (12.2.14) for $f = f_a$, i.e., that for every vector $e = (e_1, e_2)$ and $z = (x, y) \in S$ we have

$$|f_a(z) - f_a(z + h\tilde{d}_S(e, z)^{1/2} e)| \leq M|h|^\alpha, \qquad |h| \leq 1. \qquad (12.2.15)$$

Actually we can show this for $\alpha = 1$. Let

$$D_a = \{(x, y) | -\sqrt{a} \leq x \leq \sqrt{a}, x^2 \leq y \leq a\}$$

be the domain of f_a. If $z \notin D_a$, then with $z^* = (x + h\tilde{d}_S(e, z)^{1/2} e_1, y)$ (see the illustration above) we have

$$f_a(z) = f_a(z^*) = 0 \quad \text{and} \quad d(z^*, z + h\tilde{d}_S(e, z)^{1/2} e) \leq M_2 h;$$

and in this case (12.2.15) follows from the fact that f_a is a $\text{Lip}_1 1$ function in y. If $z \in D_a$, then straightforward calculation shows that

$$\tilde{d}_S(e, z)^{1/2} \leq M_3 \sqrt{a};$$

and hence (12.2.15) follows easily from the fact that on D_a

$$\left|\frac{\partial f_a}{\partial x}\right| \leq \frac{\pi}{2\sqrt{a}} \quad \text{and} \quad \left|\frac{\partial f_a}{\partial y}\right| \leq 1. \qquad \square$$

CHAPTER 13
COMPARISONS AND CONCLUSIONS

In this chapter we make some comparisons with moduli that were introduced in earlier mathematical literature. We will not discuss comparisons with $\omega^r(f,h)_p$; that was done throughout the paper. Nor will we discuss comparison with the Freud and Mhaskar modulus, as that was done in Chapter 11. We will emphasize results relating to best polynomial approximation, as that problem was the primary incentive for some authors to search for a modulus different from $\omega^r(f,h)_p$.

13.1. Comparison with Similar Expressions

For $p = \infty$ and $\varphi(x)$ satisfying Conditions I and II of Section 1.2 one can show by simple substitution that the condition

$$\sup_{0 < h \le t} \varphi(x)^\alpha |\Delta_h^r f(x)| = O(t^\alpha) \quad \text{for } t > 0 \tag{13.1.1}$$

is equivalent to $\omega_\varphi^r(f, t)_\infty = O(t^\alpha)$ (see Ditzian [1], [4] and Totik [7], [14]). However, for $1 \le p < \infty$ it was shown (Totik [8]) that

$$\sup_{0 < h \le t} \|\varphi(x)^\alpha \Delta_h^r f\|_{L_p} = O(t^\alpha)$$

is not equivalent to $\omega_\varphi^r(f, t)_p = O(t^\alpha)$. Therefore for the problem of particular operators or best polynomial approximation and $1 \le p < \infty$ an analogue of (13.1.1) would not be adequate.

13.2. The Integral Modulus of Smoothness of Ivanov and Sendov

A different modulus of smoothness given by

$$\tau_r(f;\delta)_p = \|\omega_r(f,\cdot,\delta)\|_p, \qquad \delta = \text{const}, \tag{13.2.1}$$

where

$$\omega_r(f,\cdot,t) = \sup\{|\vec{\Delta}_h^r f(t)|: [t, t+rh] \subset [x - k\delta/2, x + k\delta/2] \cap D\} \tag{13.2.2}$$

was introduced by B.L. Sendov [1] and used also by other Bulgarian mathematicians, A. Andreev, K.G. Ivanov, and V.A. Popov to mention only a few (see for references Ivanov [5]). This modulus was further extended by K.G. Ivanov to

$$\tau_r(f, w; \delta)_{p', p[a,b]} = \|w(\cdot)\omega_r(f,\cdot,\delta(\cdot))_{p'}\|_{L_p[a,b]}, \tag{13.2.3}$$

and

$$\omega_r(f,\cdot,\delta(\cdot))_{p'} = \left(\frac{1}{2\delta(x)} \int_{-\delta(x)}^{\delta(x)} |\vec{\Delta}_v^r f(x)|^{p'} dv\right)^{1/p'}. \tag{13.2.4}$$

With this modulus Ivanov solved the problem of characterizing the class of functions for which the rate of polynomial approximation is given. With weights that have zeros this requires conditions on $\tau_r(f, w_n, \delta_n)_{p', p[a,b]}$ with two sequences of weights $\delta_n(x)$ and $w_n(x)$. While we believe that our characterization and modulus is the "correct" one, as the condition is much simpler to state, and the modulus is computable (see Section 3.4 and 8.5), Ivanov's work was important to us and influenced some of our results. For $w(x) = 1$, $\delta(x) = \delta\varphi(x)$, and $p' = p$ Ivanov's modulus $\tau_r(f, 1; \delta\varphi(\cdot))_{p,p}$ can be shown to be equivalent to $\omega_\varphi^r(f,t)_p = \sup_{h \le t} \|\Delta_{h\varphi}^r f\|_p$ via the K-functional.

The result that Ivanov obtained regarding best polynomial approximation (see, e.g., Ivanov [5]) can be summarized in the following theorem.

Theorem 13.2.1 (Ivanov). *For $f^{(k)} \in L_p[-1, 1]$ and $\Delta_n(x) = n^{-1}(n^{-1} + \sqrt{1-x^2})$*

$$E_{n+k}(f)_{w,p} \le C(k)\tau_1(f^{(k)}, w\Delta_n^k; \Delta_n)_{1,p} \tag{13.2.5}$$

and

$$\tau_k(f, w; \Delta_n)_{p',p} \le \frac{C(k)}{n^k} \sum_{s=0}^{n} (s+1)^{k-1} E(f)_{w,p}, \qquad p' \in [1,p], \tag{13.2.6}$$

where w, satisfies for some fixed $s > 0$

$$w(x) \le c\lambda^s w(t), \quad \text{for } |x - t| \le \lambda\Delta_n(x), \quad \lambda \ge 1 \quad \text{and } x, t \in [-1, 1]. \tag{13.2.7}$$

One should note that w that behaves like $(1 + x)^\alpha (1 - x)^\beta$ is not allowed by (13.2.7) and in fact no zero or infinite singularity is allowed. To overcome this difficulty Ivanov uses two sequences of weights, $w_n(x)$ and $\delta_n(x)$, that depend on n. In (13.2.5) Ivanov needs a sequence of weights even for $w(x) = 1$. Another shortcoming is the more complicated form of (13.2.3). However, sometimes τ_k is defined when ω_φ^r is not, and therefore, it too has its advantages.

13.3. Moduli Generated by Multipliers and Integral Transforms

For T (the "circle" $[-\pi, \pi]$) translation is given by multipliers, i.e., $\widehat{\tau_h f}(m) = e^{imh}\hat{f}(m)$ where $\hat{f}(m) \equiv \langle f(x), e^{imx}\rangle$ and $\tau_h f(x) \equiv f(x + h)$. P.L. Butzer, R.L. Stens, and M. Wehrens [1], [2], (see also Butzer [2], Butzer and Stens [1], [2], [3], Stens and Wehrens [1], and Löfstrom [1]) used expansions in Legendre and Jacobi polynomials to define translation and then a new modulus for characterization of best polynomial approximation with and without weights.

Let $\alpha, \beta > -1$ and let $R_n^{(\alpha,\beta)}(x) = p_n^{(\alpha,\beta)}(x)/p^{(\alpha,\beta)}(1)$ be the normalized Jacobi polynomials with parameters α and β. These are orthogonal with respect to the Jacobi weight

$$w = w_{\alpha,\beta}(u) = (1-u)^\alpha (1+u)^\beta$$

and, setting

$$\hat{f}(k) = \int_{-1}^{1} f(u) R_k(u) w(u) \, du, \tag{13.3.1}$$

one can define a generalized translation operator τ_t by

$$(\tau_t f)(k) = R_k(t)\hat{f}(k). \tag{13.3.2}$$

In the Legendre case, that is $\alpha = \beta = 0$, this can be shown to be equivalent to the integral transform

$$\widehat{\tau_h f}(x) = \frac{1}{\pi} \int_{-1}^{1} f(xh + u\sqrt{(1-x^2)(1-u^2)}) \frac{du}{\sqrt{1-u^2}}. \tag{13.3.3}$$

A corresponding modulus for $B = C[-1,1]$ or $L_p[-1,1]$ can now be given by

$$\omega_1^L(\delta, f, B) = \sup_{1-\delta < h \leq 1} \|\tau_h f - f\| = \sup_{1-\delta \leq h \leq 1} \|\Delta_h^L f\|, \tag{13.3.4}$$

where L stands for Legendre. This is an analogue to the second-order modulus and an analogue for the 2r modulus was given by

$$\omega_r^L(\delta, f, B) = \sup_{1-\delta < h_j \leq 1} \|\Delta_{h_1}^L(\Delta_{h_2}^L(\ldots(\Delta_{h_r}^L f(x))\ldots))\|_B. \tag{13.3.5}$$

One should note that the treatment did not allow $\sup_{1-\delta < h \leq 1} \|(\Delta_h^L)^r f(x)\|$ instead of (13.3.5).

With these definitions Butzer, Stens, and Wehrens proved (see Butzer [2]) the following theorem.

Theorem 12.3.1 (Butzer et al). *For l, s, r integers, $0 < \alpha \leq 1$ satisfying $r + \alpha < s$ where r and s are integers we have*

$$E_n(f)_B = O(n^{-2(r+\alpha)}) \Leftrightarrow \omega_s^L(\delta, f, B) = O(\delta^{r+\alpha}). \tag{13.3.6}$$

Most of the results in this direction were announced by Butzer [2] where a detailed discussion of what was done earlier is also given. Proofs and details appeared in several papers of Butzer, Stens, and Wehrens (see also Löfstrom [1]). For Jacobi weight their modulus was defined by

$$\omega_r^J(\eta; f, B)_w = \sup_{\eta \leq h_j < 1} \|w(x)\Delta_{h_1}^J(\ldots(\Delta_{h_r}^J f(x))\ldots)\|_B, \tag{13.3.7}$$

where $\Delta_h^J f = f - \tau_h f$ and $\tau_h f$ is given by (13.3.2), and a result similar to Theorem 12.3.1 was announced by Butzer, Stens, and Wehrens [1, p. 85].

The main advantage of our modulus over those based on generalized translation is the fact that our modulus has simpler expression and is computable (see Sections 3.4 and 8.5). That is, $\Delta_{h\varphi(x)}^{2r} f(x)$ is simpler than the rth iterated integral transform, and is more naturally related to the usual concept of smoothness than the expressions constructed by (13.3.2), (13.3.3), (13.3.4), (13.3.5), and (13.3.7).

13.4. A Modulus Introduced by Potapov

In recent very intricate articles M.K. Potapov [3], [4] introduced a generalized translation using integral transforms which he utilizes to characterize weighted best polynomial approximation with Jacobi weights. Because the definition of the translation (and therefore of the modulus) is over a page long, we cannot quote them here in full but will give one of the four cases given there.

For $\mu + (1/2) > \beta + (1/2p) > 0$, $\nu - \mu > \alpha - \beta > 0$, $\nu > \mu > -1/2$, and $fw \in L_p[-1, 1]$, where $w(x) = (1-x)^\alpha (1+x)^\beta$, Potapov defines

$$\tau_t f(x) = \frac{1}{\gamma(\nu, \mu)} \int_{-1}^{1} f\left(x \cos t + yz \sin t \sqrt{1-x^2} - (1-y^2)(1-x)\sin^2 \frac{t}{2}\right)$$
$$\times (1-y^2)^{\nu-\mu-1} y^{2\mu+1} (1-z^2)^{\mu-(1/2)} \, dz \, dy,$$

where

$$\gamma(\nu, \mu) = \int_0^1 \int_{-1}^{1} (1-y^2)^{\nu-\mu-1} y^{2\mu+1} (1-z^2)^{\mu-(1/2)} \, dz \, dy.$$

For $E_n(f)_{w,p} \equiv \inf \|w(f - P_n)\|_p$ he proves that

$$E_n(f)_{w,p} \leq C\tilde{\omega}(f, 1/n)_{p,w} \equiv C \sup_{0 < t < 1/n} \|w(f - \tau_t f)\|_p$$

and

$$\tilde{\omega}(f, 1/n)_{p,w} \leq Cn^{-2} \sum_{k=0}^{n} (k+1) E_k(f)_{w,p},$$

where $\tilde{\omega}(f, t)_{p,w}$ is defined using the generalized translation τ_t.

13.5. Hoeffding's Result

In a paper about B_n^* (see (9.2.1)) Hoeffding [1] proved for $f \in L_1[0, 1]$, $f \in \text{B.V.}_{\cdot \text{loc}}$, and $\varphi(x)^2 = x(1-x)$ that

$$\|B_n^* f - f\|_1 \leq Cn^{-1/2} J_\varphi(f), \quad \text{where} \quad J_\varphi(f) \equiv \int_0^1 \varphi(x) |df(x)|. \quad (13.5.1)$$

In the context of the present book (13.5.1) is natural, as by Theorem 4.2.1, $J_\varphi(f) < \infty$ and $\omega_\varphi^1(f, h) = O(h)$ are equivalent, $\omega_\varphi^2(f, h)_1 \leq K \omega_\varphi^1(f, h)$ and by (9.3.3) $\|B_n^* f - f\|_1 \leq Mn^{-1/2}$ and $\omega_\varphi^2(f, h)_1 \leq M_1 h$ are equivalent. In an attempt to show that (13.5.1) is best possible, Hoeffding proved that for f, a step-function of finitely many jumps in any interval $[a, b] \subset (0, 1)$ satisfying $0 < a < b < 1$ (and some additional conditions),

$$\lim_{n \to \infty} n^{1/2} \|B_n^* f - f\|_1 = \left(\frac{2}{\pi}\right)^{1/2} J_\varphi(f), \quad (13.5.2)$$

irrespective of whether $J_\varphi(f)$ is finite or not.

Actually the condition $\|B_n^* f - f\|_1 = O(n^{-1/2})$ is satisfied for $f(x) = x^{-1/2}$ (for which $\omega_\varphi^2(f, h)_1 = O(h)$) but for $f(x) = x^{-1/2}$, $J_\varphi(f) = \infty$, which shows that in this direction (13.5.1) is not the best result. Moreover some of the assumptions under which Hoeffding proved (13.5.2) are redundant. For any $f \in \text{B.V.}_{\cdot \text{loc}}$ we may write $f = f_a + f_j + f_s$ where f_a is locally absolutely continuous, f_j a pure jump function, and f_s is a continuous singular function. For L_n any of the operators B_n^*, S_n^*, V_n^*, P_n, or G_n of Section 9.2 with φ and D related to them as in Definition 9.3.1 (for G_n $D = R_+$ and $\varphi(x) = x$) we can obtain with not much additional effort the following result.

Theorem 13.5.1. *If* $f \in \text{B.V.}_{\cdot \text{loc}}(D)$, $f_s \equiv 0$ *and*

$$J_\varphi(f) \equiv \int_D \varphi(x) |df(x)| < \infty,$$

then

$$\lim_{n \to \infty} n^{1/2} \|L_n(f) - f\|_1 = \left(\frac{2}{\pi}\right)^{1/2} J_\varphi(f_j).$$

We will omit the proof of this theorem. We note that $J_\varphi(f_a) < \infty$ implies $\|L_n(f_a) - f_a\|_1 = o(n^{-1/2})$. However, the really interesting question in this context remains: "Does $n^{1/2} \|B_n^*(f_s, x) - f_s(x)\|_1$ tend to $(2/\pi)^{1/2} J_\varphi(f_s)$?" Conjectures: Totik-Yes; Ditzian-No.

13.6. Conclusion

We believe that the present set of ideas and definitions is the "correct" treatment for the behavior of a modulus of smoothness near an edge of the interval discussed or near a singularity of a weight function. It is our hope that the theorems and methods here will be useful and will lead to many other applications.

APPENDIX

Two concepts that in the body of the paper were defined specifically only for $D = R_+$ will be defined in detail here for $D = (0, 1)$ or $D = R$. While this is similar to the definition on R_+, we do not like to leave a definition entirely to the imagination of the reader.

A. The Analogue of Definition 5.3.1

For $D = (0, 1)$ and $\varphi(x)$ satisfying

$$A_1 x^{\beta(0)} \leq \varphi(x) \leq A x^{\beta(0)}, \qquad x \in (0, 1/2)$$

and

$$B_1(1-x)^{\beta(1)} \leq \varphi(x) \leq B(1-x)^{\beta(1)}, \qquad x \in (1/2, 1)$$

we set

$$t_0^* = \begin{cases} (Ar)^{1/1-\beta(0)} t^{1/1-\beta(0)} & \text{if } \beta(0) < 0 \\ 0 & \text{otherwise,} \end{cases} \qquad (A.1)$$

$$t_1^* = \begin{cases} (Br)^{1/1-\beta(1)} t^{1/1-\beta(1)} & \text{if } \beta(1) < 0 \\ 0 & \text{otherwise.} \end{cases} \qquad (A.2)$$

We call φ admissible if there exist $C \geq 2$, M_0, and t_0 such that for $0 < h < t_0$

$$\text{meas}\{x: x \pm h\varphi(x) \in E\} \leq M_0 \text{ meas } E$$

for every interval $E \subset (Ch_0^*/2, 1 - Ch_1^*/2)$. The analogue of (5.3.5) is then given by

$$W_\varphi^r(f,t)_p = \|f - P_{r,t}f\|_{L_p(0,t_0^*)} + \sup_{0<h\leq 1} \|\Delta_{h\varphi}^r f\|_{L_p(Ch_0^*, 1-Ch_1^*)}$$
$$+ \|f - \tilde{P}_{r,t}f\|_{L_p(1-t_1^*, 1)}, \qquad \text{(A.3)}$$

where $P_{r,t}f$ and $\tilde{P}_{r,t}f$ denote the orthogonal projection of f onto $L_2[0, t_0^*]$ and $L_2(1 - t_1^*, 1)$ respectively.

For $D = (-\infty, \infty)$ and $\varphi(x)$ satisfying

$$A_1|x|^{\beta(-\infty)} \leq \varphi(x) \leq A|x|^{\beta(-\infty)} \quad \text{if} \quad x < -1$$
$$B_1 x^{\beta(\infty)} \leq \varphi(x) \leq B x^{\beta(\infty)} \quad \text{if} \quad x > 1$$

we set

$$t_0^* = \begin{cases} -(Ar)^{1/1-\beta(\infty)} t^{1/1-\beta(\infty)} & \text{if } \beta(-\infty) > 1 \\ -\infty & \text{otherwise} \end{cases} \qquad \text{(A.4)}$$

$$t_1^* = \begin{cases} (Br)^{1/1-\beta(\infty)} t^{1/1-\beta(\infty)} & \text{if } \beta(\infty) > 1 \\ \infty & \text{otherwise.} \end{cases} \qquad \text{(A.5)}$$

We call φ admissible (in this case) if for some $C \geq 2$, M_0, and t_0

$$\text{meas}\{x : x \pm h\varphi(x) \in E\} \leq M_0 \text{ meas } E$$

whenever $0 < h < t_0$, E is an interval, and $E \subset (2t_0^*/C, 2t_1^*/C)$. The modulus is given by

$$W_\varphi^r(f,t)_p = \|f\|_{L_p(-\infty, t_0^*)} + \sup_{0<h\leq t} \|\Delta_{h\varphi}^r f\|_{L_p(h_0^*/C, h_1^*/C)} + \|f\|_{L_p(t_1^*, \infty)}. \qquad \text{(A.6)}$$

B. The Definition of the Weighted Modulus of Smoothness on $(0, 1)$

To replace (6.1.5) and (6.1.6), for $D = (0, 1)$ the weighted modulus is given by

$$\omega_\varphi^r(f,t)_{w,p} = \sup_{0<h\leq t} \|w\Delta_{h\varphi}^r f\|_{L_p[t_0^*, 1-t_1^*]} + \sup_{0<h\leq t_0^*} \|w\vec{\Delta}_h^r f\|_{L_p[0, 12t_0^*]}$$
$$+ \sup_{0<h\leq t_1^*} \|w\overleftarrow{\Delta}_h^r f\|_{L_p[1-12t_1^*, 1]} \qquad \text{(B.1)}$$

if $\gamma(0) > 0$, $0 \leq \beta(0) < 1$, $\gamma(1) > 0$, $0 \leq \beta(1) < 1$ (t_0^* and t_1^* are defined by (A.1) and (A.2) above when we replace $\beta(i) < 0$ there by $\beta(i) < 1$);

$$\omega_\varphi^r(f,t)_{w,p} = \sup_{0<h\leq t} \|w\Delta_{h\varphi}^r f\|_{L_p[0, 1-t_1^*]} + \sup_{0<h\leq t_1^*} \|w\overleftarrow{\Delta}_h^r f\|_{L_p[1-12t_1^*, 1]} \qquad \text{(B.2)}$$

if $\gamma(0) = 0$ or $\beta(0) \geq 1$ and $\gamma(1) > 0$, $0 \leq \beta(1) < 1$; and

$$\omega_\varphi^r(f,t)_{w,p} = \sup_{0<h\leq t} \|w\Delta_{h\varphi}^r f\|_{L_p[t_0^*, 1]} + \sup_{0<h\leq t_0^*} \|w\vec{\Delta}_h^r f\|_{L_p[0, 12t_0^*]} \qquad \text{(B.3)}$$

if $\gamma(0) > 0$, $0 \leq \beta(0) < 1$ but $\gamma(1) = 0$ or $\beta(1) \geq 1$.

REFERENCES

N.I. Akhiezer
 [1] On the weighted approximation of continuous functions by polynomials on the entire number axis, *AMS Translations*, Series 2, **22**(1962), 95–137.
 [2] *The Classical Moment Problem*, Hafner Publishing Co., New York, 1965.
 [3] *Lectures on the Theory of Approximation*, 2nd edition (Russian), Isdatelstvo "Nauka", Moscow, 1965.

M. Becker
 [1] Global approximation theorems for Szász–Mirakian and Baskakov operators in polynomial weight spaces, *Indiana J. Math.*, **27**(1978), 127–142.

M. Becker and R.J. Nessel
 [1] A global approximation theorem for Meyer–König and Zeller operators, *Math. Z.*, **160**(1978), 195–206.
 [2] Inverse results via smoothing, *Constructive Function Theory*, Proc. Conf., Blagoevgrad, (conference held 1977), Publ. House Bulg. Acad. Sci., Sofia, 1980, 231–243.
 [3] On global saturation for Kantorovich polynomials, *Approximation and Function Spaces*, Proc. Conf., Gdansk, 1979, North Holland, 1981, 89–101.
 [4] On global approximation by Kantorovich polynomials, *Approximation Theory III*, Proc. Conf., Austin, Ed. E.W. Cheney, Academic Press, 1980, 207–212.

H. Berens and G.G. Lorentz
 [1] Inverse theorems for Bernstein polynomials, *Indiana J. Math.*, **21**(8)(1972), 693–708.

S.N. Bernstein
 [1] Sur l'ordre de la meilleure approximation des functions continués par de polynomes de degré donné, *Memoires publiés par la classe des sci. Acad. de Belgique* (2), **4**(1912), 1–103. (*Collected works* (Russian), Vol. **1**, 11–104, Akad. Nauk SSSR, Moscow, 1952).

Yu. Brudnyi
 [1] Generalization of a theorem of A.F. Timan (Russian), *Dokl. Akad. Nauk*, **148**(1963), 1237–1240; English transl., *Soviet Math. Dokl.*, **4**(1963), 244–247.

P.L. Butzer
 [1] Linear combinations of Bernstein polynomials, *Can. J. Math.*, **5**(1953), 559–567.
 [2] Legendre transform methods in the solution of basic problems in algebraic

approximation, *Coll. Math. Soc. János Bolyai*, **35**, Functions, Series, Operators, Budapest, 1980, 277–301.
  [3] On the extensions of Bernstein polynomials to the infinite interval, *Proc. Amer. Math. Soc.*, **5**(1954), 547–553.

P.L. Butzer and R.L. Stens
  [1] Chebyshev transform methods in the theory of best algebraic approximation, *Abh. Math. Sem. Univ. Hamburg*, **45**(1976), 165–190.
  [2] The operational properties of the Chebyshev transform II. Fractional derivatives, *Theory of Approximation of Functions*, Proc. Conf., Kaluga, 1975, Eds. S.B. Stechkin and S.A. Teljakovski, Moscow, 1977, 49–62.
  [3] Chebyshev transform methods in the solution of the fundamental theorem of best algebraic approximation in the fractional case, *Proc. Coll. on Fourier Analysis and Approximation Theory*, Budapest, 1976, 191–212.

P.L. Butzer, R.L. Stens, and M. Wehrens
  [1] Approximation by algebraic convolution integrals, *Approximation Theory and Functional Analysis*, Proc., Ed. J.B. Prolla, North Holland. Publ. Co., 1979, 71–120.
  [2] Higher order moduli of continuity based on the Jacobi translation operator and best approximation, *C.R. Math. Rep. Acad. Sci. Canada*, **2**(1980), 83–87.

R.A. DeVore
  [1] *The Approximation of Continuous Functions by Positive Linear Operators*, Lecture Notes in Mathematics, **293**, Springer–Verlag, 1972.
  [2] $L_p[-1,1]$ approximation by algebraic polynomials, *Linear Spaces and Approximation*, Eds. P.L. Butzer and B. Sz-Nagy, Birkhäuser-Verlag, Basel, 1978, 397–406.

R.A. DeVore and L.R. Scott
  [1] Error bounds for Gaussian quadrature and weighted L^1-polynomial approximation, *SIAM J. Numer. Anal.*, **2**(1984), 400–412.

Z. Ditzian
  [1] A global inverse theorem for combinations of Bernstein polynomials, *J. Approx. Theory.*, **26**(1979), 277–292.
  [2] On global inverse theorems for Szász and Baskakov operators, *Can. J. Math.*, **2**(1979), 255–263.
  [3] On interpolation of $L_p[a,b]$ and weighted Sobolev spaces, *Pacific J. Math.*, **90**(1980), 307–323.
  [4] Interpolation theorems and the rate of convergence of Bernstein polynomials, *Approximation Theory III*, Ed. E.W. Cheney, Academic Press, 1980, 341–347.
  [5] Some remarks on approximation theorems on various Banach spaces, *J. Math. Anal. Appl.*, **77**(2)(1980), 567–576.
  [6] Moduli of continuity in R^n and $D \subset R^n$, *Trans. Am. Math. Soc.*, **282**(1984), 611–623.
  [7] Rate of convergence for Bernstein polynomials revisited, *J. Approx. Theory*, **50**(1987), 40–48.
  [8] Rate of approximation of linear processes, *Acta Sci. Math.*, **48**(1985), 103–128.
  [9] Polynomials of best approximation in $C[-1,1]$, *Israel J. Math.*, **52**(1985), 341–354.

Z. Ditzian and C.P. May
  [1] L_p saturation and inverse theorems for modified Bernstein polynomials, *Indiana J. Math.*, **25**(1976), 733–751.

Z. Ditzian, D.S. Lubinsky, P. Nevai, and V. Totik
  [1] Polynomial approximation with exponential weight, *Acta Math. Hungar.*, **49** (1987).

V.K. Dzjadyk
  [1] A further strengthening of Jackson's theorem on the approximation of continuous functions by ordinary polynomials (Russian), *Dokl. Akad. Nauk SSSR*, **121**(1959), 403–406.
M.M. Dzrbasyan
  [1] On weighted best polynomial approximation on the whole real axis (Russian), *Dokl. Akad. Nauk SSSR*, **84**(1952), 1123–1126.
  [2] Some questions of the theory of weighted polynomial approximations in a complex domain (Russian), *Mat. Sb.*, **36**(78)(1955), 353–440.
M.M. Dzrbasyan and A.B. Tavadyan
  [1] On weighted uniform approximation by polynomials of functions of several variables (Russian), *Mat. Sb.*, N.S. **43**(85)(1957), 227–256.
W. Hoeffding
  [1] The L_1 norm of the approximation error for Bernstein-type polynomials, *J. Approx. Theory*, **4**(1971), 347–356.
G. Freud
  [1] Über die Approximation reeler stetiger Funktionen durch gewöhnliche Polynome, *Math. Ann.*, **137**(1959), 17–25.
  [2] On direct and converse theorems in the theory of weighted polynomial approximation, *Math. Z.*, **126**(1972), 123–134.
  [3] Markov–Bernstein type inequalities in L_p, *Approximation Theory II*, Eds. G.G. Lorentz et al, Academic Press, 1976, 369–377.
  [4] On Markov–Bernstein-type inequalities and their applications, *J. Approx. Theory.*, **19**(1977), 22–37.
G. Freud, A. Giroux, and Q.I. Rahman
  [1] Sur l'approximation polynomiale avec poids $\exp(-|x|)$, *Can. J. Math.*, **30**(1978), 358–372.
G. Freud and H.N. Mhaskar
  [1] Weighted polynomial approximation in rearrangement invariant Banach function spaces on the whole real line, *Indian J. Math.*, **22**(1980), 209–224.
  [2] K-functionals and moduli of continuity in weighted polynomial approximation, *Ark. Mat.*, **21**(1983), 145–161.
A.L. Fuksman
  [1] The approximation of functions of real variables by algebraic polynomials in a closed region (Russian), *Dokl. Akad. Nauk SSSR*, **178**(1968), 1263–1266.
G.H. Hardy, J.E. Littlewood, and G. Pólya
  [1] *Inequalities*, Cambridge University Press, Cambridge, 1934.
M. Ismail and C.P. May
  [1] On a family of approximation operators, *J. Math. Anal. Appl.*, **63**(1978), 446–462.
K.G. Ivanov
  [1] Direct and converse theorems for the best algebraic approximation in $C[-1,1]$ and $L_p[-1,1]$, *C.R. Acad. Bulgare Sci.*, **33**(10)(1980), 1309–1312.
  [2] Direct and converse theorems for the Best algebraic approximation in $C[-1,1]$ and $L_p[-1,1]$, *Coll. Math. Soc., János Bolyai*, **35**, Functions, Series, Operators, Budapest, 1980, 675–682.
  [3] Some characterizations of the best algebraic approximation in $L_p[-1,1]$ $1 \leq p \leq \infty$, *C.R. Acad. Bulgare Sci.*, **34**(9)(1981), 1229–1232.
  [4] On Bernstein polynomials, *C.R. Acad. Bulgare Sci.*, **35**(7)(1982), 893–896.
  [5] On a new characterization of functions I, *Serdica*, **8**(1982), 262–279.
  [6] A constructive characteristic of the best algebraic approximation in $L_p[-1,1]$ $1 \leq p \leq \infty$, *Constructive Function Theory*, Proc. Conf., Varna, 1981, Ed. B.L. Sendov, Publ. House Bulg. Acad. Sci., Sofia, 1983, 357–367.

[7] On a new characterization of functions II, direct and converse theorems for the best algebraic approximation in $C[-1,1]$ and $L_p[-1,1]$, *PLISKA Stud. Math. Bulgar.*, **5**(1983), 151–163.

[8] Approximation of functions of two variables by algebraic polynomials, I, *Anniversary Volume on Approximation Theory and Functional Analysis*, Eds. P.L. Butzer, R.L. Stens and B.Sz.-Nagy, ISNM 65, Birkhäuser, Basel, 1984, 249–255.

D. Jackson

[1] On the approximation by trigonometric sums and polynomials, *Trans. Amer. Math. Soc.*, **13**(1912), 491–515.

H. Johnen and K. Scherer

[1] On equivalence of K functional and moduli of continuity and some applications, *Constructive Theory of Functions of Several Variables*, Proc. Conf., Oberwolfach, 1976, *Lecture Notes in Mathematics*, **571**, Springer-Verlag, 119–140.

L. Kantorovich

[1] Sur certains developpements suivant les polynomes de la forme de S. Bernstein, I, II, *C.R. Acad. Sci. URSS*, 1930, 563–568, 595–600.

B.A. Khalilova

[1] On some estimates for polynomials (Russian), *Izv. Akad. Nauk Azerbaidzhan SSR*, **2**(1974), 46–55.

S.V. Konjagin

[1] Bounds on the derivatives of polynomials, (Russian), *Dokl. Akad. Nauk SSSR*, **243**(1978), 1116–1118; English transl., *Soviet Math. Dokl.*, **19**(1978), 1477–1480.

V.N. Konovalov

[1] On some constructive characteristics of some classes of functions of several variables (Russian), Dissertation, Kiev, 1972.

N.X. Ky

[1] On Jackson and Bernstein type approximation theorems in the case of approximation by algebraic polynomials in L_p-spaces, *Studia Sci. Math. Hungar.*, **9**(1974), 405–415.

A.L. Levin and D.S. Lubinsky

[1] Canonical products and the weights $\exp(-|x|^\alpha)$, $\alpha > 1$, with applications, *J. Approx. Theory*, **49**(1987), 149–169.

[2] Weights on the real line that admit good relative polynomial approximation with applications, *J. Approx. Theory*, **49**(1987), 170–195.

J. Löfstrom

[1] Best approximation in $L_p(w)$ by algebraic polynomials, (manuscript).

[2] Interpolation of weighted L_p and Sobolev spaces on intervals, (manuscript).

G.G. Lorentz

[1] *Bernstein Polynomials, Mathematical Expositions*, No. 8, University of Toronto Press, Toronto, 1953.

[2] *Approximation of Functions*, Holt, Rinehart and Winston, Athena series, New York, 1966.

D.S. Lubinski

[1] A weighted polynomial inequality, *Proc. Amer. Math. Soc.*, **92**(1984), 263–267.

V. Maier

[1] The L_1 saturation class of the Kantorovich operator, *J. Approx. Theory*, **22**(1978), 223–232.

[2] L_p-approximation by Kantorovic operators, *Analysis Math.*, **4**(1978), 289–295.

C.P. May
  [1] Saturation and inverse theorems for combinations of Bernstein-type operators, Thesis 1974, University of Alberta.
  [2] Saturation and inverse theorems for combinations of a class of exponential-type operators, *Canad. J. Math.*, **28**(1976), 1224–1250.
R. Martini
  [1] On the approximation of functions together with their derivatives, *Indag. Math.*, **31**(1969), 473–481.
H.N. Mhaskar
  [1] Weighted polynomial approximation, *J. Approx. Theory*, **46**(1986), 100–110.
S.M. Mazhar and V. Totik
  [1] Approximation by modified Szász operators, *Acta Sci. Math.*, **49**(1985), 257–269.
V.P. Motornii
  [1] Approximation of functions by algebraic polynomials in L_p metric (Russian), *Izv. Akad. Nauk SSSR*, **35**(1971), 874–899.
P.G. Nevai
  [1] Orthogonal Polynomials, *Memoires Amer. Math. Soc.*, **213**(1979).
  [2] Bernstein's inequality in L_p for $0 < p < 1$, *J. Approx. Theory*, **27**(1979), 239–243.
P. Nevai and V. Totik
  [1] Weighted polynomial inequalities. *Constructive Approximation*, **2**(1986), 113–127.
S.M. Nikolskii
  [1] On the best approximation by polynomials of functions which satisfy a Lipschitz condition (Russian), *Izv. Akad. Nauk SSSR*, **10**(1946), 295–318.
P. Oswald
  [1] Ungleichungen vom Jackson-Typ für die algebraische beste Approximation in L_p, *J. Approx. Theory*, **23**(1978), 113–136.
J. Peetre
  [1] On the connection between the theory of interpolation spaces and approximation theory, *Proc. Conf. Const. Theory of Functions, Budapest*, Eds. G. Alexits and S.B. Stechkin, Akad. Kiadó Budapest, 1969, 351–363.
M.K. Potapov
  [1] Some inequalities for polynomials and their derivatives (Russian), *Vestnik Moscow Univ.*, **2**(1960), 10–19.
  [2] Approximation by polynomials on a finite interval of the real axis, (Russian), *Constructive Function Theory*, Proc. Conf., Varna, 1981, Ed. B.L. Sendov, Publ. House Bulg. Acad. Sci., Sofia, 1983, 134–138.
  [3] On assumptions concerning coincidence of certain function spaces, (Russian), *Trudi Seminar I.G. Petrovskava*, **6**(1981), 223–238.
  [4] Approximation by algebraic polynomials in an integral metric with Jacobi weights (Russian), *Vestnik Moscow Univ.*, **38**(4)(1983), 43–52; English transl., *Moscow Univ. Math. Bull.*, **38**(4)(1983), 48–57.
S.D. Riemenschneider
  [1] The L_p-saturation of the Kantorovic–Bernstein Polynomials, *J. Approx. Theory*, **23**(1978), 158–162.
B.L. Sendov
  [1] Doctoral dissertation, Moscow 1968.
H.S. Shapiro
  [1] *Topics in Approximation Theory*, Lecture Notes in Mathematics, **187**, Springer-Verlag, 1970.
E.M. Stein
  [1] *Singular Integrals and Differentiability Properties of Functions*, Princeton Univ. Press, Princeton, New Jersey, 1970.

R.L. Stens
- [1] Characterization of best algebraic approximation by weighted moduli of continuity, *Collog. Math. Soc. János Bolyai*, **35**, Functions, Series, Operators, Budapest, 1980, 1125–1131.

R.L. Stens and M. Wehrens
- [1] Legendre transform methods and best algebraic approximation, *Comment. Math. Prace Mat.*, **22**(1979), 351–380.

L.I. Strukov and A.F. Timan
- [1] Mathematical expectation of continuous functions of random variables. Smoothness and variance, (Russian), *Sibirsk. Math. Zh.*, **18**(1977), 658–664; English transl., *Siberian Math. J.*, **18**(1977), 469–474.

G.I. Sunouchi
- [1] Derivatives of polynomials of best approximation, *Abstract Spaces and Approximation*, Proc. of Conf., Oberwolfach, 1968, Eds. P.L. Butzer and B. Sz-Nagy, ISNM 10, Birkhauser-Verlag, 1969, 233–241.

A.F. Timan
- [1] Strengthening of Jackson theorem on the best approximation of continuous functions on a finite segment of the real axis (Russian), *Dokl. Akad. Nauk SSSR*, **78**(1951), 17–20.
- [2] *Theory of Approximation of Functions of a Real Variable*, English translation 1963, Pergamon Press, The MacMillan Co., Russian original published in Moscow by Fizmatgiz in 1960.

V. Totik
- [1] Uniform approximation by Szász-Mirakjan operators, *Acta Math. Acad. Sci. Hungar.*, **41**(1983), 291–307.
- [2] Uniform approximation by Baskakov and Meyer-König and Zeller Operators, *Periodica Math. Hungar.*, **14**(1983), 209–228.
- [3] Approximation by Szász–Mirakjan Kantorovich operators in $L_p(p > 1)$, *Anal. Math.*, **9**(1983), 147–167.
- [4] $L_p(p > 1)$-Approximation by Kantorovich polynomials, *Analysis*, **3**(1983), 79–100.
- [5] Approximation in L^1 by Kantorovich polynomials, *Acta Sci. Math.*, **46**(1983), 211–222.
- [6] Approximation by Meyer-König and Zeller type operators, *Math. Z.*, **182**(1983), 425–446.
- [7] An interpolation theorem and its applications to positive operators, *Pacific J. Math.*, **111**(1984), 447–481.
- [8] Problems and solutions concerning Kantorovich operators, *J. Approx. Theory*, **47**(1983), 51–68.
- [9] Some properties of a new kind of modulus of smoothness, *Z. Anal. Anwendungen*, **3**(2)(1984), 167–178.
- [10] Saturation of Kantorovich type operators, *Periodica. Math. Hungar.*, **16**(1985), 115–126.
- [11] The gamma operators in L_p spaces, *Publ. Math.* (Debrecen), **32**(1985), 43–55.
- [12] Uniform approximation by positive operators on infinite intervals, *Anal. Math.*, **10**(1984), 163–183.
- [13] Approximation by exponential-type operators, *J. Math. Anal. Appl.*, (to appear).
- [14] The necessity of a new kind of modulus of smoothness, *Anniversary Volume on Approximation Theory and Functional Analysis*, Eds. P.L. Butzer, R.L. Stens and B. Sz-Nagy, ISNM 65, Birkhauser, Basel, 1984, 233–249.

E. van Wickeren
- [1] Stechkin–Marchaud-type inequalities in connection with Bernstein polynomials, *Constructive Approximation*, **2**(1986), 331–337.

[2] Weak-type inequalities for Kantorovich polynomials and related operators, (manuscript).

M. Zamansky
 [1] Classes de saturation de certains procédes d'approximation des séries de Fourier des fonctions continues et applications à quelques problèmes d'approximation. *Ann. Sci. École Norm. Sup.*, **66**(1949), 19–93.

X. Zhou
 [1] On a problem of Ditzian (Chinese), *Jour. of Hangzhou Univ.*, **12**(1985), 178–182.

LIST OF SYMBOLS

$A_{m,n}(x)$	128	$\Omega_\varphi^r(f,t)_p, \Omega_\varphi^r(C,f,t)_p$	28
$B_n(f,x)$	112	$\Omega_\varphi^r(f,t)_{w,p}, \Omega_\varphi^r(C,f,t)_{w,p}$	59
$B_n^*(f,x)$	115	$P_n(f,x)$	114
$B_{n,r}(f,x), B_{n,r}^*(f,x)$	116–117	$P_{n,r}(f,x)$	116–117
$\Delta_h^r f(x), \vec{\Delta}_h^r f(x), \overleftarrow{\Delta}_h^r f(x)$	7	Π_n	78
$\Delta_{1/t}(x)$	80	$p_{n,k}(x)$	115
$\delta_n(x)$	80	$Q(x)$	180, 184
$\sim$	8	q_n	184
$E_n(f)_p$	79	$R_m(f,u,x)$	134
$E_n(f)_{w,p}$	90	$S_n(f,x)$	113
$\varphi(x)$	8	$S_n^*(f,x)$	115
$G_n(f,x)$	114	$s_{n,k}(x)$	115
$G_{n,r}(f,x)$	116–117	$S_{n,r}(f,x), S_{n,r}^*(f,x)$	116–117
J_p^*	91	t^*	17, 28, 50, 182
$K_{r,\varphi}(f,t^r)_p$	10	$w_+(x), w_-(x)$	91
$K_{r,\varphi}(f,t^r)_{w,p}$	55	$W_\gamma(t), \overline{W_\gamma}(t)$	101
$\mathcal{K}_{r,\varphi}(f,t^r)_{w,p}$	59	$W_{\gamma,n}(t), \overline{W_{\gamma,n}}(t)$	101
$\mathcal{K}_{r,\varphi}(C,f,t^r)_{w,p}$	59	$W_\varphi^r(f,t)_p$	50
$K_r(f,t^r)_{W,p}$	181	$W_Q(x)$	181
$L_n(f,x)$	116, 117	$v_+(x), v_-(x)$	91
$L_{n,r}(f,x)$	116	$V_i(x)$	101
$L_{p,W}(a,b)$	181	$V_n(f,x)$	113
$\omega_\varphi^r(f,t)_p$	110	$V_n^*(f,x)$	115
$\omega_\varphi^r(f,t)_{w,p}$	56	$v_{n,k}(x)$	115
$\omega^r(f,t)$	1	$V_{n,r}(f,x), V_{n,r}^*(f,x)$	116–117

INDEX

Baskakov operator, 113
Baskakov–Kantorovich operator, 115
Bernstein polynomials, 112
Difference
  forward, 8
  symmetric, 8
Finite overlapping condition, 9, 47
Gamma operator, 114
Kantorovich polynomials, 115
Marchaud inequality, 43
Modulus of smoothness, 1, 8
  main-part, 28
  weighted, 56, 71
  weighted main part, 59
Polynomial approximation
  best, 78
  best weighted, 90
Polytope, 201, 202
  simple, 202
Post–Widder operator, 114
Szász–Kantorovich operator, 115
Szász–Mirakian operator, 113
Weight function, 56
  Freud, 184
  step, 9